AF830805

Magnetoelectric Composites

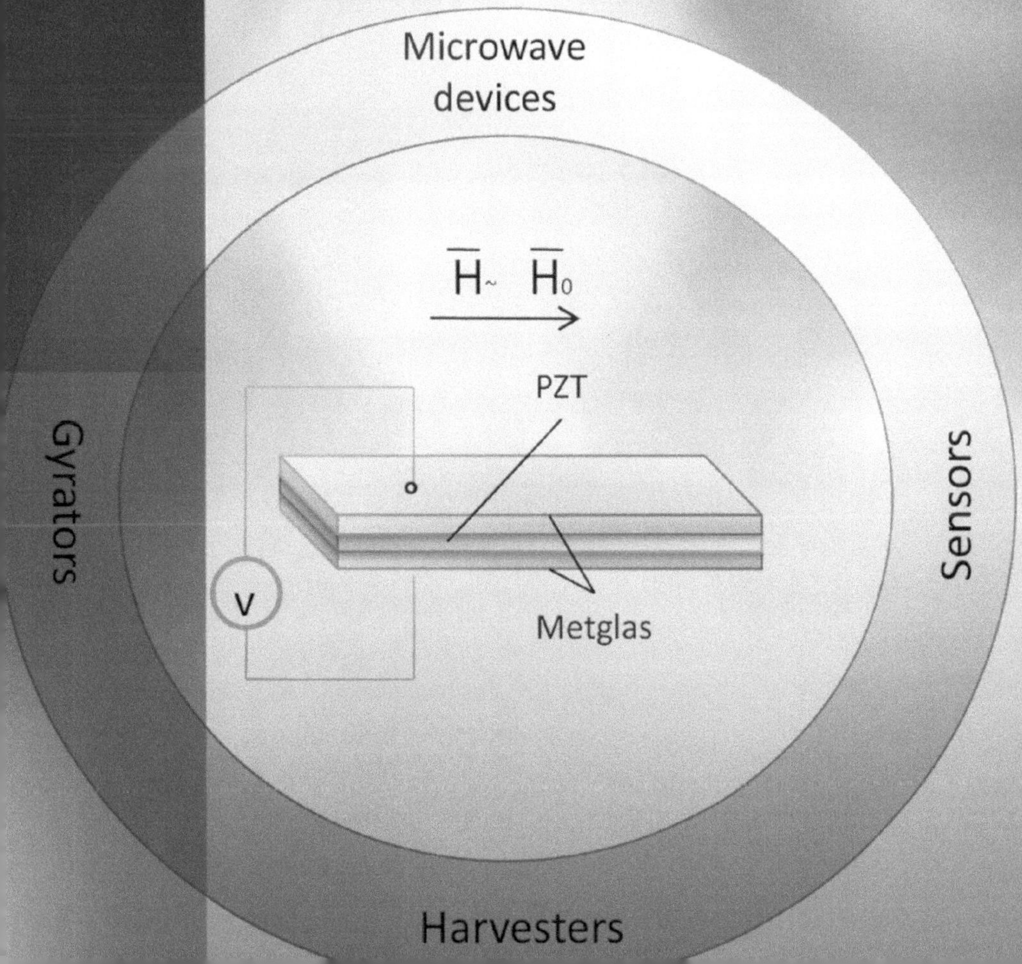

Magnetoelectric Composites

Mirza I. Bichurin | Vladimir M. Petrov
Roman V. Petrov | Alexander S. Tatarenko

PAN STANFORD PUBLISHING

Published by

Pan Stanford Publishing Pte. Ltd.
Penthouse Level, Suntec Tower 3
8 Temasek Boulevard
Singapore 038988

Email: editorial@panstanford.com
Web: www.panstanford.com

British Library Cataloguing-in-Publication Data
A catalogue record for this book is available from the British Library.

Magnetoelectric Composites

Copyright © 2019 by Pan Stanford Publishing Pte. Ltd.
All rights reserved. This book, or parts thereof, may not be reproduced in any form or by any means, electronic or mechanical, including photocopying, recording or any information storage and retrieval system now known or to be invented, without written permission from the publisher.

For photocopying of material in this volume, please pay a copying fee through the Copyright Clearance Center, Inc., 222 Rosewood Drive, Danvers, MA 01923, USA. In this case permission to photocopy is not required from the publisher.

ISBN 978-981-4800-04-4 (Hardcover)
ISBN 978-0-429-48867-2 (eBook)

Contents

Foreword

Functional materials are of crucial importance in novel electronic devices. Among functional materials, magnetoelectric multiferroics represent a class of smart materials that transforms one basic physical property into another. Magnetoelectric effect in these materials is a product property of magnetostrictive and piezoelectric components. There are many research papers on the magnetoelectric interaction due to significant technological promises. The fact is that composites have an obvious advantage over single-phase materials, since they demonstrate a giant magnetoelectric response above room temperature and are ready for technological applications. High magnetoelectric coefficients necessary for engineering applications can be achieved by the appropriate choice of components with high magnetostriction and piezoelectricity. In this book, the authors present their many years of experience in the study of magnetoelectric composites and devices. The central issue explored by the authors of this book is bringing together numerous contributions to modeling of ME composites and the creation of prototypes based on them, allowing one to consider magnetoelectric phenomena from a new point of view based on their practical applications. It should be noted that currently large number of works are devoted to the study of the possibility of constructing the magnetoelectric devices, which in a number of parameters should surpass the well-known semiconductors, ferrite and ferroelectric ones. The authors have examined several low-frequency and microwave magnetoelectric devices and discussed the observed advantages in this book.

The focus of *Magnetoelectric Composites* is on the advanced modeling techniques and state-of-the-art experimental methods that are essential to improve the magnetoelectric performance of composites and devices. The book is useful for researchers and engineers who are engaged in the study and application of new magnetoelectric multiferroics.

Prof. Yury V. Gulyaev
Full Member of Russian Academy of Sciences
President of Russian and International Union of Scientific and Engineering Associations

Preface

The book is dedicated to the modeling and application of magnetoelectric (ME) effect in layered and bulk composites based on magnetostrictive and piezoelectric materials. At present, a large number of works are devoted to theoretical and experimental studies of ME composites, it is of interest to give a summary of pioneering works in the theory and application of the ME effect. Works of Landau, Lifshitz, Dzyaloshinsky, Astrov, Rado, and others [1–4] are fundamental to the discovery of the ME effect. The first experimental works in single crystals showed that the magnitude of the effect is insufficient for practical applications [5] and the interest then shifted [6–8] toward bulk and layered composites. Here it is important to note the works [9-13] in which the theory and experiments of the ME effect in the low-frequency range were presented. Since the main purpose of the research was to increase the ME effect, the next important step was made in the works where it was proposed to investigate the ME effect in the electromechanical resonance (EMR) region [14, 15], because the energy exchange between the magnetic and electrical component in the composite occurs through mechanical oscillations. This led to an increase in the effect by 1 to 2 orders of magnitude and made it possible to proceed to the solution of practical problems. At the same time, an idea arose about observing the ME effect in the region of magnetoacoustic resonance, that is, when electromechanical and magnetic resonances are combined. In this case, preliminary estimates showed a significant increase in the ME effect [14, 16]. As for the microwave range, here from the very beginning of the research it was clear that the use of an electric field instead of a magnetic field for tuning of parameters should give a big benefit in terms of the speed and efficiency of the devices [17]. But these ideas were restrained by the large width of the ferromagnetic resonance (FMR) line in bulk and polycrystalline layered composites [18]. Therefore, the proposal to use a single-crystal magnetic structure with a narrow FMR line solved this problem [19–23]. One of the drawbacks of ME devices is the need for a bias field, which is most often created by permanent magnets. This disadvantage was corrected in [24] by using gradient magnetostrictive–piezoelectric structures, since in this case it is possible to obtain an internal bias field and an electric field for the polarization of the composite without external additional devices. The first models of ME devices [25–30] fully confirmed the calculated ratios. An important source of information on the ME effect are the *Proceedings of the MEIPIC Conference*, a number of reviews, and books [31–51]. All the works mentioned so far have served as a broad base for the development of theoretical and experimental studies of ME composites.

In this book, we mainly describe our many years of experience in the study of ME composites and creation of models of instruments based on them. Therefore, we apologize in advance to readers for the fact that it does not represent all the research directions of ME composites. The book comprises two chapters.

Section 1.1 of Chapter 1 dwells on the effective medium approach to modeling of ME coupling in the low-frequency range. The expressions for ME voltage coefficients are obtained using the material parameters of composite phases—both multilayer and bulk composites are considered. Much attention is paid to the study of the ME effect in graded structures that make it possible to create a bias field without an external magnetic field. A giant Maxwell–Wagner relaxation is predicted at ultra-low frequencies for a two-phase composite. Section 1.2 focuses on ME effects in the EMR region and presents exact solutions to equations of motion for layered structures. The obtained expressions demonstrate an enhancement of the ME voltage coefficients in the EMR region. Sections 1.3 and 1.4 are devoted to modeling of ME effect in the ferromagnetic and magnetoacoustic resonance range. The size- and shape-dependent scaling behavior of the ME effect is explained by giving an example of composite thin films. In Section 1.5, the new test method based on nomographs for ME composites is presented. This method may be useful for a primary estimation of the possible application of a particular ME composite in an electronic device. Possibilities of application of ME composites for making low-frequency and microwave devices are discussed in Chapter 2. Sections 2.1–2.3 are devoted to low-frequency ME inductance, magnetic field, and current sensors and harvesters. In Sections 2.4–2.8, the important applications of ME composites in the microwave range are considered. The basis for the use of ME composites in the microwave range is the ME effect in the FMR region. Investigations of FMR line shift by an external electric field are applied for different microwave devices. The possibility of constructing an ME microwave module for a phased-array antenna is considered.

The authors are thankful to their colleagues S. Priya, G. Srinivasan, D. Viehland, C.-W. Nan, E. Liverts, A. Zakharov, G. Semenov, D. Kovalenko, O. Sokolov, M. Havanova, K. Belicheva, D. Snisarenko, V. Leontiev, and Yu. Kiliba for important contributions to this work.

Mirza I. Bichurin
Vladimir M. Petrov
Roman V. Petrov
Alexander S. Tatarenko
Autumn 2018

---

1. L. D. Landau, E. M. Lifshitz, *Statistical Physics*, 3rd ed., Pergamon, 1980, p. 562.
2. I. E. Dzyaloshinskii, On the magneto-electrical effect in antiferromagnets, *Sov. Phys. JETP*, **10**, 628 (1960).
3. D. N. Astrov, Magnetoelectric effect in chromium oxide, *Sov. Phys. JETP*, **13**, 729 (1961).
4. V. J. Folen, G. T. Rado, E. W. Stalder, Anisotropy of the magnetoelectric effect in Cr_2O_3, *Phys. Rev. Lett.*, **6**, 607 (1961).
5. A. J. Freeman, H. Schmid (Eds.). *Magnetoelectric Interaction Phenomena in Crystals* (MEIPIC-I, Seattle, USA, 1973), London, N.-J. Paris: Gordon and Breach, 1975, p. 228.
6. J. Van Suchtelen, Product properties: a new application of composite materials, *Philips Res. Rep.*, **27**, 28 (1972).
7. J. Van den Boomgard, *et al.*, An in situ grown eutectic magnetoelectric composite materials, Part I, *J. Mater. Sci.*, **9**, 1705 (1974).
8. A. M. J. G. Van Run, *et al.*, An in situ grown eutectic magnetoelectric composite materials, Part II, *J. Mater. Sci.*, **9**, 1710 (1974).
9. G. Harshe, J. O. Dougherty, R. E. Newnham, Theoretical modelling of multilayer magnetoelectric composites, *Int. J. Appl. Electromagn. Mater.*, **4**, 145 (1993).
10. G. Harshe, J. P. Dougherty, R. E. Newnham, Theoretical modelling of 3-0, 0-3 magnetoelectric composites, *Int. J. Appl. Electromagn. Mater.*, **4**, 161 (1993).
11. I. Getman, Magnetoelectric composite materials: theoretical approach to determine their properties, *Ferroelectrics*, **162**, 45 (1994).

12. C.-W. Nan, Magnetoelectric effect in composites of piezoelectric and piezomagnetic phases, *Phys. Rev.* B, **50**, 6082 (1994).

13. M. I. Bichurin, V. M. Petrov, G. Srinivasan, Theory of low-frequency magnetoelectric coupling in magnetostrictive-piezoelectric bilayers, *Phys. Rev.* B, **68,** 054402 (2003).

14. M. I. Bichurin, D. A. Filippov, V. M. Petrov, G. Srinivasan, Magneto-piezoelectric and electro-piezomagnetic effects in composites, *Proceedings of the International Conference "Physics of Electronic Materials"*, 1–4 October, 2002, Kaluga, Russia, p. 309 (in Russian).

15. M. I. Bichurin, D. A. Filippov, V. M. Petrov, V. M. Laletsin, N. N. Paddubnaya, G. Srinivasan, Resonance magnetoelectric effects in layered magnetostrictive-piezoelectric composites, *Phys. Rev.* B, **68**, 132408 (2003).

16. M. I. Bichurin, V. M. Petrov, O. V. Ryabkov, S. V. Averkin, G. Srinivasan, Theory of magnetoelectric effects at magnetoacoustic resonance in single-crystal ferromagnetic–ferroelectric heterostructures, *Phys. Rev.* B, **72**, 060408 (2005).

17. M. I. Bichurin, E. S. Kovalenko, Electric field effect in the paramagnetic resonance of S-ions in cadmium tungstate, *Sov. Phys. FTT*, **13**, 720 (1971).

18. M. I. Bichurin, O. S. Didkovskaya, V. M. Petrov, S. E. Sofronev, Resonant magnetoelectric effect in composite materials, *Izv. vuzov., ser. Physic.*, **1**, 121 (1985) (in Russian).

19. M. I. Bichurin, V. M. Petrov, Magnetic resonance in layered ferrite–ferroelectric structures, *Sov. Phys. JETP*, **58**, 2277 (1988).

20. M. I. Bichurin, Magnetoelectrics in microwave range, *Ferroelectrics*, **161**, 53 (1994).

21. M. I. Bichurin, V. M. Petrov, Composite magnetoelectrics: their microwave properties, *Ferroelectrics*, **162**, 33 (1994).

22. M. I. Bichurin, V. M. Petrov, Influence of external electric field on magnetic resonance frequency in magnetic ferroelectrics, *Ferroelectrics,* **167**, 147 (1995).

23. M. I. Bichurin, I. A. Kornev, V. M. Petrov, I. V. Lisnevskaya. Investigation of magnetoelectric interaction in composite, *Ferroelectrics,* **204**, 289–297 (1997).

24. M. I. Bichurin, V. M. Petrov, G. A. Semenov. Magnetoelectric material for components of radioelectronic devices, Patent No. RU 2363074 C1 (submitted 11.03.2008, published 27.07.2009).

25. M. I. Bichurin, Resonance microwave magnetoelectric effect, In: *Magnetoelectric Substances*. Nauka, Moscow, Russia, pp. 53–67 (1990) (in Russian).

26. M. I. Bichurin, Yu. N. Venevtsev, O. S. Didkovskaya, V. M. Petrov, N. N. Fomich, Magnetoelectric materials: technology features and application perspectives. In: *Magnetoelectric Substances*, Nauka, Moscow, Russia, pp. 118–132 (1990) (in Russian).

27. M. I. Bichurin, R. V. Petrov, Yu. V. Kiliba, Magnetoelectric microwave phase shifters, *Ferroelectrics*, **204**, 311 (1997).

28. M. I. Bichurin, R. V. Petrov, Magnetoelectric Phasers for PAS, *Proceedings of the 2nd International Conference and Exhibition on Satellite Communications* (ICSC `96), Moscow, Russia, 1996, pp. 236–241.

29. M. I. Bichurin, V. M. Petrov, A. S. Tatarenko, et al., Magnetoelectric microwave devices, *Ferroelectrics*, **280**, 211 (2002).

30. M. I. Bichurin, V. M. Petrov, et al., Magnetoelectric sensor of magnetic field, *Ferroelectrics*, **280**, 199 (2002).

31. Proceedings of the 2nd International Conference on Magnetoelectnc Interaction Phenomena in Crystals (MEIPIC-2, Ascona) Eds.Schmid H.,Janner A.,Grimmer H.,Rivera J.-P., and Ye Z.-G. *Ferroelectrics*, **161-162**, 1993, 748p.

32. Proceedings of the 3rd International Conference on Magnetoelectric Interaction Phenomena in Crystals (MEIPIC-3, Novgorod) Ed. M.I Bichurin. *Ferroelectrics*, **204**, (1997), 356p.

33. Proceedings of the Fourth Conference on Magnetoelectric Interaction Phenomena in Crystals (MEIPIC-4, Veliky Novgorod) Ed. M.I.Bichurin. *Ferroelectrics*, **279-280**, 2002, 386p.

34. M. Fiebig, V. V. Eremenko, I. E. Chupis (Eds.), *Proceedings of the Fifth Conference on Magnetoelectric Interaction Phenomena in Crystals* (MEIPIC-5, Sudak), Kluwer Academic Publishers, NATO Sciences Series, 2004, p. 334.

35. M. Fiebig, Revival of the magnetoelectric effect, *J. Phys. D: Appl. Phys.*, **38**, R1 (2005).

36. W. Eerenstein, N. D. Mathur, J. F. Scott, Multiferroic and magnetoelectric materials, *Nature*, **442**, 759 (2006).

37. R. Ramesh, N. A. Spaldin, Multiferroics: progress and prospects in thin films, *Nature Materials*, **6**, 21 (2007).

38. S. Priya, R. Islam, S. Dong, D. Viehland, Recent advancements in magnetoelectric particulate and laminate composites, *J. Electroceram*, **19**, 149 (2007).

39. M. I. Bichurin, D. Viehland, G. Srinivasan, Magnetoelectric interactions in ferromagnetic–piezoelectric layered structures: phenomena and devices, *J. Electroceram.*, **19**, 243 (2007).

40. C.-W. Nan, M. I. Bichurin, S. Dong, D. Viehland, G. Srinivasan, Multiferroic magnetoelectric composites: historical perspectives, status, and future directions, *J. Appl. Phys.*, **103**, 031101 (2008).

41. J. Zhai, Z. Xing, S. Dong, J. Li, D. Viehland, Magnetoelectric laminate composites: An overview, *J. Am. Ceram. Soc.*, **91**, 351 (2008).

42. C. A. F. Vaz, J. Hoffman, C. Ahn, R. Ramesh, Magnetoelectric coupling effects in multiferroic complex oxide composite structures, *Adv. Mater.*, **22**, 2900 (2010).

43. G. Srinivasan, Magnetoelectric composites, *Annu. Rev. Mater. Res.*, **40**, 153 (2010).

44. Y. Wang, J. Hu, Y. Lin, C-W. Nan, Multiferroic magnetoelectric composite nanostructures, *NPG Asia Mater.*, **2**, 61 (2010).

45. J. Ma, J. Hu, Z. Li, C.-W. Nan, Recent progress in multiferroic magnetoelectric composites: from bulk to thin films, *Adv. Mater.*, **23**, 1062 (2011).
46. A. S. Tatarenko, M. I. Bichurin, Microwave magnetoelectric devices, *Adv. Cond. Matter Phys.*, **2012**, 286562 (2012).
47. M. I. Bichurin, D. Viehland (Eds.), *Magnetoelectricity in Composites*, Pan Stanford Publshing, Singapore, 2012, p. 273.
48. M. Bichurin, V. Petrov, *Modeling of Magnetoelectric Effects in Composites* (Springer Series in Materials Science), **201**, 2014, p. 108.
49. G. Srinivasan, S. Priya, N. X. Sun (Eds.), *Composite Magnetoelectrics*, Woodhead Pulishing, Cambridge, UK, 2015, p. 381.
50. H. Palneedi, V. Annapureddy, S. Priya, J. Ryu, Status and perspectives of multiferroic magnetoelectric composite materials and applications, *Actuators*, **5**, 9 (2016).
51. G. Srinivasan, C.-W. Nan, M. S. R. Rao, N. Sun (Eds.), Special issue on magnetoelectrics and their applications, *J. Phys. D: Appl. Phys.*, **51**, 2018.

Chapter 1

Modeling of Magnetoelectric Composites

The magnetoelectric (ME) effect in a material consists in inducing an electric polarization by an applied external magnetic field, or vice versa in inducing a magnetization by external electric field. Linear state functions that define these cross couplings can be given respectively as

$$P_i = \alpha_{ij} H_j , \tag{1.1}$$

$$M_i = \alpha_{ij} / \mu_0 E_j , \tag{1.2}$$

where P_i is the electric polarization, M_i the magnetization, E_j and H_j the electric and magnetic fields, α_{ij} the ME susceptibility tensor, and μ_0 the permeability of vacuum. The ME effect in solids was first theoretically predicted by Landau and Lifshitz [1], calculated by Dzyaloshinskii in Cr_2O_3 [2], and experimentally observed by Astrov [3] and Rado *et al.* [4].

When uniform magnetic and/or electric fields are applied to a material, the change in the Gibbs free energy density can be expressed as [1]

$$dF = -P_i \, dE_i - \mu_0 \, M_i \, dH_i \tag{1.3}$$

Magnetoelectric Composites
Mirza I. Bichurin, Vladimir M. Petrov, Roman V. Petrov, and Alexander S. Tatarenko
Copyright © 2019 Pan Stanford Publishing Pte. Ltd.
ISBN 978-981-4800-04-4 (Hardcover), 978-0-429-48867-2 (eBook)
www.panstanford.com

Using Eq. 1.3 enables one to obtain the following thermodynamic relationships for the dielectric polarization and magnetization:

$$P_i = -(\partial F/\partial E_i)_{\mathrm{H,T}}, \tag{1.4}$$

$$\mu_0 M_i = -(\partial F/\partial H_i)_{\mathrm{E,T}}, \tag{1.5}$$

where T is the absolute temperature. If we assume that the electric susceptibility (χ^{E}) and magnetic susceptibility (χ^{M}) are, respectively, independent of their primary ordering fields E and H, we can obtain the following free energy expression for a linear dielectric and magnetic system that has an ME exchange between the subsystems:

$$F = -1/2\, \chi^{\mathrm{E}}{}_{ij}\, E_i\, E_j - 1/2\, \chi^{\mathrm{M}}{}_{ij}\, H_i\, H_j - \alpha_{ij}\, E_i H_j. \tag{1.6}$$

The first term on the right is the electrical energy stored on application of E, the second is the magnetic energy stored on application of H, and the third is the bilinear coupling between the magnetic and polar subsystems on simultaneous application of E and H. From Eq. 1.6, we can then obtain the following expressions for the total polarization and magnetization induced by simultaneous application of E and H:

$$P_i = \chi^{\mathrm{E}}{}_{ij}\, E_j + \alpha_{ij}\, H_j, \tag{1.7}$$

$$M_i = \chi^{\mathrm{M}}{}_{ij}\, H_j + \alpha_{ji}/\mu_0\, E_j. \tag{1.8}$$

The ME susceptibility is a second-rank axial tensor, involving an exchange between polar (E_j) and axial (H_j) vectors. It is unique in these regards to the dielectric susceptibility and magnetic permeability, which are both second-rank polar tensors. This is an important point, because as a consequence, the values of the components of the ME susceptibility tensor will be dependent on the magnetic point group symmetry, rather than merely on the crystallographic one [5–7]. By now, ME coupling in single crystals have been studied in details [8–19].

Composite materials offer the opportunity to engineer properties that are not available within any of its constituent phases. First, composites have the conventional group of colligative properties which includes, for example, density and stiffness, where composite quantitative adjectives are determined by individual component adjectives, and their volume or weight fractions.

However, composites can also have a (more interesting) second group of properties, which are not intrinsic to its constitutive phases: these are the so-called product tensor properties, first proposed by Van Suchtelen [20, 21]. The appearance of new properties in a composite, which were not present in any individual phase that went into the composite, can be explained as follows. If one of the phases facilitates transformation of an applied independent variable A into an effect B, the inter-relationship of A and B can be characterized by the parameter $X = \partial B/\partial A$. Then, if the second phases converts the variable B to an effect C, the inter-relationship of B and C can be characterized by $Y = \partial C/\partial B$. Then, A must be able to be transformed into C, whose transformation can be characterized by a parameter that is the product of the component parameters: $\partial C/\partial A = (\partial C/\partial B)(\partial B/\partial A) = YX$. Thus, a composite of the two materials must possess a new property which can convert A into C: this property only results from the action of one phase on the other, and is absent in each phase when physically separated from the other.

It is fair to note that B. D. H. Tellegen in 1948 had already proposed a device based on ME composites, which was later named the Tellegen's gyrator [22]. Composites with strong ME effects were not known 60 years ago, and therefore Tellegen's conjecture ME device was not practically realized.

Finally, with regards to any composite materials, there are many possible ways to vary their physical properties, simply by changes and optimizations in the construction and dimensionality of the composite design.

The ME effect in composites is a result of elastically coupled piezoelectric and piezomagnetic effects. The mechanism of ME effects in hybrid composites is as follows: the piezomagnetic material is deformed under an applied magnetic field. This deformation results in a mechanical voltage that acts on the piezoelectric component, and hence induces an electric polarization change in the material via piezoelectricity. Obviously, the converse effect is also possible: an applied electric field causes the piezoelectric component to deform, resulting in mechanical voltage acting on the piezomagnetic material, generating a change in magnetization. Either directly or conversely, the net effect is a new tensor property of the composite—the ME effect—which consists of an electric polarization induced by an external magnetic field, and a magnetization induced by an external electric field.

Most of the known magnetically ordered materials have some measurable magnetostrictive effect: as it is a fourth-rank polar tensor property, and thus must be exhibited by all crystal classes. However, the piezomagnetic effect is not required by symmetry to be present in all crystal classes; since it is a third-rank axial tensor, its property matrix can become null by action of specific symmetry operations. As magnetostriction is the most prevalent form of magnetoelastic coupling, deformations ε_{ij} induced by external magnetic fields (H) most frequently depend quadratically on field strength, rather than linearly: this is the definition of magnetostriction, that is, $\varepsilon_{ij} = Q_{ijkl}H_kH_l$, where Q_{ijkl} is the magnetostriction coefficient. This fact makes use of composites difficult in device applications which require linearity over large field ranges. Linearization can only be achieved under applied dc bias magnetic biases. In this case, the ME effect will be close to linear as long as the range of ac magnetic fields remains small in comparisons to that of the superimposed dc magnetic bias.

ME composites were prepared for the first time by van den Boomgard and coworkers. They used unidirectional solidification of a eutectic composition in the quinary system Fe–Co–Ti–Ba–O [23, 24]. This process promotes formation of alternating layers of magnetic spinel and piezoelectric perovskite phases. The unidirectional solidification process requires careful control of composition, in particular, when one of the components (oxygen) is a gas. Investigations of said composites have shown that excess TiO_2 (1.5 weight %) allowed obtainment of a ME voltage coefficient of $\alpha_E = dE/dH$ = 50 mV/(cm · Oe): which was considered quite high at that time. However, other compositions showed a much lower ME voltage coefficient in the range of 1–4 mV/(cm · Oe). In a subsequent work, the authors reported a ME voltage coefficient of α_E = 30 mV/(cm · Oe) for the eutectic composition $BaTiO_3$–$CoFe_2O_4$, again prepared by unidirectional solidification [35]. This value was nearly an order of magnitude greater than that for the best single-phase response previously reported, which was α_E = 20 mV/(cm · Oe) for single crystal Cr_2O_3. Using the ME voltage coefficient, it is possible to obtain other relevant ME parameters, such as $\alpha = dP/dH = \alpha_E K \varepsilon_o$, where K is the average relative dielectric constant of the composite and ε_0 the permittivity of free space. Using dielectric constants of $K = 500$ for the composite and $K = 11.9$ for Cr_2O_3, we can obtain an estimated value of $\alpha = 7.22 \times 10^{-10}$ s/m for the composite, which is approximately in 270 times higher than that of $\alpha = 2.67 \times 10^{-12}$ s/m for Cr_2O_3.

In ceramic composites, the value for the ME effect of $BaTiO_3$ and $NiFe_2O_4$ alloyed by cobalt and manganese was reduced relative to that prepared by unidirectional solidification. The maximum value of the ME voltage coefficient was 24 mV/(cm · Oe). The authors reported an unusual polarization behavior in which the field polarity was reversed at a Curie temperature [25]. Using the field created by volume charges in the composite allowed one to obtain high degrees of electrical poling. For ceramic composites of $BaTiO_3$ and Ni(Co,Mn) Fe_2O_4, it was found possible to obtain ME voltage coefficients of 80 mV/(cm · Oe).

There are other early studies of ceramic composites. Bunget and Raetchi observed an ME effect in Ni–Zn ferrite–PZT composites, and studied its dependence of applied magnetic field [26, 27]. The magnitude of the ME voltage coefficient was found to be 3.1 mV/(cm · Oe). Discussions on the potential for broadband sensors based on $BaTiO_3$–$NiFe_2O_4$ composites with working frequencies up to 650 kHz should be noted. Composites of Mg–Mn ferrite–$BaTiO_3$ are simultaneously ferroelectric and ferrimagnetic: ferrite/$BaTiO_3$ composites with weight fractions of 30:70, 50:50, 70:30, and 90:10 have all revealed ferroelectric and ferromagnetic hysteresis. Subsequent measurements of the piezoelectric properties of ME composites have shown that the piezoelectric resonance frequency depends on applied magnetic field, where the maximum variation of resonance frequency was 0.2% under $H = 875$ kA/m.

Magnetostrictive metals have, more recently, been investigated as alternatives to ferrites in ME laminated composites. Efforts have focused on use of permendur, Terfenol, and Metglas [14–15]. These (Fe, Ni, Co) magnetostrictive alloys offer much larger magnetostrictions. The ME effect in layered composites based on these magnetostrictive alloys and piezoelectric PZT ceramics is significantly larger than those based on ferrites, which is important for engineering applications. The maximum ME voltage coefficient reported was about 4 V/(cm · Oe), which was obtained in multilayer structures of Terfenol and PZT [28–29]. This opens real possibilities for practical devices.

It is also relevant to note that these composites have been considered by our scientific group across a wide frequency bandwidth ranging from near dc to millimeter. Research on microwave ME effects in layered ferrite–piezoelectric composites has been performed [30]. Application of external electric field to the piezoelectric phase is known to induce a shift in the FMR line frequency of the ferrite phase. An analogous effect was reported earlier in bulk ferrite–piezoelectric composites [31]. Detailed analysis of the resonance ME effects in paramagnetic and magnetic ordered materials were performed [32]. Microscopic theories of ME effects were presented for the magnetic resonance frequency range in magnetically ordered crystals with 3d-electrons. In addition, a theory for low-frequency ME effects and subsequent dispersion with increasing frequency have also been reported.

Finally, there are other important books and review articles that deal with fundamental properties of ME composites [39–43]. There are some special considerations for the modeling of ME composites that need to be mentioned. First, piezoelectric layers of a composite can be electrically connected in series or parallel. Depending upon which piezoelectric d_{33} or d_{31} and magnetostriction $\lambda_{//}$ or $\lambda_{\perp}$ coefficients that one wishes to take advantage of, different multilayer ME composite configurations can be created [45]. Investigations have been devoted to longitudinal ME effects in bilayer composite based on Terfenol-D [47], in which case an ME voltage coefficient of 1.43 V/(cm · Oe) was reported. Second, optimally poled piezoelectric layer effects can be achieved in multilayer structures where ferrite layers are paralleled by intermediate electrodes. In this case, we get a composite with a mechanical connectivity type of 2-2, but with an electrical connectivity type of 3-0. This can be considered as a composite with a mixed connectivity type.

It is often required to predict the effective composite parameters from their constitutive components. Such effective parameters can be determined by the Maxwell Garnett Eq. 1.9. For example, in the case of a 0-3 composite, the effective permittivity of a continuous dielectric matrix having a permittivity ε_1 containing isolated second phase particles with ε_2 can be given as follows:

$$\varepsilon_{эфф} = \varepsilon_1[2\varepsilon_1 + \varepsilon_2 + 2y(\varepsilon_1 - \varepsilon_2)]/[2\varepsilon_1 + \varepsilon_2 - y(\varepsilon_1 - \varepsilon_2)]; \qquad (1.9)$$

where y is the volume fraction of any additional component.

Calculations of the ME properties of composites have some discrepancy with measured values. Measurement of the ME effect in sintered composites of $NiFe_2O_4$ or $CoFe_2O_4$ and $BaTiO_3$ have been reported [23]. Samples in the form of thin disks were polarized by an electric field applied perpendicular to the sample plane. The ME coefficient was then measured for two cases: (i) transverse fields, where dc and ac magnetic fields were parallel to each other and to the plane of the disk (along directions 1 and 2), which were perpendicular to an ac electric field (along direction 3) and (ii) longitudinal fields, where all three fields (dc and ac magnetic, and ac electric) were parallel to each other and perpendicular to the sample's plane. In general, bulk ceramic composites exhibit values of the ME voltage coefficient that are notably lower than those predicted theoretically by continuum mechanics [49]. One reason for this is the low resistivity of the ferrite phase that was aforementioned, which (i) decreases the electric field that can be applied during poling of the piezoelectric phase, resulting in insufficient polarization and (ii) leaks current across the electrodes of the composites, that results in an inability of the composite to maintain charge after it has been induced by applied magnetic fields via the piezoelectric effect. The essential advantage of bulk composites is the possibility of reaching the required values of the given parameters using combinations of the constituent phases that have the necessary values of electric and magnetic materials parameters, and alternatively by tuning the effective composite parameters by adjusting the relative phase volume fractions.

Cubic models of ferrite–piezoelectric ME composites with a 3-0 and 0-3 connectivity types have been considered in [46], which allow for numerical computation of ME coefficients. However, this early theoretical model was inadequate: as evidenced by experimental observations that have shown ME voltage coefficients [8×10^{-3} V/(cm · Oe)] more than two orders of magnitude lower than theoretical predicted one [3.9 V/(cm · Oe)]. Nan *et al.* [48] developed a computational approach to ME effects in bulk composites based on Green's functional approach and perturbation theory. The work offered the perspective of a three-phase composite with high mechanical and ME properties.

The ME effect could find wide applications in various types of electronic devices. Possible applications were considered earlier in Refs. [17, 30, 33, 34 49, 50], microwave phase shifters were considered in Ref. [51], and ME magnetic field and microwave power sensors in Ref. [52]. For such applications, ME composites are necessary (rather than single-phase crystals) due to their high values of ME coefficients and higher working temperature range. However, application of ME composites is hindered by bad reproducibility of effective composite parameters. For example, good mechanical connection across layer interface between phases is necessary to achieve good ME coupling. Furthermore, the constituent phases of a composite should not react with each other. This is a concern particularly for sintered composites based on ferrite–piezoelectric ceramics, as very high temperatures are used in the densification process: avoidance of chemical reactions at interfacial areas can then complicate processing of sintered composites.

Magnetoelectric susceptibility is known to be the fundamental parameter describing the ME coupling [53]. Recently, it was shown that direct and converse ME susceptibilities are equivalent for strain-coupled two-phase systems [54], that is,

$$\alpha = \partial P/\partial H = \mu_0(\partial M/\partial E) \qquad (1.10)$$

ME coefficients for direct and converse ME effects can be expressed in terms of ME susceptibility. Thus, ME voltage coefficient can be defined as $\alpha_E = \partial E/\partial H$ under open electric circuit condition and can be expressed as $\alpha_{\mathrm{E}} = \alpha/\varepsilon$ where ε is effective permittivity. Similarly, electrically induced magnetic field can be determined as follows: $\alpha_{\mathrm{H}} = \partial H/\partial E = \alpha/\mu$ under open magnetic circuit condition where μ is effective permeability.

Direct and converse coupling strengths may be simply related to two-phase systems via the Maxwell relation, provided it is rewritten as $dp/dH = \mu_0\ dm/dE$, in order to describe the entire system via effective magnetic dipole moment m and effective electrical dipole moment p.

For describing the ME response due to mechanical-electric-magnetic coupling, the constitutive equation for the piezoelectric effect can be written in the following form:

$$^{p}S_i = {}^{p}s_{ij}pT_j + {}^{p}d_{ki}E_k, \tag{1.11}$$

$$D_k = {}^{p}d_{ki}\,{}^{p}T_i + {}^{p}\varepsilon_{kn}E_n, \tag{1.12}$$

where $^{p}S_i$ is a strain tensor component of the piezoelectric phase; E_k is a vector component of the electric field; D_k is a vector component of the electric displacement; $^{p}T_i$ is a stress tensor component of the piezoelectric phase; $^{p}s_{ij}$ is a compliance coefficient of the piezoelectric phase; $^{p}d_{ki}$ is a piezoelectric coefficient of the piezoelectric phase; and $^{p}\varepsilon_{kn}$ is a permittivity matrix of the piezoelectric phase.

Analogously, the strain and magnetic induction tensors of the magnetostrictive phase are given as:

$$^{m}S_i = {}^{m}s_{ij}\,{}^{m}T_j + {}^{m}q_{ki}H_k, \tag{1.13}$$

$$B_k = {}^{m}q_{ki}\,{}^{m}T_i + {}^{m}\mu_{kn}H_n, \tag{1.14}$$

where $^{m}S_i$ is a strain tensor component of the magnetostrictive phase; $^{m}T_j$ is a stress tensor component of the magnetostrictive phase; $^{m}s_{ij}$ is a compliance coefficient of the magnetostrictive phase; H_k is a vector component of magnetic field; B_k is a vector component of magnetic induction; $^{m}q_{ki}$ is a piezomagnetic coefficient; and $^{m}\mu_{kn}$ is a permeability matrix.

In what follows, we consider the trilayer magnetoelectric structure (Metglas/PZT/Metglas) in the form of thin and narrow strip so that 1D theoretical model can be used. In-plane magnetic and out-of-plane electric fields are supposed to be directed along 1 and 3 axes, correspondingly. The following boundary conditions can be used to derive expressions for ME coefficients.

$$^{p}S_1 = {}^{m}S_1;$$

$$^{p}T_1\,{}^{p}t = -{}^{m}T_1\,{}^{m}t; \tag{1.15}$$

where ^{p}t and ^{m}t denote thicknesses of the piezoelectric and magnetostrictive phases, respectively. Stress components can be found from Eqs. 1.20 by taking into account Eqs. 1.16 and 1.18. Substituting the found expressions for stress components into Eqs. 1.17 and 1.19 enables one to express the electric and magnetic inductions in terms of electric and magnetic fields. Then ME susceptibility can be found as $\partial D_3/\partial H_1$ or $\partial B_1/\partial E_3$.

$$\alpha_{13} = \frac{{}^{m}q_{11}\,{}^{p}d_{31}}{{}^{p}s_{11} + r\,{}^{m}s_{11}} \tag{1.16}$$

where $r = {}^{\mathrm{p}}t/{}^{\mathrm{m}}t$ with ${}^{\mathrm{p}}t$ and ${}^{\mathrm{m}}t$ denoting the thicknesses of piezoelectric and piezomagnetic layers, correspondingly.

For DME, the transverse ME voltage coefficient can be found as the ratio of induced electric field to applied magnetic field under open electric circuit condition, $D_3 = 0$.

$$\alpha_{\mathrm{E31}} = -\frac{{}^{\mathrm{m}}q_{11}\,{}^{\mathrm{p}}d_{31}}{[(1-{}^{\mathrm{p}}K_{31}^2)^{\mathrm{p}}s_{11} + r^{\mathrm{m}}s_{11}]^{\mathrm{p}}\varepsilon_{33}}, \tag{1.17}$$

For CME at transverse field orientation, the relevant ME coefficient is defined as the ratio of induced magnetic field to applied electric field at open magnetic circuit condition, $B_1 = 0$.

$$\alpha_{\mathrm{H13}} = -\frac{{}^{\mathrm{m}}q_{11}\,{}^{\mathrm{p}}d_{31}}{[(1-{}^{\mathrm{m}}K_{11}^2)^{\mathrm{m}}s_{11} + p^{\mathrm{p}}s_{11}]^{\mathrm{m}}\mu_{11}}, \tag{1.18}$$

where $p = {}^{\mathrm{m}}t/{}^{\mathrm{p}}t$.

Experimentally, ME susceptibility is obtained under closed electric circuit condition for DME (output voltage equals zero and output current should be measured) and under closed magnetic circuit condition for CME. As opposed to ME susceptibility, ME voltage coefficient can be measured at open electric circuit condition. Similarly, α_{H} should be measured at open magnetic circuit condition.

Since the ME effect in layered structures is due to mechanically coupled piezoelectric and magnetostrictive subsystems, it sharply increases in the vicinity of the EMR. Mechanical oscillations of an ME composite can be induced by alternating either the magnetic or the electric field. Theoretical modeling and measuring the ME coupling at EMR was first carried out by our group [55–57, 74]. The equation of medium's axial motion can be written as

$$\bar{\rho}\frac{\partial^2 u_i}{\partial t^2} = V\frac{\partial^{\mathrm{p}}T_{ij}}{\partial x_j} + (1-V)\frac{\partial^{\mathrm{m}}T_{ij}}{\partial x_j}, \tag{1.19}$$

where μ_i is the displacement vector component, $\bar{\rho} = ({}^{\mathrm{p}}t\,{}^{\mathrm{p}}\rho + {}^{\mathrm{m}}t\,{}^{\mathrm{m}}\rho)/({}^{\mathrm{p}}t + {}^{\mathrm{m}}t)$ is the average mass density, ${}^{\mathrm{p}}\rho$ and ${}^{\mathrm{m}}\rho$ are the densities of the ferroelectric and ferromagnet. If the length of the electromagnetic wave exceeds the spatial size of the composite by some orders of magnitude, then it is possible to neglect gradients of the electric and magnetic fields within the sample volume. Therefore, based on elastodynamics, magnetostatics, and electrostatics, the equation of medium's axial motion for harmonic vibrations are governed by:

$$\frac{\partial^2 u_1}{\partial x^2} = -k^2 u_1 , \tag{1.20}$$

$$k = \omega \sqrt{\left({}^{\mathrm{p}}\rho\, {}^{\mathrm{p}}t + {}^{\mathrm{m}}\rho\, {}^{\mathrm{m}}t \right) \left(\frac{{}^{\mathrm{p}}t}{{}^{\mathrm{p}}s} + \frac{{}^{\mathrm{m}}t}{{}^{\mathrm{m}}s^{\mathrm{B}}} \right)^{-1}} ,$$

where ${}^{\mathrm{p}}s$ and ${}^{\mathrm{m}}s^{\mathrm{B}}$ denote the thicknesses of piezoelectric layer at constant electric field and piezomagnetic layer at constant magnetic induction, respectively. Eq. 1.25 should be solved for the boundary conditions that consist of vanishing total force at the ends of the trilayer.

Solving Eq. 1.25 for u_1 and substituting the solution into Eq. 1.16 enables us to obtain the stress component. Once the stress components are determined, expression for the ME susceptibility can be obtained as $\partial \mathrm{D}_3/\partial H_1$ with D_3 denoting the average electric induction:

$$\alpha_{13} = \frac{2\, {}^{\mathrm{m}}q_{11}\, {}^{\mathrm{p}}d_{31} \tan\left({kL}/{2} \right)}{\left[\left(1 + {}^{\mathrm{m}}K_{11}^2 \right) \left(p\, {}^{\mathrm{p}}s_{11} + {}^{\mathrm{m}}s_{11}^{\mathrm{B}} \right) \right] kL - 2\, {}^{\mathrm{m}}k_{11}^2\, p\, {}^{\mathrm{p}}s_{11} \tan\left({kL}/{2} \right)} . \tag{1.21}$$

where L is the sample length.

It should be noted that an increase in ME susceptibility occurs at resonance frequency for both DME and CME. To obtain the expression for the ME voltage coefficient for DME, we use the open circuit condition

$$\int_{-\frac{L}{2}}^{\frac{L}{2}} D_3\, dx = 0$$

$$\alpha_{\mathrm{E31}} = \frac{2\, {}^{\mathrm{m}}q_{11}\, {}^{\mathrm{p}}d_{31} / {}^{\mathrm{p}}\varepsilon_{33} \tan\left({kL}/{2} \right)}{\left[2\, {}^{\mathrm{m}}K_{11}^2\, {}^{\mathrm{p}}s_{11} - 2\, {}^{p}K^2\, {}^{\mathrm{m}}s_{11}^{\mathrm{B}} r \right] \tan\left({kL}/{2} \right) - \left({}^{\mathrm{p}}s_{11} + r\, {}^{\mathrm{m}}s_{11}^{\mathrm{B}} \right) kL} . \tag{1.22}$$

Increase in ME voltage coefficient is obtained at antiresonance frequency.

In this chapter, we discuss the ME properties of ferrite–piezoelectric composites, to create new ME composites with enhanced ME couplings that would enable them for application in functional electronics devices. To address this important scientific and technical goal, a generalization of various theoretical and experimental studies of ME composites has been given. One of the main tasks according to the formulated approach is a comparative analysis of ME composites that have different connectivity types. The relative simplicity of manufacturing multilayer composites with a 2-2 type connectivity having giant ME responses is an important benefit. In addition, composites with 3-0 and 0-3 connectivity types can also be made in considerable quantity by a minimum monitoring of the synthesis process.

Any material with a high piezoelectric constant is a good choice for the piezoelectric phase in ME composites. The most suitable ones are $Pb(Zr,Ti)O_3$, $BaTiO_3$, and $Pb(Mg_{1/3}Nb_{2/3})O_3$–$PbTiO_3$, which is due to their large piezoelectric coefficients. Analogously, any material with a high piezomagnetic coefficient at low magnetic biases is a good candidate for the magnetostrictive phase. The most suitable choices are ferrites and ferromagnetic metals (such as Terfenol-D and Metglas).

Availability of theoretical models for composite properties is necessary to interpret experimental data and to restrict oneself among the multitude of composite configurations. Theoretical estimations of the ME voltage coefficient for series and parallel composite models, and also a cubic model for composites with a 3-0 connectivity type [52, 55], are known. However, as already mentioned, ME voltage coefficient was computed as the ratio of an electric field induced in the piezoelectric phase to the magnetic field applied to the magnetic one: that is, $\alpha_E = {}^{p}E/{}^{m}H$. But, in reality, the internal fields in composite components can be significantly different from the external fields. In particular, formulas [45, 46] predicted a maximum ME voltage coefficient in a pure piezomagnetic phase (i.e., ${}^{p}v = 0$), which distinctly mismatches reality. In addition, generalized models for composite based on an effective medium method were presented. These offer determination of the effective composite parameters with phase connectivity types of 2-2, 3-0, and 0-3 that are based on exact solutions. The ultimate purpose of any theoretical work must be to predict the ME parameters—both susceptibility and voltage coefficients—as these are the most basic parameters of magnetoelectricity.

It is important to realize that ME effects occur over a broad frequency bandwidth, extending from the quasi-static to millimeter ranges. This offers important opportunities in potential device applications. It makes possible new concepts in sensing, gyrators, microwave communications, phase shifters, just to name a few. It also complicates the understanding of magnetoelectricity, as there are significant changes in its spectra with frequency. There are strong enhancements in the ME coefficients near both the electromechanical and magnetomechanical resonances. In addition, there is the important problem of studying dispersion in the ME parameters over a broad frequency range of $10^{-3} < f < 10^{10}$ Hz. Relaxation parameters depend on connectivity type, composite geometry and structure, and volume fraction of constituent phases.

This book surveys the ME effect in ME composites over a wide frequency range, offering suggestions for making new ME materials with sufficient exchange to enable practical applications. Generalized theoretical and experimental studies will be presented that try to gain advantage by comparing existing solutions with existing data.

1.1 Low-Frequency Range

In this chapter, we discuss the theoretical modeling of low-frequency ME effect in layered and bulk composites based on magnetostrictive and piezoelectric materials. Our analysis rests on the effective-medium approach and exact calculation based on elastostatic, electrostatic and magnetostatic equations. The expressions for effective parameters including ME susceptibilities and ME voltage coefficients as functions of material parameters and volume fractions of components are obtained. Longitudinal, transverse and in-plane field orientations are considered. The use of the offered model has allowed to estimate the ME effect in ferrite cobalt–barium titanate, ferrite cobalt–PZT, ferrite nickel–PZT, lanthanum–strontium manganite–PZT composites adequately.

1.1.1 Symmetric Layered Structures

Using layered structures enables one to overcome a series of difficulties that are characteristic for bulk composites. The reasons for the giant ME effects in layered composites are: (a) high piezoelectric and piezomagnetic coefficients in individual layers, (b) effective stress transfer between layers, (c) ease of poling and subsequent achievement of a full piezoelectric effect, and (d) ability to hold charge due to suppression of leakage currents across composites with a 2-2 connectivity.

Prior theoretical models based on mechanics and constitutive relationships by Harshe *et al.* [45] were restricted to account for longitudinal ME voltage coefficient in laminates having ideal mechanical connection at the interfaces between layers. Principal disadvantages of this earlier approach were as follows: (i) For the case of longitudinally oriented fields, the effect of the magnetic permeability of the ferrite phase was ignored. Diminution of interior (local) magnetic fields results in a weakening of ME interactions via demagnetization fields. (ii) The case of fields applied in cross orientations to the ME layer connectivity was not considered, which later experimental investigations revealed as large ME responses.

In this work, we present a summary of a more recent theory of ME laminate composites, which are free from the disadvantages mentioned just above. The approach is based on continuum mechanics and considers the composite as a homogeneous medium having piezoelectric and magnetostrictive subsystems. To derive the effective material parameters of composites, an averaging method consisting of two steps [55–59] should be used. In the first step, the composite is considered as a structure whose magnetostrictive and piezoelectric phases are distinct and separable. ME composites are characterized by the presence of magnetic and electric subsystems interacting with each other.

The constitutive equation for the piezoelectric effect can be given in the following form:

$$ {}^{p}S_i = {}^{p}s_{ij}\,{}^{p}T_j + {}^{p}d_{ki}\,{}^{p}E_k, \tag{1.23}$$

$$ {}^{p}D_k = {}^{p}d_{ki}\,{}^{p}T_i + {}^{p}\varepsilon_{kn}\,{}^{p}E_n; \tag{1.24}$$

where ${}^{p}S_i$ is a strain tensor component of the piezoelectric phase; ${}^{p}E_k$ is a vector component of the electric field; ${}^{p}D_k$ is a vector component of the electric displacement; ${}^{p}T_i$ is a stress tensor component of the piezoelectric phase; ${}^{p}s_{ij}$ is a compliance coefficient of the piezoelectric phase; ${}^{p}d_{ki}$ is a piezoelectric coefficient of the piezoelectric phase; and ${}^{p}\varepsilon_{kn}$ is a permittivity matrix of the piezoelectric phase.

Analogously, the strain and magnetic induction tensors of the magnetostrictive phase are respectively

$$ {}^{m}S_i = {}^{m}s_{ij}\,{}^{m}T_j + {}^{m}q_{ki}\,{}^{m}H_k, \tag{1.25}$$

$$ {}^{m}B_k = {}^{m}q_{ki}\,{}^{m}T_i + {}^{m}\mu_{kn}\,{}^{m}H_n, \tag{1.26}$$

where ${}^{m}S_i$ is a strain tensor component of the magnetostrictive phase; ${}^{m}T_j$ is a stress tensor component of the magnetostrictive phase; ${}^{m}s_{ij}$ is a compliance coefficient of the magnetostrictive phase; ${}^{m}H_k$ is a vector component of magnetic field; ${}^{m}B_k$ is a vector component of magnetic induction; ${}^{m}q_{ki}$ is a piezomagnetic coefficient; and ${}^{m}\mu_{kn}$ is a permeability matrix.

Prior models assumed that the connection at interfaces between layers was ideal. However, in this chapter, we assume that there is a coupling parameter $k = ({}^{p}S_i - {}^{p}S_{i0})/({}^{m}S_i - {}^{p}S_{i0})$ $(i = 1, 2)$, where ${}^{p}S_{i0}$ is a strain tensor component assuming no friction between layers [57]. This interphase–interface elastic–elastic coupling parameter depends on interface quality, and is a measure of a differential deformation between piezoelectric and magnetostrictive layers. The coupling parameter is $k = 1$ for the case of an ideal interface, and is $k = 0$ for the case of no friction.

In the second step of the averaging method to derive the effective materials parameters, the bilayer composite is considered as a homogeneous solid, whose behavior can be described by the following coupled sets of linear algebraic equations:

$$\begin{aligned} S_i &= s_{ij}T_j + d_{ki}E_k + q_{ki}H_k,\\ D_k &= d_{ki}T_i + \varepsilon_{kn}E_n + \alpha_{kn}H_n,\\ B_k &= q_{ki}T_i + \alpha_{kn}E_n + \mu_{kn}H_n; \end{aligned} \tag{1.27}$$

where, S_i is a strain tensor component; T_j is a stress tensor component; E_k is a vector component of the electric field; D_k is a vector component of the electric displacement; H_k is a vector component of the magnetic field; B_k is a vector component of the magnetic induction; s_{ij} is an effective compliance coefficient; d_{ki} is a piezoelectric coefficient; q_{ki} is a piezomagnetic coefficient; ε_{kn} is an effective permittivity; μ_{kn} is a permeability coefficient; and α_{kn} is an ME coefficient.

The simultaneous solution of the coupled sets of linear algebraic equations allows one to find the effective parameters of a composite.

Let us consider that the layers of a composite are oriented along the planes (X_1, X_2), and that the direction X_3 is perpendicular to the same plane. In this case, the direction of polarization in a sample coincides with the X_3 axis. If we by choice apply a constant magnetic bias and variable magnetic field along the same direction coincidental with that of the polarization, then any resultant electric field will also be parallel to the X_3 axis, as shown in Fig. 1.1. This summarization supposes that the symmetry of the piezoelectric phase is $\infty\, m$, and that of the magnetic phase is cubic. The following boundary conditions can then be used to derive expressions for ME coefficients.

$$ {}^{p}S_i = k\, {}^{m}S_i + (1-k)\, {}^{p}S_{i0};\ (i = 1,2) $$

$$ {}^{p}T_i = -\, {}^{m}T_i (1-v)/\, v;\ (I = 1,2) \qquad (1.28) $$

where $v = {}^{p}v/({}^{p}v + {}^{m}v)$ and ${}^{p}v$ and ${}^{m}v$ denote the Poisson's ratio of the piezoelectric and magnetostrictive phases, respectively; and ${}^{p}S_{10}$ and ${}^{p}S_{20}$ are the strain tensor components for $k = 0$.

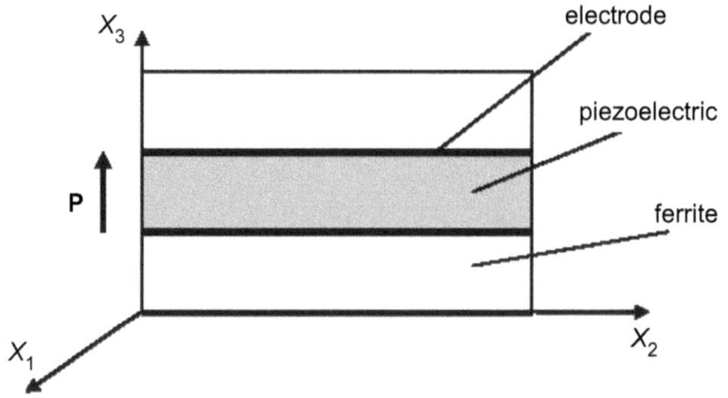

Figure 1.1 Schematic of the layered composite structure.

Using continuity conditions for magnetic and electric fields, and using open circuit condition, one can then obtain the following expressions for the ME susceptibility, and longitudinal ME voltage coefficient.

$$'\alpha_{33} = 2\frac{k\mu_0(\nu-1)\ {}^{\mathrm{p}}d_{31}\ {}^{\mathrm{m}}q_{31}}{[\mu_0(\nu-1) - {}^{\mathrm{m}}\mu_{33}\nu][k\nu({}^{\mathrm{m}}s_{12} + {}^{\mathrm{m}}s_{11}) - ({}^{\mathrm{p}}s_{11} + {}^{\mathrm{p}}s_{12})(\nu-1)] + 2{}^{\mathrm{m}}q_{31}^2 k\nu^2}$$

$$\alpha_{\mathrm{E},33} = \frac{\mathrm{E}_3}{\mathrm{H}_3} = 2\frac{\mu_0 k\nu(1-\nu)\ {}^{\mathrm{p}}d_{31}\ {}^{m}q_{31}}{\{2{}^{\mathrm{p}}d_{31}^2(1-\nu) + {}^{\mathrm{p}}\varepsilon_{33}[({}^{\mathrm{p}}s_{11} + {}^{\mathrm{p}}s_{12})(\nu-1) - \nu({}^{\mathrm{m}}s_{11} + {}^{\mathrm{m}}s_{12})]\}}$$
$$\times\frac{[({}^{\mathrm{p}}s_{11} + {}^{\mathrm{p}}s_{12})(\nu-1) - k\nu({}^{\mathrm{m}}s_{11} + {}^{\mathrm{m}}s_{12})]}{\{[\mu_0(\nu-1) - {}^{\mathrm{m}}\mu_{33}\nu][k\nu({}^{\mathrm{m}}s_{12} + {}^{\mathrm{m}}s_{11}) - ({}^{\mathrm{p}}s_{11} + {}^{\mathrm{p}}s_{12})(\nu-1)] + 2{}^{\mathrm{m}}q_{31}^2 k\nu^2\}} \tag{1.29}$$

The earlier expression obtained by Harshe *et al.* [45] matched our theory for the special case of $k = 1$, provided that the magnetic field is applied only to the ferrite phase.

The model presented above allows for the determination of the longitudinal ME coefficients as functions of volume fractions, physical parameters of phases, and elastic–elastic interfacial coupling parameter k.

Next we consider the transverse fields' orientation that corresponds to E and δE being applied along the X_3 direction, and H and δH along the X_1 direction (in the sample plane). The expressions for the ME susceptibility and transverse ME voltage coefficient are then, respectively,

$$\alpha_{31} = \frac{(\nu-1)\nu({}^{\mathrm{n}}q_{11} + {}^{\mathrm{n}}q_{21}){}^{\mathrm{p}}q_{31}k}{(\nu-1)({}^{\mathrm{p}}s_{11} + {}^{\mathrm{p}}s_{12}) - k\nu({}^{\mathrm{m}}s_{11} + {}^{\mathrm{m}}s_{12})}$$

$$\alpha_{\mathrm{E},31} = \frac{E_3}{H_1} = \frac{-k\nu(1-\nu)({}^{\mathrm{m}}q_{11} + {}^{\mathrm{m}}q_{21}){}^{\mathrm{p}}q_{31}}{p_{\varepsilon_{33}}({}^{\mathrm{m}}s_{12} + {}^{\mathrm{m}}s_{11})k\nu + p_{\varepsilon_{33}}({}^{\mathrm{p}}s_{11} + {}^{\mathrm{p}}s_{12})(1-\nu) - 2k{}^{p}d_{31}^2(1-\nu)} \tag{1.30}$$

Finally, we consider a bilayer laminate that is poled with an electric field E in the plane of the sample. We suppose that the in-plane fields H and δH are parallel and that the induced electric field δE is measured in the same direction (i.e., along the c-axis). Expression for α can be obtained in the following form:

$$\alpha_{E,31} = \{{}^{m}q_{11}({}^{p}s_{33}{}^{p}d_{11} - {}^{p}s_{12}{}^{p}d_{12}) + {}^{m}q_{12}({}^{p}s_{11}{}^{p}d_{12} - {}^{p}s_{12}{}^{p}d_{11})\}(1-v) +$$
$$({}^{m}q_{11}({}^{m}s_{11}{}^{p}d_{11} - {}^{m}s_{12}{}^{p}d_{12}) + {}^{m}q_{12}({}^{m}s_{11}{}^{p}d_{12} - {}^{m}s_{12}{}^{p}d_{11}))vk)(1-v)/$$
$$(((1-p)^{m}\varepsilon_{11} + v^{p}\varepsilon_{11})((1-v)^2({}^{p}s_{11}{}^{p}s_{33} - {}^{p}s_{12}^2) + (1-v)vk({}^{m}s_{11}{}^{p}s_{11} +$$
$${}^{p}s_{33}{}^{m}s_{11} - 2{}^{p}s_{12}{}^{m}s_{12}) + k^2v^2({}^{m}s_{11}^2 - {}^{m}s_{12}^2)) - kv(1-p)^2(2{}^{p}s_{12}{}^{p}d_{11}{}^{p}d_{12}$$
$$-{}^{p}s_{33}{}^{p}d_{11}^2 - {}^{p}s_{11}{}^{p}d_{12}^2 + k^2v^2(1-v)({}^{m}s_{11}{}^{p}d_{12}^2 + {}^{m}s_{11}{}^{p}d_{11}^2 - 2{}^{m}s_{12}{}^{p}d_{11}{}^{p}d_{11}). \quad (1.31)$$

Among all the cases presented so far, the in-plane ME coefficient is expected to be the largest. This is due to availability of magnetostrictive and piezoelectric phases with high q- and d-values, respectively, and to the absence of demagnetization fields. We will further use these outcomes later in the estimation of ME parameters for some specific examples.

1.1.2 Bilayer Structure

Theoretical modeling of low-frequency ME effect described above is based on the homogeneous longitudinal strain approach. However, configurational asymmetry of a bilayer implies bending the sample in applied magnetic or electric field and variation in ME response. ME effect at bending modes was first studied by our scientific group [60]. One of the principal objective to present this section is to discuss modeling of the ME interaction in a magnetostrictive–piezoelectric bilayer taking into account the flexural strains. We calculated ME voltage coefficients α_E for transverse field orientations to provide minimum demagnetizing fields and maximum α_E.

The thickness of the plate is assumed small compared to remaining dimensions. We assume the longitudinal axial strains of each layer to be linear functions of the vertical coordinate z_i to take into account bending of the sample. To preserve force equilibrium, the axial forces in the three layers add up to zero, that is,

$$F_{p1} + F_{m1} = 0 \quad (1.32)$$

where

$$F_{i1} = \int_{-i_{t/2}}^{i_{t/2}} {}^{i}T_1 d\!:$$

The moment equilibrium condition has the form:

$$F_{m1}h_m = M_{m1} + M_{p1}, \tag{1.33}$$

where $M_{i1} = \int_{-i_{t/2}}^{i_{t/2}} z_i\, {}^iT_1 dz$

Solving Eqs. 1.30 and 1.32 simultaneously enables finding the axial stress components in the piezoelectric layer ${}^pT_1$ and ${}^pT_2$. Then the expression for ME voltage coefficient can be expressed using the open circuit condition

$$\alpha_{E31} = \frac{E_3}{H_1} = -\frac{{}^pd_{31} \int_{-{}^pt/2}^{{}^pt/2} ({}^pT_1 + {}^pT_2)dz}{tH_1 {}^p\varepsilon_{33}}$$

where $t = {}^mt + {}^pt + {}^st$ is the total thickness of the considered structure.

Using the 1-D approximations of Eqs. 1.23–1.32 enables one to obtain an explicit form of expression for ME voltage coefficient:

$$\alpha_{E31} = \frac{[(1-{}^pK_{31}^2){}^ps_{11} + {}^ms_{11}\, {}^p\frac{t^3}{{}^mt^3}]{}^mq_{11}\frac{{}^pd_{11}}{{}^p\varepsilon_{11}}}{{}^ps_{11}(1-{}^pK_{31}^2)[2\frac{{}^pt}{{}^mt}\,{}^ms_{11}(2+3\frac{{}^pt}{{}^mt}+2^p\frac{t^2}{{}^mt^2})+(1-{}^pK_{31}^2){}^ps_{11}]+{}^ms_{11}^2\frac{{}^pt^4}{{}^mt^4}} \tag{1.34}$$

In case of neglecting the flexural strains, it is easily shown that expression for ME voltage coefficient reduces to the well-known expression in Section 1.1.1, which was obtained with the assumption of homogeneous longitudinal strains.

1.1.3 Examples of Multilayer Structures

The preceding comprehensive theoretical treatment resulted in expressions of the ME voltage coefficients for three different orientations of fields, which were the ones of most importance, including longitudinal, transverse, and in-plane longitudinal. The most significant features of the model are as follows: (i) Consideration of three different field configurations. This allows for the determination of a single-valued interface parameter k, facilitating quantitative characterization of the bilayer interface; (ii) consideration of a new field configuration, that is, in-plane longitudinal fields, which has very strong ME coupling; and, (iii) consideration of the effect of a finite magnetic permeability on the magnetostriction of the magnetic subsystem, which was ignored in prior investigations.

Next, we apply the theory for the calculation of ME coupling in layered composites. Consider the materials couple cobalt ferrite and lead zirconate titanate (CFO–PZT), which is a system that has been of significant prior interests. Since the value of α_E depends notably on the concentration of the two phases, the ME voltage coefficient has been determined as a function of the volume fraction v of the piezoelectric phase in composite. Material parameters used for estimates are given in Table 1.1. Results of calculations using the model are illustrated in Fig. 1.2, which were obtained by assuming an ideal interface coupling (k = 1).

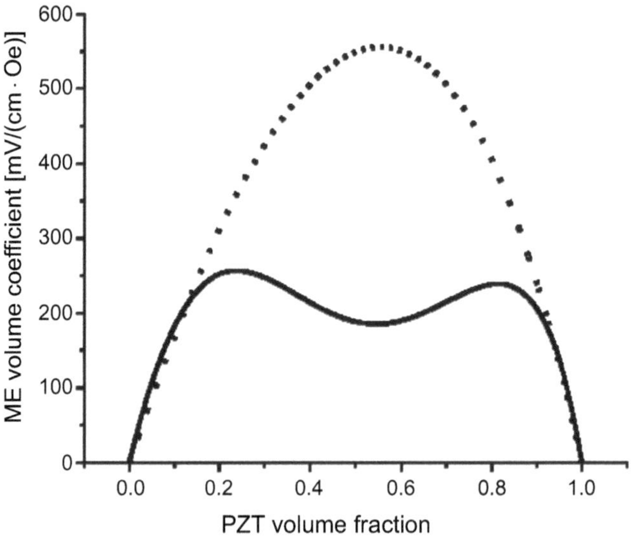

Figure 1.2 PZT volume fraction dependence of transverse ME voltage coefficient $\alpha_{E,31} = \delta E_3/\delta H_1$ for a perfectly bonded (k = 1) bilayer (solid line) and symmetric structure (dot line) of CFO and PZT.

Table 1.1 Material parameters (compliance coefficient *s*, piezomagnetic coupling *q*, piezoelectric coefficient *d*, and petmittivity ε) for lead zirconate titanate (PZT), cobalt ferrite (CFO), and lanthanum strontium manganite used for theoretical values

Material	s_{11} ($10^{-12}\,m^2/N$)	s_{12} ($10^{-12}\,m^2/N$)	s_{13} ($10^{-12}\,m^2/N$)	s_{33} ($10^{-12}\,m^2/N$)	q_{33} ($10^{-12}\,m/A$)	q_{31} ($10^{-12}\,m/A$)	d_{31} ($10^{-12}\,m/V$)	d_{33} ($10^{-12}\,m/V$)	$\varepsilon_{33}/\varepsilon_0$
PZT	15.3	−5	−7.22	17.3	-	-	−175	400	1750
CFO	6.5	−2.4			−1880	556	-	-	10
LSMO	15	−5			250	−120	-	-	10

Results of $\alpha_{E,31}$ versus PZT volume fraction reveals a double maximum that is due to fact that the strain produced by the ferrite consists of two components: longitudinal and flexural. For a symmetric structure such as a trilayer, there are no flexural strains and the maximum ME coefficient occurs for $V = 0.6$ [57]. Since the flexural strain is opposite in sign compared to the longitudinal one and reaches its maximum value for $V = 0.6$, the two types of strains combine to produce suppression of $\alpha_{E,31}$ at $V = 0.6$ and a double maximum in the ME coefficient as in Fig. 1.2. In what follows, we consider theoretical models of low frequency ME coupling for symmetric structures taking into account that such structures result in higher values of ME coefficients.

The variation of $\alpha_{E,33}$ with v for various values of coupling parameter k is shown in Fig. 1.3a for a symmetric structure which excludes the flexural deformations. The magnitude of $\alpha_{E,33}$ decreases with decreasing k, and v_{max} shifts to PZT-rich compositions. Figure 1.3b shows the dependence of the maximum value in $\alpha_{E,33}$ on k, where calculations are illustrated for various values of v_{max}. With increasing k, a near-linear increase was found in the maximum value of $\alpha_{E,33}$. For transverse fields, the maximum α_E is 40% higher than that of $\alpha_{E,33}$. This is due to the strong parallel piezomagnetic coupling q_{11} which determines α_E, relative to that of q_{31} which determines $\alpha_{E,33}$.

Next, we consider the ME effect in CFO–PZT for the in-plane longitudinal field orientation.

The most significant prediction of the present model is that the strongest ME coupling should occur for in-plane longitudinal fields, as shown in Fig. 1.4. One can easily see in Fig. 1.4 that when the field is switched from longitudinal to in-plane longitudinal, the maximum value of the relevant ME coefficient increases by nearly an order of magnitude: $\alpha_{E,max} = 325$ mV/(cm · Oe) for the longitudinal orientation, whereas $\alpha_{E,11} = 3600$ mV/(cm · Oe) for the in-plane longitudinal. The v-dependence of $\alpha_{E,11}$ reveals a rapid increase in the ME coefficient to a maximum value of $\alpha_{E,11} = 3600$ mV/(cm · Oe) for $v = 0.11$, which is followed by a near-linear decrease with further increase of v. Such an enhancement in the in-plane longitudinal coefficient relative to the longitudinal one is understandable due to (i) the absence of demagnetizing fields in the in-plane configuration, and (ii) increased piezoelectric and piezomagnetic coupling coefficients compared to longitudinal fields. The down-shift in the value of v_{max} (from 0.5 to 0.6 for longitudinal and transverse fields to a much smaller value of 0.1) is due to the concentration dependence of the effective permittivity.

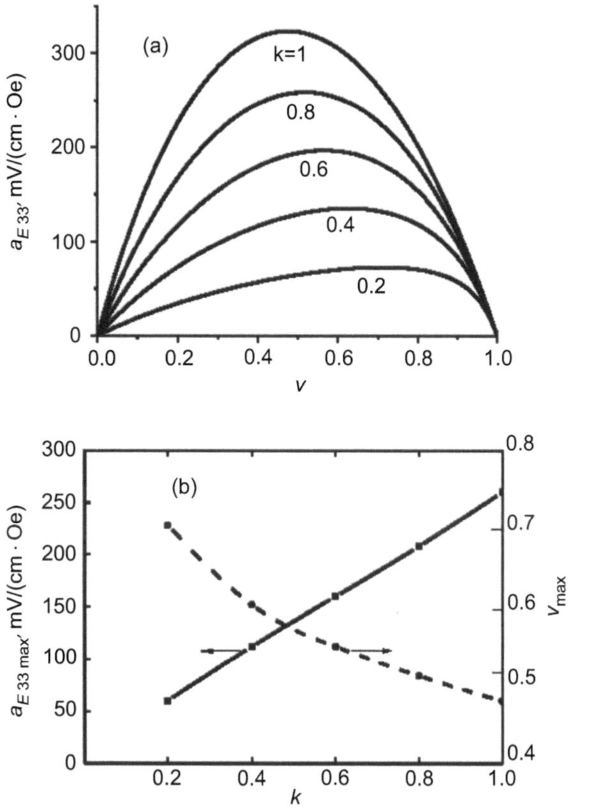

Figure 1.3 (a) Estimated dependence of longitudinal ME voltage coefficient on interface coupling *k* and volume fraction *v* for symmetric structure of CFO and PZT. (b) Variation with *k* of maximum $\alpha_{E,33}$ and the corresponding v_{max}.

Another layered structure of importance is nickel ferrite (NFO)–PZT. Although NFO is a soft ferrite with a much smaller anisotropy and magnetostriction than CFO, efficient magnetomechanical coupling in NFO–PZT gives rise to ME voltage coefficients comparable to those of CFO–PZT. Using the model presented in this chapter, we can estimate α_E for NFO–PZT for different field orientations and conditions, similar to that for CFO–PZT.

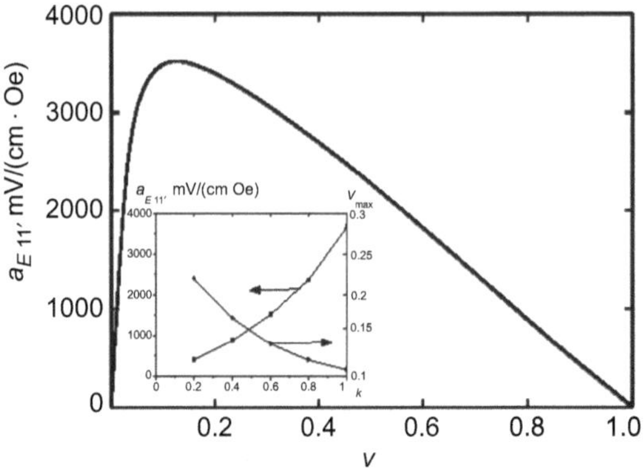

Figure 1.4 ME voltage coefficient for a perfectly bonded (k = 1) symmetric structure of CFO and PZT for in-plane longitudinal field orientation. The poling field and dc and ac magnetic and electric fields are in the sample plane and parallel to each other. Inset shows variation of maximum $\alpha_{E,33}$ and the corresponding v_{max} with k.

Finally, we consider composites that have lanthanum strontium manganites for the magnetostrictive phase. Lanthanum manganites with divalent substitutions have attracted considerable interest in recent years due to double exchange-mediated ferromagnetism, metallic conductivity and giant magnetoresistance. The manganites are potential candidates for ME composites because of (i) high magnetostriction and (ii) metallic conductivity that eliminates the need for a foreign electrode at the interface. Figure 1.5 shows the longitudinal and transverse ME voltage coefficients for unclamped $La_{0.3}Sr_{0.7}MnO_3$ (LSMO)–PZT laminate that assumes ideal coupling at the interface and no bending strain. In this case, the values of the ME coefficients are quite small compared to that of ferrite–PZT: this is due to weak piezomagnetic coefficients and compliances parameters for LSMO. The ME coefficient for in-plane longitudinal fields and the effects of clamping for different field orientations were similar in nature to those for ferrite–PZT layered structure and thus, are not discussed in any detail here.

It is important to compare the theoretical predictions, illustrated above, with experimental data. Let us consider first a bilayer of CFO–PZT taking into account the flexural deformations. Figure 1.6 shows α_E as a function of v. These data were obtained at low frequencies (100–1000 Hz). The desired volume fractions v were achieved by careful control of the layer thickness. Data show an increase in α_E with v until a maximum is reached. However, these data clearly demonstrated that the actual experimental value is an order of magnitude smaller than that predicted in Figs. 1.2 and 1.3 (assuming $k = 1$). It is, therefore, reasonable to compare the data with calculated values of α_E as a function of v using a reduced interface coupling parameters of $k = 0.1$: in this case, agreement between theory and experiment can be seen, as shown in Fig. 1.6. The key inference that can be made concerns the inherently poor interface coupling for CFO–PZT, irrespective of sample synthesis techniques. We address possible causes for this poor coupling later in this section.

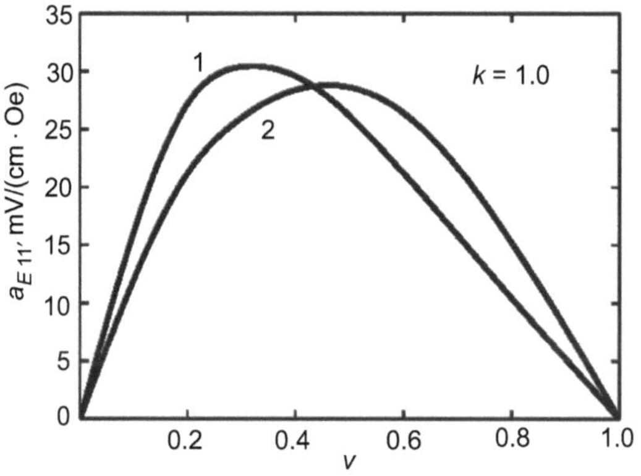

Figure 1.5 (1) Longitudinal and (2) transverse ME voltage coefficients as functions of PZT volume fraction for symmetric layered structure of $La_{0.3}Sr_{0.7}MnO_3$ (LSMO) and PZT for interface coupling parameter $k = 1$.

A third materials couples, LSMO–PZT, is considered in Fig. 1.7, which shows α_E as a function of v for longitudinal and transverse fields. The α_E values are the smallest amongst the three systems considered here. Calculated values assuming $k = 1$ were found to be quite high compared to the data, rather it was found that nonideal values of $k = 0.2$ gave reasonable agreement with the data. Thus, one can readily infer poor interfacial coupling in LSMO–PZT, similar to that for CFO–PZT.

Finally, we should comment on a possible cause of poor interfacial coupling for CFO–PZT and LSMO–PZT and ideal coupling for NFO–PZT. The parameter k can be expected to be sensitive to the mechanical, structural, chemical, and electromagnetic parameters at the interface. We attribute unfavorable interface conditions in CFO–PZT and LSMO–PZT to inefficient magnetomechanical coupling. The magnetomechanical coupling k_m is given by $k_m = (4\pi\lambda'\mu_r/E)^{1/2}$, where λ' is the dynamic magnetostrictive constant and μ_r is the reversible permeability, and E is Young's modulus. In ferrites, under the influence of a dc magnetic bias H and ac magnetic field δH, domain wall motion and domain rotation contribute to the Joule magnetostriction and consequently to the effective linear piezomagnetic coupling. A key requirement for strong coupling is unimpeded domain wall motion and domain rotation. A soft ferrite with a high initial permeability (i.e., low anisotropy), such as NFO, will have key materials parameters favoring a high k_m, and consequently, strong ME effects. Measurements have shown that NFO has an initial permeability of 20, whereas that of LSMO and CFO is 2–3. Thus, one can infer that a plausible simple explanation of the near interfacial parameter for NFO–PZT is (in part) the favorable domain motion.

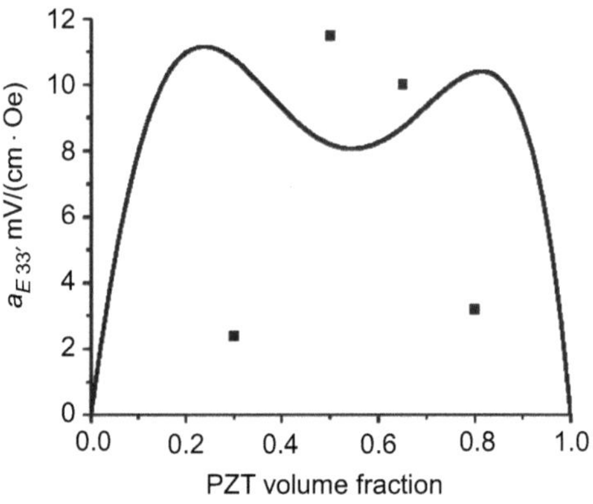

Figure 1.6 PZT volume fraction dependence of longitudinal ME voltage coefficient of CFO–PZT bilayer. Solid line are theory for k = 0.1 and points are experiment.

In deriving the above expression, we assumed the electric field to be zero in magnetic phase since magnetostrictive materials that are used in the case under study have a small resistance compared to piezoelectric phase. Estimate of ME voltage coefficient for CFO-PZT layered structure gives $\alpha_{E,33} = 325$ mV/(cm · Oe) provided the bending strains are ignored. However, considering CFO as a dielectric results in $\alpha_{E,33} = 140$ mV/(cm · Oe) [61] while the experimental value doesn't exceed 74 mV/(cm · Oe) [45]. We believe CFO should be considered as a conducting medium compared to dielectric PZT in the low-frequency region in accordance with our model. The discrepancy between theoretical estimates and data can be accounted for by features of piezomagnetic coupling in CFO and interface coupling of bilayer [57].

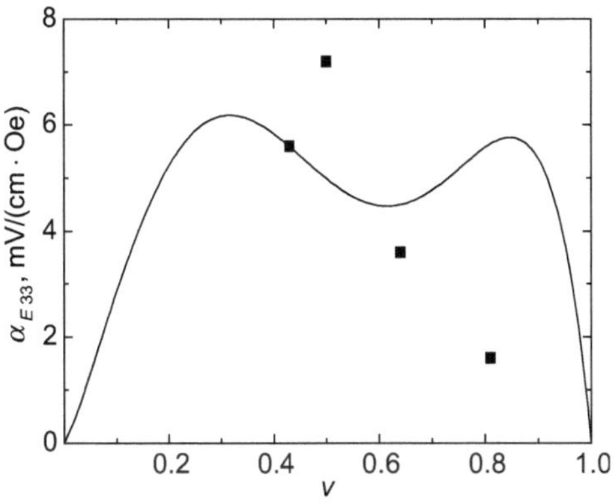

Figure 1.7 PZT volume fraction dependence of longitudinal ME voltage coefficients for LSMO–PZT bilayer: Solid lines are the theoretical values for k = 0.2 and points are the experimental values.

1.1.4 Bulk Composites

The design of new ME composites assumes the use of reliable theoretical models, allowing prediction of properties for various materials couples and over a range of laminate parameters. Manufacturing methods of all-ceramic composites are based on an initial mixing of starting powders batched in proportion to the composite volume fraction, followed by pressing and densification/sintering to a netshape. Clearly, if the concentration of one of the constituent phases is small, then that phase will consist of isolated particles in a matrix. Following accepted classification nomenclature [44], this composite should be referred to as a 0-3 type because one phase is isolated (i.e., connected in zero dimensions) and the second is interconnected in three dimensions. If the volume fraction of the secondary phase in the matrix is increased, and a percolation limit is reached, then it is classified as a 1-3 type composite. If the secondary phase then crosses that initial percolation limit, and subsequently begins to be interconnected in two dimensions, the composite connectivity is known as the 2-3 type. We mention these things at this time to make the point that the same ceramic manufacturing technology allow the fabrication of a wide range of relative volume fractions of the different phases in an all-ceramic composite, and consequently to various possible types of dimensional interconnectivities. Accordingly, it is very important to choose the correct method of calculation for effective constants of a composite at various relative volume fractions of components.

Unfortunately, exact solutions of three-dimensional problems related to the calculation of effective constants of inhomogeneous systems are unknown. Therefore, there is presently no precise structural classification of composites. Within the limited theory of heterogeneous systems of two-phase composites, there are two principle approaches to approximate solutions: matrix systems and two-component mixtures, for which behavior of effective parameters depend continuously on the concentration.

In the case of matrix systems, modification of the concentration from 0 to 1 does not change the qualitative structure of the composite: at any concentration, one of the components must form a coherent matrix that contains isolated particles of the second component. The system always remains essentially noncentral, and matching formulas for an evaluation of effective constants give their continuous dependence on concentration in the entire range from 0 to 1. We should note that the application of these formulas to the calculation of effective constants of composites is not always justified.

The case of two-component mixtures is characterized by a qualitative modification of the structure of the composite because the concentration is changed. Such systems are characterized, as is well known, by critical concentrations at which point there are important property changes such as metal–insulator or rigid–plastic transformations. The metal–insulator transformation occurs in a composite consisting of insulating and conductive phases. Assume that the insulating phase is initially the matrix and that the conductive one consists of isolated particles. In this case, initially the composite is insulating; however, when the percolation limit is crossed, the conducting particles form an interconnected conduction pathway, dramatically lowering the resistivity near a critical volume fraction. In the second example (rigid–plastic transformation), it is supposed that the composite is a mixture in which the elastic compliance of one of the constituent phases tends to infinity (for example, a porous composite). This composite type posses a critical concentration of the second phase, above which the rigid framework of the composite loses its stability. It should be straightforward to see that bulk composites will have numerous effective materials properties, all of which change with relative phase volume fraction in a manner that is independent of other properties.

As an example, we consider a composite with a 3-0 type connectivity. Cubic models for ferrite–ferroelectric composites with a connectivity of 3-0 and 0-3 have been considered by Harshe [46]. Numerically, the ME coefficient is equal to the ratio of the electric field induced on the composite by an applied magnetic field: the ME coefficient is equal to E_3/H_3. It is necessary to realize that the magnetic field was applied only to the ferrite phase: that is, $E_3/{}^{\mathrm{m}}H_3$ where ${}^{\mathrm{m}}H_3$ is the local magnetic field on the ferrite phases that may exceed that applied to the entire composite. Harshe's study only considered the case of free cubic cells, and effective parameters of the composite in known model systems were not determined. However, in real composites, we must consider the case of nonfree cells. It is also very important to use any such model to predict the effective composite parameters. In the following section, we present a generalized model for ferrite–piezoelectric composites that allows one to define and predict the effective parameters of said composite using given conditions.

The properties of this ME composite will depend on the parameters of the corpuscles, and also on the terminal conditions. Now, let us suppose that the geometrical model for ME composites in this figure is miniaturized to fine scales. If the given cubic model ME composite is considered as a material consisting of consecutive and parallel connections of cubic cells with legs of unit length, then it is obvious by the definition of properties of a composite that it is possible to consider only one cubic cell rather than the entire ensemble of cells. The magnetostriction phase is enclosed by piezoelectric ones along different directions.

$$S_1 = -s_{c1}\, s_{11}\, T_1,$$
$$S_2 = -s_{c2}\, s_{22}\, T_2, \tag{1.35}$$
$$S_3 = -s_{c3}\, s_{33}\, T_3,$$

where $s_{ci} = {}^{c}s_{ii}/s_{ii}$ is the relative compliance of surroundings and ${}^{c}s_{ii}$ is effective compliance of composite.

PZT volume fraction dependence of effective ME susceptibility is shown in Fig. 1.8 [62, 63].

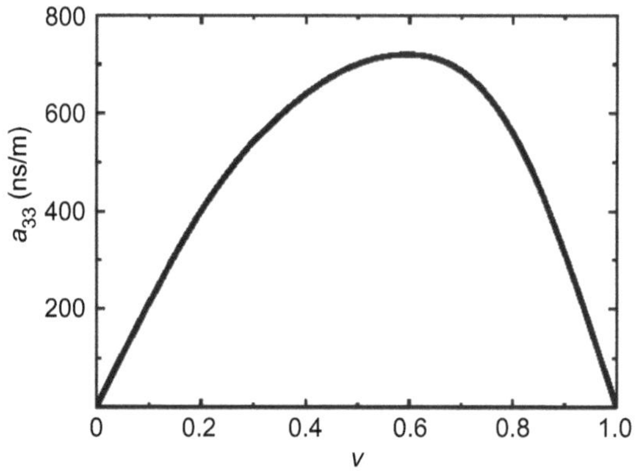

Figure 1.8 PZT volume fraction dependence of ME susceptibility for CFO-PZT composite with connectivity 3-0.

The dependence of the effective ME voltage coefficient, defined as $\alpha_{E,33} = -\alpha_{33}/\varepsilon_{33}$, on the piezoelectric phase volume fraction can then be easily obtained, as shown in Fig. 1.9. These graphical solutions then allow one to determine the piezoelectric and magnetostrictive phase volume fractions that yield maximum values for the effective ME susceptibility (Fig. 1.8) and ME voltage coefficient (Fig. 1.9).

The values of the ME voltage coefficient in Fig. 1.10 coincide with previously published data [46], demonstrating the usefulness of the predictions. As follows from Fig. 1.10, the ME voltage coefficient was approximately 20% greater than that calculated from the experimental data using the model. This is explained by the fact that the internal (local) magnetic field in the ferrite component is considerably different from the externally applied magnetic field.

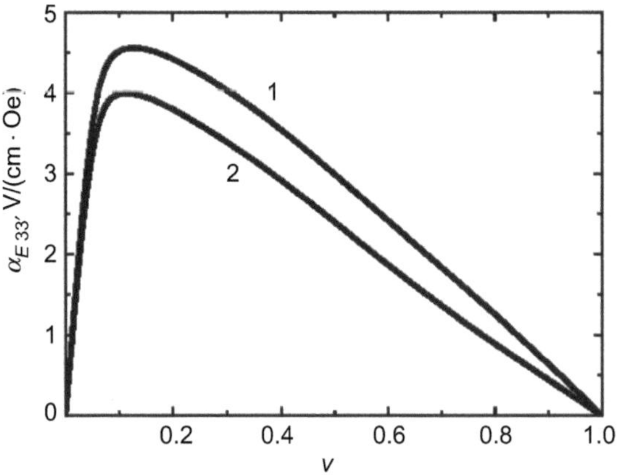

Figure 1.9 PZT volume fraction dependence of ME voltage coefficient for CFO-PZT composite with connectivity 3-0 according to model [62] (1) and model [46] (2) for $\alpha_{E33} = E_3/{}^mH_3$ and material parameters from Table 1.1.

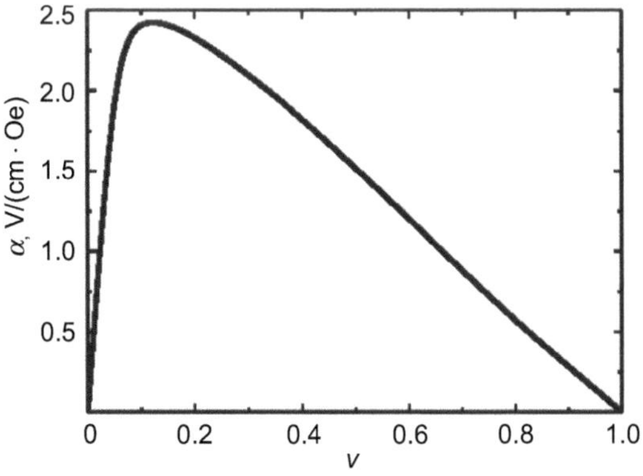

Figure 1.10 Concentration dependence of $\alpha_{E33} = E_3/{}^mH_3$ for composite with material parameters from [62].

Measurements of the ME voltage coefficient have been performed for bulk composites of NFO–PZT, using the experimental methodology mentioned above. Data are shown in Fig. 1.11 for the ME voltage coefficient as a function of the piezoelectric phase volume fraction. These measured values are much lower in magnitude than the theoretical ones predicted for the free composite condition. However, considering a clamped condition defined by the matching $s_{c11} = s_{c22} = s_{c33} = 0.3s_{33}$, agreement between theory and experiment was found, as illustrated in Fig. 1.11. These results indicate that there are in reality 0-3 ferrite–piezoelectric ceramic composites mixtures in which the component phase grains are mechanically clamped by neighboring grains and by environmental boundary conditions.

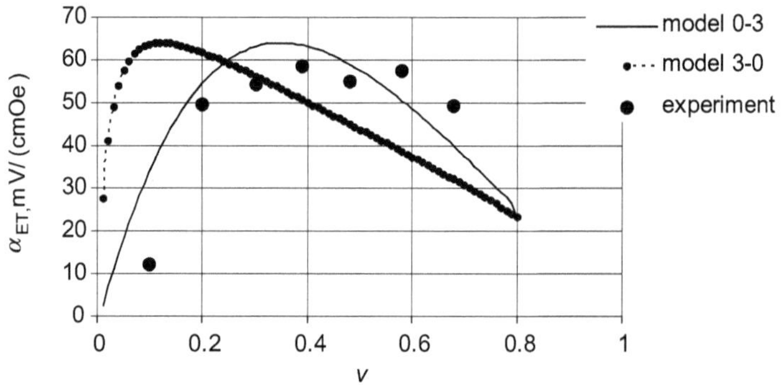

Figure 1.11 ME effect in bulk composite of NFO and PZT.

1.1.5 Magnetoelectric Effects in Compositionally Graded Layered Structures

A key drawback of functional devices based on ME materials is the need of prior polarization of ferroelectric phase and magnetic bias field of magnetic phase. Poling consists in heating the sample to Curie temperature and slow cooling in a static electric field. The bias magnetic field magnitude should correspond to maximal piezomagnetic coupling coefficient which provides the maximal ME coefficient. Thus supplementary constructional elements are necessary to provide poling and biasing the sample.

Here we prepare a composite based on magnetization-graded magnetic phase from compositionally graded ferromagnetic and polarization-graded piezoelectric phase from compositionally graded ferroelectric. Using the polarization-graded piezoelectric phase is known to result in a built-in static electric field. The amplitude of this field is determined by polarization gradient:

$$E_i = \frac{1}{\varepsilon_0}\int_0^L \frac{\partial P}{\partial x}dx.$$

where L is the sample length. Analogously, magnetization gradient gives rise to a built-in static magnetic field [76]:

$$H_i = \frac{1}{\mu_0}\int_0^L \frac{\partial M}{\partial x}dx.$$

The above expressions show that the proper choice of polarization and magnetization gradients and sample length enables one to get the built-in electric and magnetic fields needed for obtaining the maximal piezoelectric and piezomagnetic coupling coefficients. It should be noted that it is not needed in preliminary poling and magnetic biasing of the sample [64, 65].

As an example, we consider a magnetostrictive–piezoelectric bilayer of nickel–zinc ferrite and PZT. Bilayer includes the magnetic layer that is compositionally graded along the sample plane and the piezoelectric layer that is compositionally graded perpendicular to the sample plane. The output voltage induced by ME coupling is measured across the sample thickness. Estimated bias field dependence of ME voltage coefficient is shown in Fig. 1.12 for equal volume fractions of ferrite and PZT.

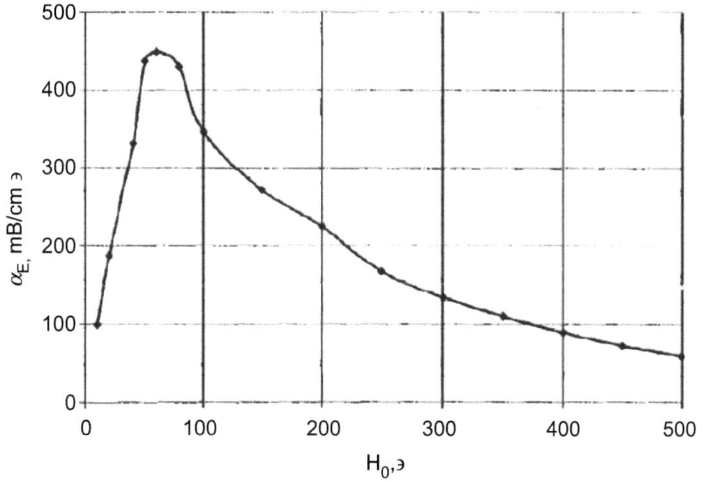

Figure 1.12 Bias field dependence of ME voltage coefficient for bilayer of compositionally graded nickel–zinc ferrite and PZT.

Variation of zinc content from 0.3 to 0.5 gives rise to a magnetization gradient in the sample plane that results in a built-in magnetic field of 44 Oe. Estimates show that variation of zinc content from 0.3 to 0.5 enables one to increase the built-in magnetic field up to 60 Oe. Fig. 1.12 shows that bias field of 60 Oe provides obtaining ME voltage coefficient of 450 mV/(cm · Oe).

As an example, we consider theoretical model of ME coupling in a bilayer of nickel zinc ferrite, $Ni_{1-x}Zn_xFe_2O_4$ (NZFO), and $Pb_{1-x}Zr_xTiO_3$ (PZT). Although ferrites of homogeneous compositions are not piezomagnetic, one can achieve a pseudo piezomagnetic effect ($q = d\lambda/dH$, where λ is the magnetostriction) by subjecting the sample to a bias magnetic field H_0 and ac field H_1. Studies show that when Zn is substituted in nickel ferrite, the room temperature q varies linearly with increasing Zn concentration x for $x < 0.4$ [64]. Thus it is possible to achieve a linear grading in q with a compositionally graded NZFO. Similar grading of the piezoelectric coefficient d is possible by compositional grading of PZT.

The analysis described here is based on the equations for the strain, electric displacement, and magnetic induction of piezoelectric and magnetostrictive phases (1.11–1.13):

For finding the low-frequency ME voltage coefficient, we solve magnetostatic and elastostatic equations in NZFO and elastostatic and electrostatic equations in PZT, taking into account boundary conditions. The longitudinal axial strains of each layer can be considered as linear functions of the vertical coordinate z_i to take into account the bending deformations of bilayer:

$$^{m}S_1 = {}^{m}S_{10} + z_m/R_1;$$

$$^{p}S_1 = {}^{p}S_{10} + z_p/R_1; \tag{1.36}$$

where $^{i}S_{10}$ is the centroidal strains at $z_i = 0$, R_1 is the radius of curvature and z_i is measured relative to centroidal plane of i-layer.

We limit the analysis to field orientation in which the bias magnetic field H_0 and ac magnetic field H_1 are parallel to each other and the samples plane. The PZT is polarized along z and the ac electric field is measured perpendicular to the sample plane. The assumed field orientation provides minimum demagnetizing fields and maximum ME coefficient.

The axial forces in the three layers must add up to zero to preserve force equilibrium,

$$ {}^{m}F_1 + {}^{p}F_1 = 0, \tag{1.37}$$

where $F_{i1} = \int_{-^{i}t/2}^{^{i}t/2} {}^{i}T_1 dz_1$, ${}^{m}t$, ${}^{p}t$, and ${}^{s}t$ are the thicknesses of piezomagnetic, piezoelectric, and substrate layers.

Using Eq. 1.37 and taking into account Eqs. 1.11–1.13 enables finding ${}^{m}S_{10}$:

$$ {}^{m}S_{10} = \frac{s_1}{t}\left[{}^{m}Y \int_{-^{m}t/2}^{^{m}t/2} {}^{m}q_{11}\, {}^{m}H_1 dz_1 + {}^{p}Y \left(\int_{-^{p}t/2}^{^{p}t/2} {}^{p}d_{31}\, {}^{p}E_3 dz_2 + \frac{{}^{p}th_{m}}{R_1} \right) \right] \tag{1.38}$$

where $s_1 = t({}^{m}t\,{}^{m}Y + {}^{p}t\,{}^{p}Y)^{-1}$, $t = {}^{m}t + {}^{p}t$, and ${}^{m}Y$ and ${}^{p}Y$ are the modulus of elasticity of piezomagnetic and piezoelectric components, correspondingly. To conserve moment, the rotating moments of axial forces in the three layers are counteracted by resultant bending moments M_{mj}, M_{pj} induced in piezomagnetic and piezoelectric. That is,

$$ F_{m1}h_m + F_{p1}(h_m + h_p) = M_{m1} + M_{p1}, \tag{1.39}$$

where $M_{i1} = \int_{-^{i}t/2}^{^{i}t/2} z_i\, {}^{i}T_1 dz_i$ and

$$ M_{i2} = \int_{-^{i}t/2}^{^{i}t/2} z_i\, {}^{i}T_2 dz_i \cdot $$

To obtain the expression for ME voltage coefficient, we use the open-circuit condition on the boundary:

$$ D_3 = 0. \tag{1.40}$$

Since electric induction is divergence free and has only one component D_3, it is evident that D_3 is equal to zero for any z. In this case, Eqs. 1.38 and 1.39 result in the expression for ME voltage coefficient

$$ \alpha_{E31} = \frac{E_3}{H_1} = -\int_{-^{p}t/2}^{^{p}t/2} \frac{{}^{p}d_{31}\, {}^{p}T_1}{tH_1\, {}^{p}\varepsilon_{33}} dz; \tag{1.41}$$

where E_3 and H_1 are the average electric field induced across the piezoelectric layer and applied ac magnetic field.

As an example, numerical estimations of ME voltage coefficient are considered for a bilayer of NZFO-PZT. First we consider grading only in $Ni_{1-x}Zn_xFe_2O_4$ (NZFO) in which the piezomagnetic coefficient linearly varies with z:

$$^{m}q_{11} = {}^{m}q_{110}\,(1+2kz_1/{}^{m}t), \qquad -{}^{m}t/2 \le z_1 \le {}^{m}t/2 \tag{1.42}$$

$$^{m}q_{12} = {}^{m}q_{120}\,(1+2kz_1/{}^{m}t),$$

where $^{m}q_0$ is the average value of this parameter and factor k specifies the grading strength and direction: $k > 0$ and $k < 0$ correspond to "positive" and "negative" magnetostriction grading, respectively. According to data on magnetostriction versus H of NZFO[21], $^{m}q_{11}$ varies from −680 pm/A for pure nickel ferrite to −1156 pm/A for NZFO with $x = 0.4$, with an average of −918 pm/V and a similar variation for $^{m}q_{12}$. In Eq. 1.42, we assume $^{m}q_{110} = -918$ pm/A and $^{m}q_{120} = 169$ pm/A and $|k| = 0.26$.

Similarly, the composition of PZT can be tailored to obtain linear grading of the piezoelectric coefficient and permittivity:

$$^{p}d_{31} = {}^{p}d_{310}\,(1 + 2kz_2/{}^{p}t), \qquad -{}^{p}t/2 \le z_2 \le {}^{p}t/2 \tag{1.43}$$

$$^{p}\varepsilon_{33} = {}^{p}\varepsilon_{330}\,(1 + 2kz_2/{}^{p}t),$$

with $^{p}d_{310} = -175$ pm/V, $^{p}\varepsilon_{330}/\varepsilon_0 = 1750$, and $|k| = 0.35$.

Next we apply the theory to estimate the ME coefficients for (a) graded NZFO and homogeneous PZT, (b) homogeneous NZFO and graded PZT, and (c) grading of both NZFO and PZT.

The ME coefficient versus PZT volume fraction V is shown in Fig. 1.13 in the case of homogeneous PZT and homogeneous, and positively and negatively graded NZFO. Results of $\alpha_{E,31}$ versus V reveals a double maximum for all cases and is due to the fact that the strain produced by the ferrite consisting of two components: longitudinal and flexural. In the absence of a flexural strain, the maximum ME coefficient occurs for $V = 0.6$. Since the flexural strain is of opposite sign relative to longitudinal strain and reaches its maximum value for $V = 0.6$, the two types of strains combine to produce suppression of $\alpha_{E,31}$ at $V = 0.6$ and a double maximum in the ME coefficient as in Fig. 1.13. Negative grading of NZFO leads to a maximum in $\alpha_{E,31}$ for low V whereas positive grading gives rise to a maximum for high V. The ME coefficient shows a 50% increase in $\alpha_{E,31}$ for negative grading of the piezomagnetic coefficient compared to the homogeneous case. Such dependence arises due to a decrease in the rotational moment of the ferrite layer for negative grading of q and an increase in the radius of curvature and longitudinal strain of PZT. It should be noted that the ME voltage according to Eq. 1.41 is determined by average longitudinal strains of the PZT layer.

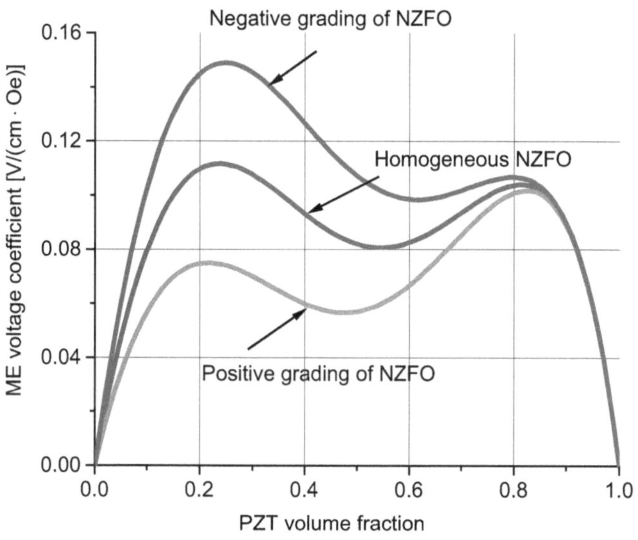

Figure 1.13 Low-frequency ME voltage coefficient as a function of PZT volume fraction for a free-standing PZT-NZFO bilayer. The results are for homogeneous PZT and homogeneous, and positively and negatively graded NZFO.

The PZT volume fraction dependence of $\alpha_{E,31}$ is shown in Fig. 1.14 for the case of homogeneous NZFO and homogeneous or graded PZT. Simultaneous variations in ${}^{p}d_{31}$ and ${}^{p}\varepsilon_{33}$ leads to a constant value for the ratio ${}^{p}d_{31}$ / ${}^{p}\varepsilon_{33}$ and the grading therefore cannot influence the ME voltage coefficient. But grading of the piezoelectric coefficient results in an additional variation in the volume average of the stress ${}^{p}T_1$. Thus z-dependence of ${}^{p}d_{31}$ induces an additional flexural moment which gives rise to an increase in ME coefficient for positive grading and a decrease for negative grading, as shown in Fig. 1.14.

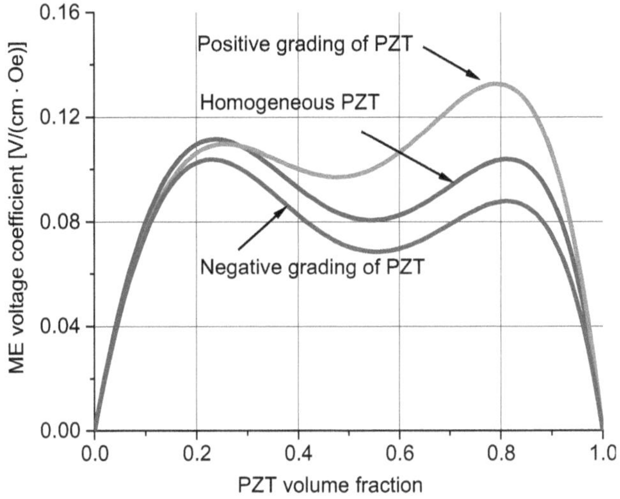

Figure 1.14 Similar results as in Fig. 2.13, but for bilayers of homogeneous NZFO and homogeneous, and positively and negatively graded PZT.

1.1.6 Magnetoelectric Effect at Zero Bias Field

The subject of the current section is ME effects that can be observed at zero bias field in a bilayer of piezoelectric and stepped graded magnetostrictive components.

We consider a trilayer of $Pb_{1-x}Zr_xTiO_3$ (PZT), Ni, and nickel ferrite (NFO) as in Fig. 1.15. Although magnetic metals or alloys of homogeneous compositions are not piezomagnetic, one can achieve a pseudo piezomagnetic effect ($q = d\lambda/dH$, where λ is the magnetostriction) by subjecting the sample to a bias magnetic field H_0 and ac field H_1. Magnetostriction of magnetic components and piezoelectric coupling of PZT are known to result in a magnetoelectric coupling. Theoretical modeling rests on the assumption that the sample undergoes the superposition of axial and bending strain. These strains can be calculated by using the elastostatic equations for axial and flexural deformations and then they can be expressed in terms of stress components with the aid of elasticity law. To obtain the expression for the ME voltage coefficient, we substitute the found stress components with the open circuit condition. The ME voltage coefficients have been calculated for the transverse field orientation to provide minimum demagnetizing fields and maximum ME output.

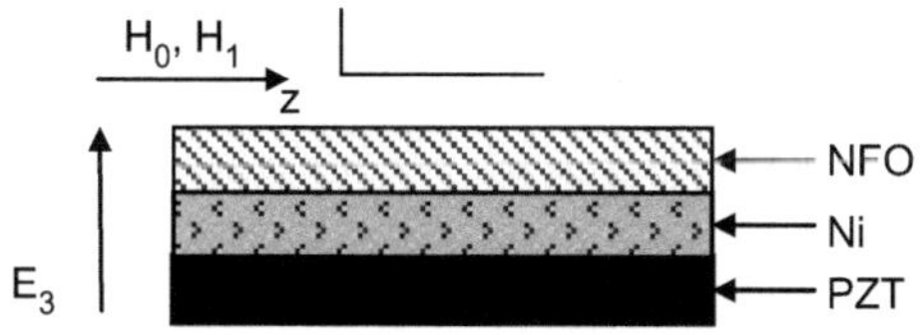

Figure 1.15 A trilayer of PZT, Ni, and NFO. The bias magnetic field H_0 and the ac magnetic field H_1 are parallel to x-axis. The piezoelectric layer is poled along z and the ac electric field E_3 is measured across the layer.

The analysis described here is based on the following equations for the strain, electric displacement, and magnetic induction of piezoelectric and magnetostrictive phases and the substrate:

$$\begin{aligned}
&{}^{\mathrm{p}}S_i = {}^{\mathrm{p}}s_{ij}\,{}^{\mathrm{p}}T_j + {}^{\mathrm{p}}d_{ki}\,{}^{\mathrm{p}}E_k;\\
&{}^{\mathrm{p}}D_k = {}^{\mathrm{p}}d_{ki}\,{}^{\mathrm{p}}T_i + {}^{\mathrm{p}}\varepsilon_{kn}\,{}^{\mathrm{p}}E_{\mathrm{n}};\\
&{}^{\mathrm{m1}}S_i = {}^{\mathrm{m1}}s_{ij}\,{}^{\mathrm{m1}}T_j + {}^{\mathrm{m1}}q_{ki}\,{}^{\mathrm{m}}H_k;\\
&{}^{\mathrm{m1}}B_k = {}^{\mathrm{m1}}q_{ki}\,{}^{\mathrm{m1}}T_i + {}^{\mathrm{m1}}\mu_{kn}\,{}^{\mathrm{m}}H_{\mathrm{n}};\\
&{}^{\mathrm{m2}}S_i = {}^{\mathrm{m2}}s_{ij}\,{}^{\mathrm{m2}}T_j + {}^{\mathrm{m2}}q_{ki}\,{}^{\mathrm{m}}H_k;\\
&{}^{\mathrm{m2}}B_k = {}^{\mathrm{m2}}q_{ki}\,{}^{\mathrm{m2}}T_i + {}^{\mathrm{m2}}\mu_{kn}\,{}^{\mathrm{m}}H_{\mathrm{n}},
\end{aligned} \tag{1.44}$$

where S_{i} and T_{j} are strain and stress tensor components, E_{k} and D_{k} are the vector components of electric field and electric displacement, H_{k} and B_{k} are the vector components of magnetic field and magnetic induction, s_{ij}, q_{ki}, and d_{ki} are compliance, piezomagnetic, and piezoelectric coefficients, $\varepsilon_{\mathrm{kn}}$ is the permittivity matrix, and μ_{kn} is the permeability matrix. The superscripts p, m1, and m2 correspond to piezoelectric and two piezomagnetic layers, respectively. We assume the symmetry of the piezoelectric to be ∞m and the magnetic to be cubic.

For finding the low-frequency ME voltage coefficient, we solve magnetostatic and elastostatic equations in piezomagnetic layers, and elastostatic and electrostatic equations in PZT, taking into account boundary conditions. In this case, the theoretical modeling is similar to recent studies on ME coupling in a bilayer of compositionally graded components on a substrate [38]. To adapt that model to the structure in Fig. 1.15, we assume the longitudinal axial strains of each layer to be linear functions of the vertical coordinate z_i to take into account the cylindrical bending of trilayer: ${}^{p}S_1 = {}^{p}S_{10} + z_p/R_1$; ${}^{m1}S_1 = {}^{m1}S_{10} + z_{m1}/R_1$; ${}^{m2}S_1 = {}^{m2}S_{10} + z_{m2}/R_1$; where ${}^{i}S_{10}$ are the centroidal strains along x-axis at $z_i = 0$, R_1 is the radius of curvature, and z_i is measured relative to centroidal plane of i-layer. It can be shown that centroidal strains obey the following conditions: ${}^{m2}S_{10} - {}^{m1}S_{10} = h_2/R_1$, ${}^{m1}S_{10} - {}^{p}S_{10} = h_1/R_1$, where $h_1 = ({}^{m1}t + {}^{p}t)/2$ and $h_2 = ({}^{m1}t + {}^{m2}t)/2$ are distances between the centroidal planes of piezoelectric and first magnetic layers and between that of two magnetic layers, ${}^{p}t$, ${}^{m1}t$, and ${}^{m1}t$ are thicknesses of three layers.

The axial forces in the three layers must add up to zero and the rotating moments of axial forces in the three layers must be counteracted by resultant bending moments ${}^{p}M_1$, ${}^{m1}M_1$, and ${}^{m2}M_1$, induced in piezoelectric and two piezomagnetic layers to preserve force and moment equilibrium,

$$ {}^{m1}F_1 + {}^{m2}F_1 + {}^{p}F_1 = 0, \tag{1.45} $$

$$ {}^{m1}F_1 h_1 + {}^{m2}F_1(h_1 + h_2) = {}^{p}M_1 + {}^{m1}M_1 + {}^{m2}M_1, \tag{1.46} $$

where

$$ {}^{i}F_1 = \int_{-{}^{i}t/2}^{{}^{i}t/2} {}^{i}T_1 dz_i, \quad {}^{i}M_1 = \int_{-{}^{i}t/2}^{{}^{i}t/2} z_i {}^{i}T_1. $$

To obtain the expression for ME voltage coefficient, we use the open-circuit condition

$$ \alpha_{E31} = \frac{E_3}{H_1} = -\int_{-{}^{p}t/2}^{{}^{p}t/2} \frac{{}^{p}d_{31}{}^{p}T_1}{{}^{p}tH_1{}^{p}\varepsilon_{33}} dz; \tag{1.47} $$

where E_3 and H_1 are the average electric field induced across the piezoelectric layer and applied ac magnetic field.

Solving Eqs. 1.45 and 1.46 for R_1 and ${}^{p}S_{10}$ enables one to find the axial stress ${}^{p}T_1$ from Eq. 1.44. Substituting obtained value of stress into Eq. 1.47 yields:

$$ \alpha_{E31} = \frac{{}^{p}d_{31}{}^{p}Y_1(r_1{}^{m1}t^{m1}q_{11} + r_2{}^{m2}t^{m2}q_{11})}{(c_1{}^{p}K_{31}^2 + c_2){}^{p}\varepsilon_{33}}; \tag{1.48} $$

where

$$c_1 = {}^{m1}Y^{m1}t(h_2b_3 - b_1) + {}^{m2}Y^{m2}t(b_3h - b_1);$$

and $c_2 = b_1Y^*t$;

$$b_1 = \frac{-{}^{m2}th({}^{m1}Y^{m1}th_1 + {}^{p}Y^{p}th) + {}^{m1}Y^{m1}th_2(h_1{}^{m2}t^{m2}Y - {}^{p}Y^{p}th_2)}{Y^*t}$$

$$-{}^{m1}D - {}^{m2}D - \frac{{}^{p}D}{(1 - {}^{p}K^2)}$$

$$r_1 = {}^{m1}Y\frac{{}^{m1}Y^{m1}th_2({}^{m2}Y^{m2}th_1 - {}^{p}Y^{p}th_2) + {}^{m2}Y^{m2}t(h_1{}^{m2}t^{m2}Y - {}^{p}Y^{p}th_2)h - b_1Y^*t}{Y^*t}$$

$$r_1 = -{}^{m2}Y\frac{{}^{m2}Y^{m2}th({}^{m1}Y^{m1}th_1 - {}^{p}Y^{p}th) + [{}^{m1}Y^{m1}t(h_1h_2{}^{m1}t^{m1}Y + {}^{p}Y^{p}th_2h) + b_1Y^*}{Y^*t}$$

$$Y^* = \frac{{}^{m1}Y^{m1}t + {}^{m2}Y^{m2}t + {}^{p}Y^{p}t}{t}$$

$$t = {}^{m1}t + {}^{m2}t + {}^{p}t; \quad h = h_1 + h_2;$$

${}^{m1}D$, ${}^{m2}D$, and ${}^{p}D$ are the cylindrical stiffnesses and ${}^{m1}Y$, ${}^{m2}Y$, and ${}^{p}Y$ are Young modulus of layers.

One can see from Eq. 1.48 that ME voltage coefficient is substantially determined by piezoelectric coefficient of piezoelectric phase and combination of piezomagnetic coefficients of magnetic phases. The weight coefficient for ${}^{m1}q_{11}$ and ${}^{m2}q_{11}$ depend on sample geometry and elastic properties of composite components. It should be noted that the weight coefficient optimization and using the most suitable magnetic phases is a new path for increasing the ME coupling. Using the stepped magnetic layer leads to a dependence of total torque moment on magnetic layers' sequence order. Beyond that point, the torque moment is a function of bias magnetic field due to piezomagnetic coupling. In addition, coercive force of a stepped magnetic bilayer is known to increase in comparison to homogeneous layer. As a result, the bias field dependence of ME voltage coefficient becomes strongly hysteretic and depends on magnetic layer composition.

Estimate of the coercive force for stepped magnetic layer is based on thermodynamic calculation of nucleation field for the decreasing branch of the loop. Free energy density includes the Zeeman F_H, crystalline anisotropy F_A, magnetoelastic coupling F_{ME}, magnetostatic interaction F_M, and elastic F_E energy of magnetic sublayers:

$$F = F_H + F_A + F_{ME} + F_E + F_M, \tag{1.49}$$

where

$$F_H = -\mu_0(v_1{}^1M_s\cos\beta_1 + v_2{}^2M_s\cos\beta_2)H_0,$$
$$F_A = v_1{}^1K_1(\sin^2\beta_1 - \sin^4\beta_1) + v_2{}^2K_1(\sin^2\beta_2 - \sin^4\beta_2),$$
$$F_{ME} = -v_1{}^1\lambda_{11}{}^1M_s{}^2(\cos^2\beta_1{}^{m1}T_{10} + \sin^2\beta_1{}^{m1}T_{20}) - v_2{}^2\lambda_{11}{}^2M_s{}^2(\cos^2\beta_2{}^{m2}T_{10} + \sin^2\beta_2{}^{m2}T_{20}),$$
$$F_E = v_1({}^{m1}T_{10}\,{}^{m1}S_{10} + {}^{m1}T_{20}\,{}^{m1}S_{20}) + v_2({}^{m2}T_{10}\,{}^{m2}S_{10} + {}^{m2}T_{20}\,{}^{m2}S_{20}),$$
$$F_M = -\mu_0 v_1 v_2{}^1M_s{}^2M_s(\cos\beta_1\cos\beta_2 + \sin\beta_1\sin\beta_2),$$

μ_0 is magnetic constant, β_1 and β_2 are angles between ${}^1M_s$ and x-axis and ${}^2M_s$ and x-axis, v_1 and v_2 are the volume fractions of magnetic layers, ${}^1K_1$ and ${}^2K_1$ are the magnetic crystalline anisotropy constants of layers, ${}^1\lambda_{11}$ and ${}^2\lambda_{11}$ are the magnetostriction constants of layers, . Magnetic field is supposed to be applied along x-axis. ${}^{m1}T_{i0}$, ${}^{m2}T_{i0}$, ${}^{m1}S_{i0}$, and ${}^{m2}S_{i0}$ are the stress and strain components induced by applied dc magnetic field H_0.

Eq. 1.49 implies preliminary finding of the equilibrium strain components from boundary mechanical conditions and substituting them into Eq. 1.49. However, we neglect magnetoelastic energy in what follows since it is small compared to other terms. Equilibrium magnetizations direction can be found by minimizing the free energy. Thus we have

$$\frac{\partial F}{\partial \beta_1} = 0; \qquad \frac{\partial^2 F}{\partial \beta_1^2}$$

$$\frac{\partial F}{\partial \beta_2} = 0; \qquad \frac{\partial^2 F}{\partial \beta_2^2} \tag{1.50}$$

The locus of critical points is determined by

$$\frac{\partial F}{\partial \beta_1} = 0; \qquad \frac{\partial^2 F}{\partial \beta_1^2} \text{ and}$$

$$\frac{\partial F}{\partial \beta_2} = 0; \qquad \frac{\partial^2 F}{\partial \beta_2^2}$$

Estimate of the coercive force for stepped magnetic layer is based on calculation of nucleation field H_n for decreasing branch of the loop. So, we get

$$^1H_n = -\,^1H_a - v_2\,^2M_s;$$

$$^2H_n = -\,^2H_a - v_1\,^1M_s; \qquad (1.51)$$

where $^1H_a = {}^1K_1/(\mu_0\,^1M_s)$ and $^2H_a = {}^2K_1/(\mu_0\,^2M_s)$.

Next, we apply the model to the composites considered here. The following material parameters were used:

$$\text{PZT}: \varepsilon_{33}/\varepsilon_0 = 3250;\ d_{31} = -270\ \text{pm/V};\ s_{33} = 17.3\times10^{-12}\ \text{m}^2/\text{N},$$

$$\text{NZFO}: s_{33} = 4.3\times10^{-12}\ \text{m}^2/\text{N},$$

$$\text{Ni}: s_{33} = 4.9\times10^{-12}\ \text{m}^2/\text{N}.$$

Notwithstanding the fact that Eqs. 1.51 are obtained in single domain approximation, they enable one to estimate the external magnetic fields on the assumption that magnetization curves of magnetic layers take the form of approximately rectangular hysteresis loop. Piezomagnetic coupling coefficients of magnetic layers are calculated from data on magnetostriction. Figure 1.16 shows calculated ME voltage coefficient versus H for PZT-Ni-NZFO for ferrite-to-Ni thickness ratio $R = 1$.

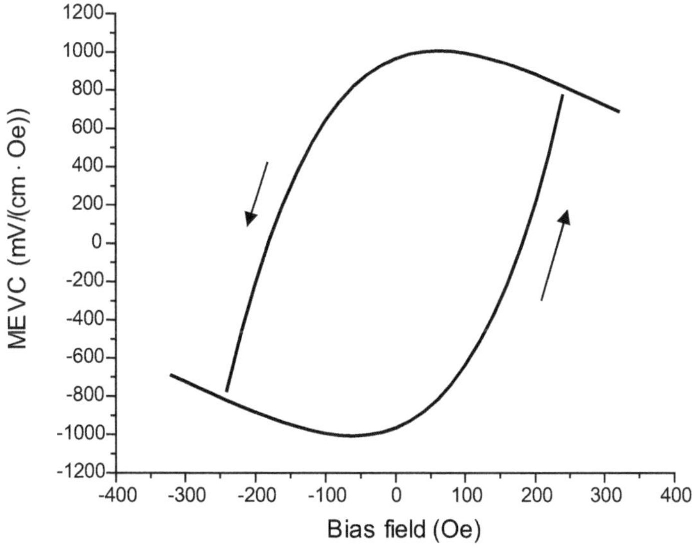

Figure 1.16 Theoretical ME voltage coefficient *vs.* H for PZT-Ni-NZFO for $R = 1$.

For theoretical values of remanent ME voltage coefficient, we first calculated the built-in field at the sample interface due to grading. The magnitude and direction of this field depends on the thickness of the Ni and NZFO layers. Next, the coercive field H_c was determined. Following this, ME voltage coefficient versus. H was estimated. The bias field dependence of MEVC becomes strongly hysteretic due to the grading-induced built-in field at the interface and its dependence on thickness of magnetic layers. The observed hysteresis and remnant ME voltage coefficient are attributed to a grading-related magnetization at the Ni-NZFO interface that results in the ME coupling for zero external magnetic bias field.

1.1.7 Magnetoelectric Effect in Dimensionally Graded Laminate Composites

Dimensionally gradient piezoelectric plate with thickness of 1 mm was fabricated by mechanical polishing and dicing technique, as shown in Fig. 1.17(a) [66]. Piezoelectric plates with composition $Pb(Zn_{1/3}Nb_{2/3})_{0.2}(Zr_{0.5}Ti_{0.5})_{0.8}O_3$ [PZNT] were synthesized by conventional mixed oxide method. Piezoelectric constant of poled PZNT plates was found to be 500 pC/N and the dielectric constant was 2219 at 1 kHz. The piezoelectric voltage constant (g_{33}) was of the order of 23.41×10^{-3} Vm/N. On this PZNT plate, 25 μm-thick Metglas (2605SA1, Metglas Inc, USA) sheets of desired dimensions were attached using epoxy (West System, USA) with a curing temperature of 80°C. Impedance spectrum of the composites was measured by LCR meter (HP 4194A). For ME voltage coefficient measurement in longitudinal–transversal (L–T) mode configuration, an electromagnet was used to apply the DC magnetic field and the samples were placed in the center of the Helmholtz coil under an AC magnetic field (H_{ac}). The voltage induced on the laminate was monitored by using a lock-in amplifier. The magnetostriction was evaluated by using the strain gauge and Wheatstone bridge.

Impedance spectrum of asymmetric piezoelectric plate was measured to identify the EMR range. The first, second, and third resonances were found to be 42, 44, and 94.3 kHz, respectively. In comparison, measured impedance spectrums for the asymmetric piezoelectric plate exhibited resonance peaks at 42, 44, and 96 kHz, as shown in Fig. 1.17(b). Thus, the results between simulation and measurement were in good agreement. The first mode at 42 kHz was associated with biaxial bending of the large rectangular section and the intermediate rectangular section, while the second mode at 44 kHz was related to the biaxial bending of the large rectangular section and the small rectangular section. The third mode at 99 kHz came from the lateral displacement of the large rectangular section, as shown in Fig. 1.17(c).

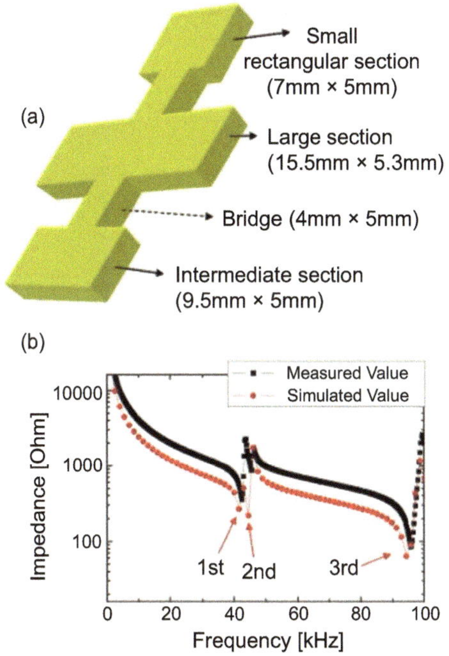

Figure 1.17 (a) Schematic diagram of asymmetric piezoelectric plate, (b) Impedance spectrums from simulation and measurementat 42, 44, and 94.3 kHz.

Figure 1.18(a) shows the fabricated ME laminate composite based on Fig. 1.17(a). On top of the PZNT layer, four layers of Metglas with an area of 13 × 7 mm^2 were attached at the middle rectangular section (Section A), thirty layers of Metglas with area an of 20 × 7 mm^2 were attached at the larger rectangular section

(Section B), and five layers of Metglas with an area of 7 × 7 mm^2 were attached at the intermediate rectangular section (Section C), as shown in Fig. 1.18(a). There are two variables which could be adjusted to achieve a wideband ME response. First, if the rectangular area of two sections in piezoelectric plate is different then the one with smaller number of Metglas layers will show higher ME coefficient. Second, if the numbers of Metglas layers are same, the rectangular section with the smaller area will show smaller ME coefficient. Thus, by adjusting the ratio of Metglas layers on various rectangular sections, a composite ME response with flat behavior can be obtained.

Figure 1.18(b) shows the magnetostriction (S_{ij}) and piezomagnetic (q_{ij}) coefficient for varying dimensions and stack configurations of Metglas. In this figure, S_{11} corresponds to longitudinal in-plane magnetostriction parallel to H_{DC} and q_{11} is the longitudinal in-plain piezomagnetic coefficient corresponding to the differential of S_{11}. The maximum in-plane magnetostriction was found to be 28 ppm regardless of Metglas stack configurations and dimensions; however, the strain behavior was strongly dependent on the stack configuration and dimensions. Four layers of Metglas with an area of 13 × 7 mm^2, thirty layers of Metglas with an area of 20 × 7 mm^2, and five layers of Metglas with an area of 7 × 7 mm^2 showed piezomagnetic coefficient corresponding to 0.38, 0.21, and 0.16 ppm/Oe at 70, 150, and 204 Oe of H_{DC}, respectively. These piezomagnetic behaviors will result in strong elastic coupling with the piezoelectric sections.

Figure 1.18(c) shows the measured ME response from the composite structure as a function of magnetic DC bias under the condition of H_{AC} = 1 Oe at f = 1 kHz. The peak at 70 Oe was associated with Section A. The second peak of the ME coefficient at 150 Oe was associated with Section C. The third peak of the ME coefficient at 209 Oe was associated with Section B. The value of DC bias at the peak in piezomagnetic coefficient corresponds to that for the peak in ME coefficient. Further, it was found that not only piezomagnetic coefficient but also piezoelectric dimensions affected elastic coupling between the Metglas and piezoelectric sections compared to Fig. 1.18(b and c). Furthermore, if only Section A and Section B were combined, there will be a valley in the intermediate range. By inserting Section C, the formation of flat ME band was facilitated. The overall ME response as a function of magnetic DC bias was a composite of three individual responses.

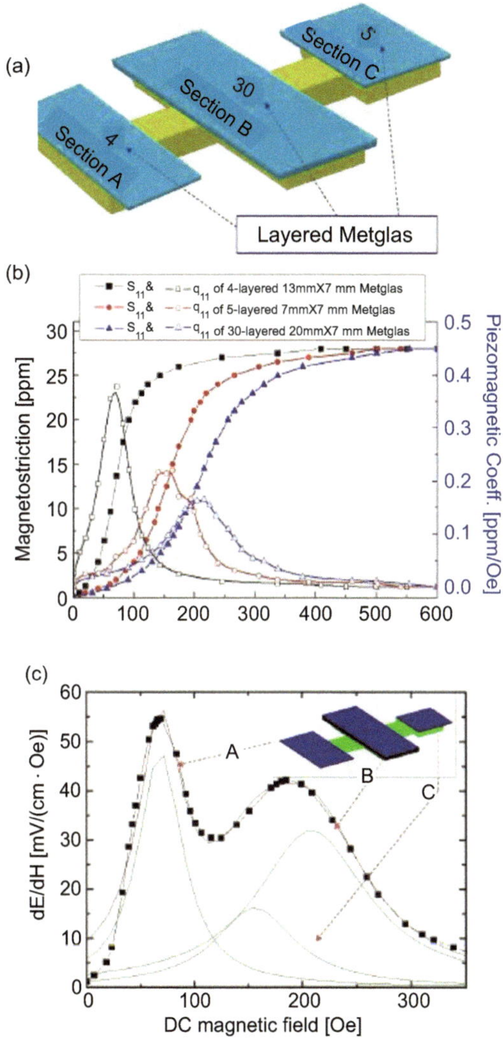

Figure 1.18 (a) Schematic diagram of laminated ME composite, (b) Magnetostriction (S_{11}) and piezomagnetic coefficient (q_{11}) for varying dimensions and stack configurations of Metglas, (c) ME response as a function of DC magnetic field under the constant condition of H_{ac} = 1 Oe at f = 1 kHz.

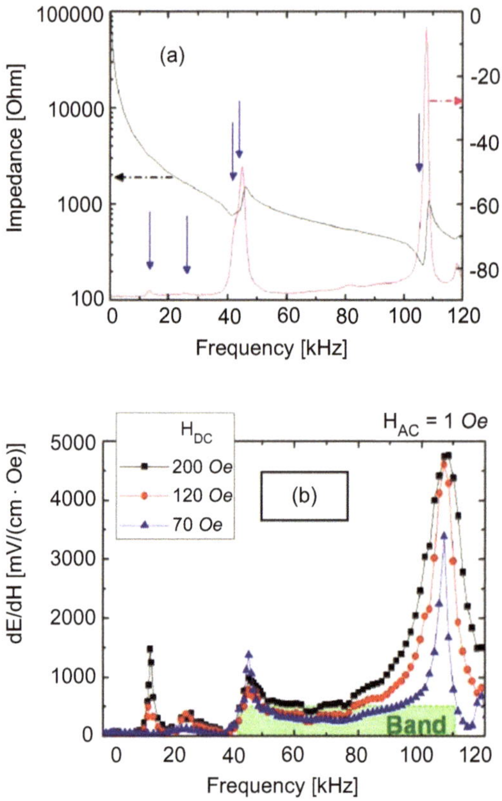

Figure 1.19 (a) Impedance and phase angle spectrums of the ME composite, (b) ME response as a function of frequency.

The impedance and phase angle spectrums for this laminate are shown in Fig. 1.19(a). After Metglas was attached on the PZNT plate, multiple resonances occurred at 14, 28, 41, 44, and 107 kHz as marked with arrows in Fig. 1.19(a). Resonant modes at 14 and 28 kHz were found to be related to the bending vibrations of all the sections. The middle section dominated the vibration mode at 14 kHz while the small rectangular section dominated the vibration mode at 28 kHz. The resonance at 41 kHz was found to be a combination of the bending motions of large rectangular section (Section B) and middle rectangular section (Section A). The resonance at 44 kHz was related to combined bending response from the bridge and small rectangular section (Section C). The resonant modes of ME composite at 41 and 44 kHz were correlated to the 1st and 2nd resonance modes of PZNT layer as determined in Fig. 1.17(b). The resonant mode at 107 kHz had a similar vibration mode as that at 44 kHz.

The ME output voltage from composite was measured as a function of frequency under constant H_{AC} = 1 Oe but varying H_{DC} = 70, 120, and 200 Oe, as shown in Fig. 1.17(b). Interestingly, ME response as a function of frequency had similar behavior as that of phase angle shift shown in Fig. 1.19(a and b). The ME peaks are shown at 14, 28, 45, and 107 kHz corresponding to the resonance frequencies observed in the impedance measurement. The peak ME response at 45 kHz was associated with resonances at 41 and 44 kHz. The maximum ME coefficient was found to be 4740 mV/(cm · Oe) at f = 107 kHz under H_{DC} = 200 Oe. Interestingly, the bands were successfully formed in the range of 41–110 kHz. The composite showed high ME coefficient of 260 mV/(cm · Oe) after the resonance peak of 41 kHz, regardless of the applied DC magnetic field. These widely extended bands were derived from a combination of the resonances at 41, 44, and 107 kHz.

The maximum ME coefficients of 1400 and 4740 mV/(cm · Oe) were found at 45 and 107 kHz, respectively. The band at 107 kHz exhibited ME coefficient higher than 3000 mV/(cm · Oe) from 52 to 242 Oe of H_{DC} while the band at 45 kHz exhibited ME coefficient higher than 780 mV/(cm · Oe) from 40 to 230 Oe of H_{DC}. The bands were obtained regardless of applied DC and AC magnetic fields. The wideband was formed in both conditions of the frequency range of 41–110 kHz and DC magnetic ranges of 40–230 Oe and 52–242 Oe at f = 45 and 107 kHz, as shown in Figs. 1.19(b).

Interestingly, there are two noticeable states in the ME frequency response in Fig. 1.19(b). Under the constant DC magnetic field condition of 70 Oe, the composite showed flat ME responses in both frequency ranges of 1–11 kHz (State I) and 41–110 kHz (State II) as shown in Fig. 1.18. Figure 1.20 shows the schematic diagram of resulting ME response as a function of frequency and the applied DC magnetic field conditions. State I shows a band lower than 129 mV/(cm · Oe) before f = 41 kHz while State II shows a band higher than 260 mV/(cm · Oe) after f = 41 kHz under the constant condition of H_{DC} = 70 Oe. The ME coefficient of State II was 2× higher than that of State I. Consequently, ΔME (=$ME_{State\ II}$ − $ME_{State\ I}$) was 131 mV/(cm · Oe) under the constant H_{DC} = 70 Oe. These states were found to be adjustable with changing H_{DC} conditions. The maximum value of the $ME_{State\ I}$ can be reduced by decreasing the DC magnetic field and ranges between 0 and 129 mV/(cm · Oe). On the other hand, the minimum value of the $ME_{State\ II}$ can be elevated by increasing DC magnetic field condition in the range of 260 to 406 mV/(cm · Oe). Thus by tuning the frequency and H_{DC}, ΔME in the range of 131 and 406 mV/(cm · Oe) can be achieved which clearly shows the tunability of this device. These signals are strong enough to allow two distinguishable states. The clear sensing margin was seen between State I and State II: the ME values in State II were at least two times higher than that in State I. This interesting and promising ME behavior can be exploited in several applications. The wideband behavior can be a candidate for magnetic field-controlled switches as well as ME harvesters. In the case of the magnetic field-controlled switches, the resonant frequency becomes the cut-off condition and State I and State II can be considered as "off" and "on," respectively.

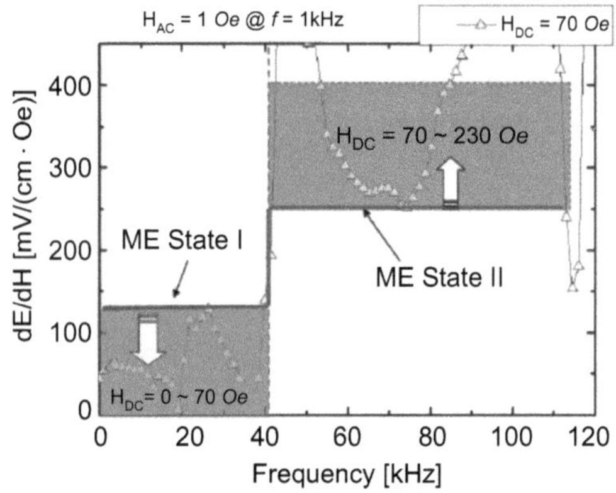

Figure 1.20 Schematic of switchable ME response dependency on the frequency and DC magnetic field conditions.

We attempt to model the response of the laminate structure by calculating the in-plane strain and stress components for all sections and at the end combining them together to find the overall solution. The total vibration spectrum of the laminate composite consists of several contributions. The first mode is supposed to be associated with simultaneous bending vibrations of the small rectangular section with its bridge and the middle rectangular section. The equation of bending motion of *i*th-area (1 and 2 areas correspond to the bridge and the section) has the form:

$$\nabla^2\nabla^2 w_{\mathrm{i}} + \frac{\rho_{\mathrm{i}} t_{\mathrm{i}}}{D_{\mathrm{i}}}\frac{\partial^2 w_{\mathrm{i}}}{\partial t^2} > 0, \tag{1.52}$$

where $\nabla^2\nabla^2$ is a biharmonic operator, w_i is the displacement in z direction, t_i is thickness, ρ_i is average density of i-area, and D_i is cylindrical stiffness. The strain's components can be expressed in terms of displacement as $S_{1i} = -z\partial^2 w_i/\partial x^2$ and $S_{2i} = -z\partial^2 w_i/\partial y^2$. The stress components can be expressed in terms of strains as:

$$({}^{\mathrm{p}}S_k)_i = {}^{\mathrm{p}}s_{kj}\,({}^{\mathrm{p}}T_j)_i + {}^{\mathrm{p}}d_{31}\,{}^{\mathrm{p}}E_{3i};$$

$$({}^{\mathrm{m}}S_k)_i = {}^{\mathrm{m}}s_{kj}\,({}^{\mathrm{m}}T_j)_i + ({}^{\mathrm{m}}g_{k1})_i\,{}^{\mathrm{m}}B_{1i}; \tag{1.53}$$

where S_{1i} and T_{1i} are strain and stress components for i-area, E_{3i} is the electric field, H_{1i} is the magnetic field, s_{kj} is compliance at constant electric field for piezoelectric component and at constant magnetic induction for magnetic component, g_{k1} and d_{31} are the piezomagnetic and piezoelectric coefficients correspondingly. The superscripts p and m correspond to piezoelectric and piezomagnetic layers. Solving Eq. 1.52 for displacement of each section by using the boundary conditions given below provides the dynamic solution: $w_1 = 0$ and $\partial w_1/\partial x = 0$ at $x = 0$, $w_1 = w_2$, $\partial w_1/\partial x = \partial w_2/\partial x$, $(M_1)_1 = (M_1)_2$, and $(V_1)_1 = (V_1)_2$ on the boundary of 1 and 2 areas; $(M_1)_2 = 0$ and $(V_1)_2 = 0$ at $x = L$ (L is the total length of section and bridge); $(M_2)_i = 0$ and $(V_2)_i = 0$ at $y = \pm b_i/2$ (b_i is the width of i-area), where $(M_j)_i$ is the moment of rotation and $(V_j)_i$ is the transverse force with respect to j-axis.

The computed displacements were used to determine the strain components and then the stress components from Eqs. 1.23 and 1.25. Substituting the stress components into open circuit condition enables the calculation of the ME voltage coefficient by taking into account the condition that average electric field induced across the piezoelectric layer is estimated an integral of internal electric field taken over the piezoelectric thickness. These calculations were carried out numerically. The resonance frequencies for the small rectangular section and the middle rectangular section were found to be approximately equal. The second harmonic of this mode can be seen at 97 kHz. It should be noted that theoretical estimate for the next resonance frequency (at 44 kHz) as the bending mode of large rectangular section was made using similar procedure. The bridge was not included into the vibrating system since its displacement (in z direction) was negligible. Finally, the mode at 120 kHz was calculated under the assumption that it came from the axial mode of the large rectangular section. The estimate for this case was found using the computing method known heretofore.

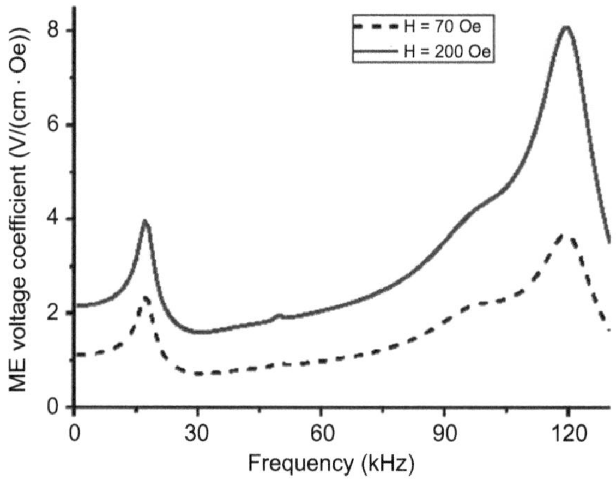

Figure 1.21 Frequency dependence of ME voltage coefficient for Metglas–PZNT composite for bias field of 200 and 70 Oe.

Fig. 1.20 shows that the first mode at 17 kHz can be associated with simultaneous bending vibrations of the small rectangular section and the middle rectangular section, while the second mode at 44 kHz can be related to the bending of the large rectangular section. The third peak corresponds to second harmonic of first mode and occurs at 97 kHz. The fourth mode at 120 kHz came from the axial mode of the large rectangular section. Fig. 1.18 reveals a frequency band of 90 to 130 kHz which is dependent on the dc magnetic field. This band arises from a combination of the resonances. One can see two states in the ME frequency response at 70 Oe. The ME response in the frequency ranges of 25–90 kHz (State I) and 90–130 kHz (State II) is shown to be a function of frequency and the applied DC magnetic field. State I corresponds to a band lower than 1.8 V/(cm·Oe) while State II shows a band higher than 1.8 V/(cm·Oe) at bias field of 70 Oe. These states are adjustable with the changing H_{DC}. Some disagreement between estimates and data can be accounted for by distinctions between the geometry of actual composite structure and that of simple model used for obtaining the estimates.

Thus, we demonstrate a ME resonator exhibiting wideband behavior by fabricating a dimensionally gradient structure and combining with laminate configuration. We were able to obtain a flat ME response in the DC magnetic bias range of 52–242 Oe where the ME coefficient was higher than 3000 mV/(cm · Oe) under resonant condition. The promising wideband behavior as a function of frequency occurred in the range of 41–110 kHz where the ME coefficient was higher than 260 mV/(cm · Oe) independent of applied H_{DC}. Under low H_{DC} condition, two different ME states were clearly obtained.

In this section, a generalized theoretical model for low-frequency ME effects in layered composites was discussed. To describe the composite's physical properties, the exact solution of elastostatic and electrostatic equations were obtained. Expressions for the ME susceptibility and ME voltage coefficient were derived as functions of an interface coupling parameter, constituent phase material parameters, and relative volume fractions of phases. Longitudinal, transverse, and in-plane cases were all considered. For a bilayer that is an asymmetric structure, the influence of flexural deformations of sample on ME output was estimated.

Predictions of the ME effect for various model composite systems were given including CFO–PZT and lanthanum strontium manganite–PZT. It was shown that ME effect in ferrite–PZT systems is maximum for in-plane magnetic and electric fields. The theoretical estimates of ME parameters were compared with experimental data.

The generalized theory allows for modeling of the low-frequency ME effect in bulk composites. To describe these low-frequency composite properties, an effective medium method was used. Calculation of the ME susceptibility, and ME voltage coefficient were performed as functions of volume fractions and component parameters. Composites with connectivity types 3-0 and 0-3 were considered. Larger ME coefficients were found for 3-0 composites with magnetic and/or electric fields applied along the longitudinal direction. For composites of CFO–PZT, values as high as 4 V/(cm · Oe) were predicted for the longitudinal ME voltage coefficient. For the transverse fields orientation, ME effect was found to be in 2–3.5 times smaller than that for longitudinal orientation. Furthermore, clamping was shown to significantly reduce the ME effect.

We presented a theory for the resonance enhancement of ME interactions at frequencies corresponding to EMR. Frequency dependence for ME voltage coefficients are obtained using the simultaneous solution of electrostatic, magnetostatic, and elastodynamic equations. The ME effect at bending mode in a bilayer is shown to be dependent on boundary conditions. A giant ME interaction at the lowest frequency is predicted for a bilayer fixed at one end and is free to vibrate at the other end. The ME voltage coefficients are estimated from known material parameters (piezoelectric coupling, magnetostriction, elastic constants, etc.) of composite components. It is shown that the ME coupling in the EMR region exceeds the low-frequency value by more than an order of magnitude.

1.1.8 Maxwell–Wagner Relaxation in ME Composites

Ferrite–piezoelectric composites can be considered as a thermodynamic system that is capable of responding to an infinitesimal externally applied field. Relaxation can then be considered as the self-adjustment of the thermodynamic system in response to an external force [67] that maintains equilibrium. It reflects an adjustment of internal parameters to the balanced values. Specifically, dielectric relaxation is a time-dependent induced polarization P response to an applied electric field E. The process is characterized by dispersion in the internal parameter $\delta P/\delta E$. Polar dielectrics with low resistivities often have a strongly enhanced permittivity at low frequencies due to charge separation. Such space charge polarization contributions to the permittivity are characterized by low-frequency dispersion [68, 69]. Investigations have shown that the effects of space charge polarization relaxation reveals itself not only in the dielectric permittivity but also in the elastic constants of layered polar dielectrics.

In the case of dielectric relaxation in the quasistatic frequency range, when the applied electric field duration period may exceed times of seconds or more, space charge polarization contributions to the dielectric permittivity have sufficient time to respond to the drive: consequently, the permittivity has a constant value. However, as the frequency is increased, this polarization mechanism cannot respond efficiently to the drive, rather there is a phase angle between drive and response. The tangent of this phase angle is the dielectric loss factor. Thus, the material is characterized by the complex dielectric permittivity that depends on frequency

$$\varepsilon_{33} = \varepsilon - i\,\gamma/\omega \tag{1.54}$$

where $\varepsilon'(\omega)$ is the real part (i.e., the permittivity of storage compliance) and $\varepsilon''(\omega)$ is the imaginary part (i.e., the phase angle or energy loss). Dynamic experiments in which the electric field applied to the material is periodically varied with a circular frequency are standard methods for measuring the dielectric constant. There are many impedance analyzers capable of measuring the complex frequency-dependent dielectric constant over frequencies extending from micro to mega Hertz.

Since dynamic response functions allow for the description of frequency-dependent materials parameters, they are suitable for use in the analysis of the ME susceptibility and voltage coefficient of ferrite–piezoelectric composites, both of which are complex functions of frequency. Investigations of the relaxation phenomenon of ferrite–piezoelectric composites have previously been reported.

The relations that we obtained for the effective parameters of composite in the last section did not consider their frequency dependence. Calculations were done using the ferroelectric polarization, which will provide a fair estimate of the high-frequency polarization. Accumulation of electric charges at the boundaries of composite components will result in additional contributions to the polarization: albeit low-frequency ones. Said accumulation of free charges at the interfaces between component phases will result in dielectric dispersion and losses under low-frequency alternating fields, which is known as Maxwell–Wagner relaxation. The purpose of the following section is an analysis of Maxwell–Wagner relaxation of the effective parameters in ferrite–piezoelectric composites, in particular, the ME susceptibility and voltage coefficient.

1.1.8.1 Layered composites

Now, let us consider Maxwell–Wagner relaxation of the ME parameters for a multilayer composite with a 2-2 type connectivity [68, 70]. Following from Eq. 1.38, the specific equations, in view of finite electrical conductivities of component phases, for the complex permittivity of the components is

$$^{\mathrm{p}}\varepsilon_{33} = {}^{\mathrm{p}}\varepsilon - i\,{}^{\mathrm{p}}\gamma/\omega,$$

$$^{\mathrm{m}}\varepsilon_{33} = {}^{\mathrm{m}}\varepsilon - i\,{}^{\mathrm{m}}\gamma/\omega; \tag{1.55}$$

where $^{\mathrm{p}}\gamma$ and $^{\mathrm{m}}\gamma$ are the conductivities of the piezoelectric and magnetostrictive phases; and ω the circular frequency, where the frequency is less than that of electromechanical resonance. The effective parameters of the composite can be then determined by an average of expressions for the components of strain, electric, and magnetic inductions. Furthermore, assume that the component phase layers are thin and arranged in the plane OX_1X_2; that the piezoelectric component is polarized along the OX_3 axis, along the same axis on which an electric field with a circular frequency ω is applied; and that a magnetic bias and variable magnetic fields are applied along the OX_1 axis.

The general characteristics of the frequency-dependent ME susceptibility will fulfill the Debye formula:

$$\alpha_{13} = \alpha'_{13} - i\alpha''_{13}, \tag{1.56}$$

$$\alpha'_{13} = \alpha_{13\infty} + \Delta\alpha_{13}/(1 + \omega^2\tau_\alpha^{\,2});\ \alpha''_{13} = \Delta\alpha_{13}\omega\tau_\alpha/(1 + \omega^2\tau_\alpha^{\,2});$$

where $\Delta\alpha_{13} = \alpha_{130} - \alpha_{13\infty}$ is the relaxation strength, α_{130} and $\alpha_{13\infty}$ are the static ($\omega\to 0$) and high-frequency ($\omega\to\infty$) ME susceptibilities, and τ_α the relaxation time. The transverse static and high-frequency ME susceptibilities, and in addition the relaxation time, can be found from the solutions of Eqs. 1.53. Then, by supposing that the symmetry of the piezoelectric phase is ∞m, and that the magnetic phase possesses cubic symmetry, we obtain the following expressions:

$$\alpha_{130} = \frac{{}^{\mathrm{m}}\gamma v k(1-v)({}^{\mathrm{m}}q_{12} + {}^{\mathrm{m}}q_{11}){}^{\mathrm{p}}d_{31}}{\left[[{}^{\mathrm{p}}\gamma(1-v) + {}^{\mathrm{m}}\gamma v][({}^{\mathrm{m}}s_{12} + {}^{\mathrm{m}}s_{11})v + ({}^{\mathrm{p}}s_{11} + {}^{\mathrm{p}}s_{12})(1-v)]\right]},$$

$$\alpha_{13\infty} = \frac{{}^{m}\varepsilon v k(1-v)({}^{m}q_{12} + {}^{m}q_{11})}{\begin{bmatrix} [v^{m}\varepsilon + {}^{p}\varepsilon(1-v)][({}^{m}s_{12} + {}^{m}s_{11})v + ({}^{p}s_{11} + {}^{p}s_{12}) \\ (1-v)] - 2^{p}d_{31}{}^{2}(1-v)^{2} \end{bmatrix}}. \quad (1.57)$$

As an example, we shall consider the composite to consist of polarized ferroelectric PZT ceramics and nickel ferro-spinel. In numerical calculations, the following values of the materials parameters for composite phase components can be used:

${}^{p}s_{11} = 15.3 \times 10^{-12}$ m^2/N, ${}^{p}s_{12} = -5 \times 10^{-12}$ m^2/N,
${}^{p}s_{13} = -7.22 \times 10^{-12}$ m^2/N, ${}^{p}s_{33} = 17.3 \times 10^{-12}$ m^2/N,
${}^{m}s_{11} = 15.3 \times 10^{-12}$ m^2/N, ${}^{m}s_{12} = -5 \times 10^{-12}$ m^2/N,
${}^{m}q_{33} = -1880 \times 10^{-12}$ m/A, ${}^{m}q_{31} = 556 \times 10^{-12}$ m/A,
${}^{p}d_{31} = -175 \times 10^{-12}$ m/V, ${}^{p}d_{33} = -400 \times 10^{-12}$ m/V,
${}^{m}\mu_{33}/\mu_0 = 3$, ${}^{p}\varepsilon/\varepsilon_0 = 1750$, ${}^{m}\varepsilon/\varepsilon_0 = 10$,
${}^{m}\gamma = 10^{-5}$ (Ohm·m)$^{-1}$, ${}^{p}\gamma = 10^{-13}$(Ohm·m)$^{-1}$.

Relaxation of the effective permittivity of the composite was then calculated, and is shown in Fig. 1.22. Under the conditions of ${}^{p}\gamma/{}^{m}\gamma \ll 1$, ${}^{p}\varepsilon/{}^{m}\varepsilon \gg 1$, and $v \ll 1$, it is possible to obtain a giant increase in the static dielectric permittivity. The results are in agreement with prior reports for ferroelectric–polymer composites. The strength of the relaxation was found to decrease with increase of this interphase–interface parameter.

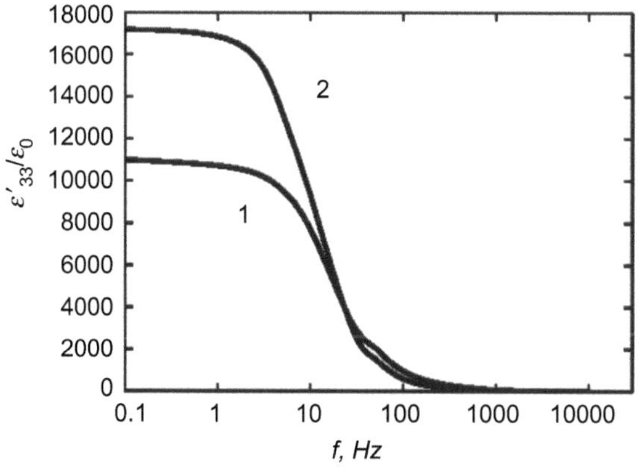

Figure 1.22 Dielectric relaxation in composite for $v = 0.1$: (1) $k = 1$, (2) $k = 0$.

Figure 1.23 illustrates the dependence of the relaxation of the ME susceptibility on piezoelectric phase volume fraction depth. A large relaxation strength is characteristic for a composite whose piezoelectric phase component has a big permittivity and whose ferrite one is electrically conductive. In the case of ${}^{m}\varepsilon/{}^{p}\varepsilon << 1$ and ${}^{p}\gamma/{}^{m}\gamma << 1$, a maximum value of the relaxation strength is observed at a piezoelectric phase volume fraction of $v_1 \approx [({}^{p}s_{11} + {}^{p}s_{12})/({}^{m}s_{11} + {}^{m}s_{12})]^{1/2} ({}^{p}\gamma/{}^{m}\gamma)^{1/2}$. If the compliances of both composite phase components are taken as equal, then one gets $v_1 \approx ({}^{p}\gamma/{}^{m}\gamma)^{1/2}$.

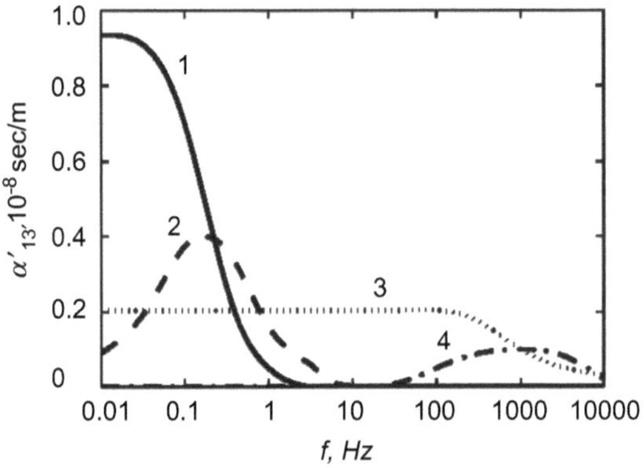

Figure 1.23 Frequency dependence real (1, 3) and imaginary (2, 4) parts of a ME susceptibility: (1, 2) $v = 0.001$, (3, 4) $v = 0.9$.

Relaxation of the ME susceptibility is shown in Fig. 1.24. For ${}^{p}\gamma/{}^{m}\gamma << v << 1$, the static ME susceptibility approaches a value equal to $({}^{m}q_{12} + {}^{m}q_{11})\, {}^{p}d_{31}/({}^{p}s_{11} + {}^{p}s_{12})$. For a composite with a composition dictated by the condition $v_1 = ({}^{p}h/{}^{m}h)^{1/2} \approx 10^{-4}$ where ${}^{p}h$ and ${}^{m}h$ are the thicknesses of the piezoelectric and ferrite phase components, the value of the ME susceptibility is equal 0.94×10^{-8} s/m. This large magnitude of the static ME susceptibility is due to a large local electric field in the thin piezoelectric phase component, to the significant electrical conductance of the ferrite layer, and to the large internal mechanical stresses ${}^{p}T_j$ and ${}^{m}T_j$ ($j = 1, 2$) that are induced by the electrical field in the piezoelectric component. The predicted maximum value for the ME susceptibility of this composite exceeds the experimentally observed one for known materials parameters.

In the case of a weak piezoelectric effect considered here as $({}^{p}d_{31}{}^{2}/[({}^{p}s_{11} + {}^{p}s_{12})\, {}^{p}\varepsilon] << 1)$, the relaxation time that follows from Eq. 1.57 at ${}^{p}\gamma/{}^{m}\gamma << v/(1 - v)$ is controlled mainly by the charging time of the piezoelectric layer capacitance, across the resistance of the ferrite layer: $\tau_\alpha \approx ({}^{p}\varepsilon/{}^{m}\gamma)(1 - v)/v$, as shown in Fig. 1.25.

With an increase of the piezoelectric phase volume fraction, the relaxation time decreases rapidly, whereas with an increase of the piezoelectric constant of the ferroelectric phase, the relaxation time of the composite also increases. The stresses ${}^{p}T_1$ and ${}^{p}T_2$ (caused by a transverse piezoelectric effect) induce additional charge on the piezoelectric layer, increase the charging time of its capacity, and decrease the relaxation time of the ME susceptibility.

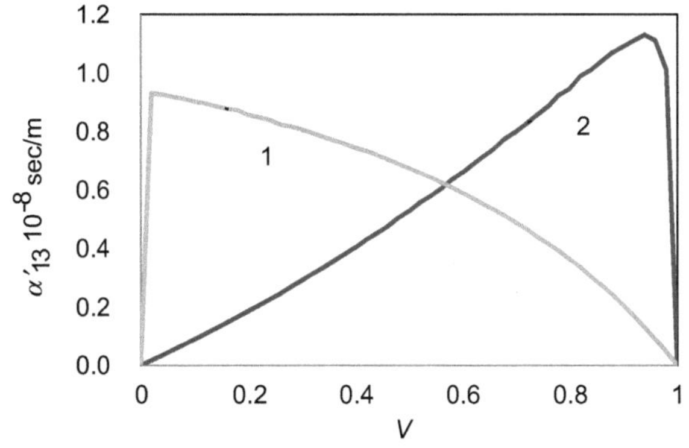

Figure 1.24 Concentration dependence of static (1) and high-frequency (2) magnetoelectric susceptibilities

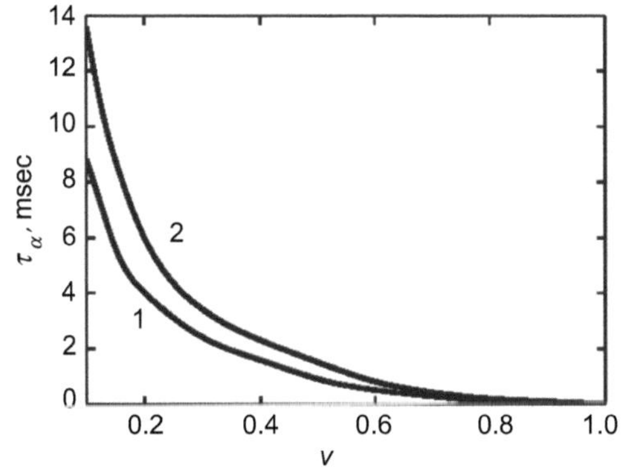

Figure 1.25 Concentration dependence of relaxation time of ME susceptibility: (1) $k = 1$, (2) $k = 0.1$.

Next, we consider the effect that the interfacial parameter k has on the relaxation time. As follows from Eq. 1.57, the relaxation time

τ_α and relaxation frequency $\omega_r = 1/\tau_\alpha$ (determined by the frequency at which maximum imaginary ME susceptibility occurs) can be predicted for a wide range of volume fractions of the component phases and also for various choices of materials for the component phases. Since the frequency dependence of the transverse ME voltage coefficient is defined by the Debye formulas, we can arrive at relations analogous to Eq. 1.57, which provides us the dependence of the relaxation time on various parameters, as follows:

$$\alpha_{E,T} = \alpha'_{E,T} - i\alpha''_{E,T},$$
$$\alpha'_{E,T} = \alpha_{E,T\infty} + \Delta\alpha_{E,T}/(1 + \omega^2\tau_{\alpha T}^2); \tag{1.58}$$
$$\alpha''_{E,T} = \Delta\alpha_{E,T}\omega\tau_{\alpha T}/(1 + \omega^2\tau_{\alpha T}^2);$$

where

$$\Delta a_{E,T} = -a_{E,T\infty} = \frac{v\,{}^{p}d_{31}k(1-v)\,({}^{m}q_{12} + {}^{m}q_{11})}{[{}^{m}s'v + {}^{m}s'(1-v)] - 2\,{}^{p}d_{31}^2(1-v)}$$

$$\tau_{\alpha T} = \frac{{}^{p}\varepsilon}{{}^{p}\gamma} - \frac{2k(1-v)\,{}^{p}d_{31}^2}{{}^{p}\gamma[v\,{}^{m}s' + {}^{p}s'(1-v)]}$$

with ${}^{m}s' = {}^{m}s_{11} + {}^{m}s_{12}$ and ${}^{p}s' = {}^{p}s_{11} + {}^{p}s_{12}$.

Following this relationship, we can see in Fig. 1.25 that increasing k has the same effect on relaxation time as previously shown for increasing piezoelectric constant: it increases notably the value of τ_α.

Unlike the ME susceptibility, the real part of the ME voltage coefficient increases with increasing frequency: that is, inverse relaxation occurs, as shown in Fig. 1.26. The relaxation strength was maximum for $v \approx 0.5$. In the case of a weak piezoelectric effect (defined as $d_{31}^2/[({}^{p}s_{11} + {}^{p}s_{12})\,{}^{p}\varepsilon] << 1$), the relaxation time defined is controlled by the discharge time of the piezoelectric layer capacitance through its own resistance $\tau_{\alpha T} \approx {}^{p}\varepsilon/{}^{p}\gamma$, and does not depend on volume fraction of the component phases. In this case, it can be shown that $\tau_{\alpha T} >> \tau_\alpha$, as illustrated in Fig. 1.27. This figure also shows that increasing piezoelectric phase volume fraction decreases $\tau_{\alpha T}$. This diminution is due to the appearance of additional charge on the piezoelectric layer capacitance under effect of stresses ${}^{p}T_1$ and ${}^{p}T_2$, which is induced by external magnetic fields applied to the composite.

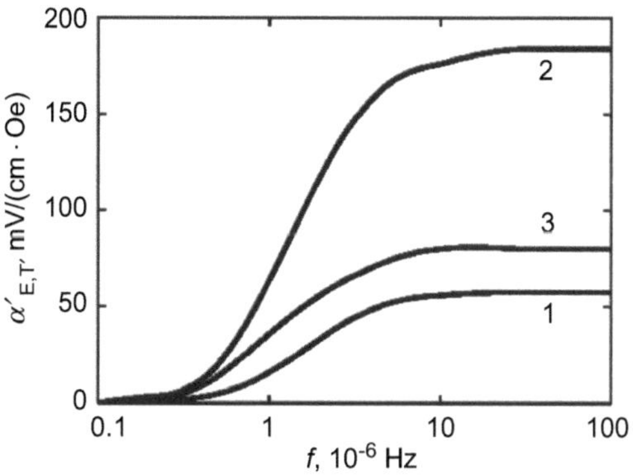

Figure 1.26 Frequency dependence of real part of ME voltage coefficient: (1) $v = 0.1$, (2) $v = 0.5$, (3) $v = 0.9$.

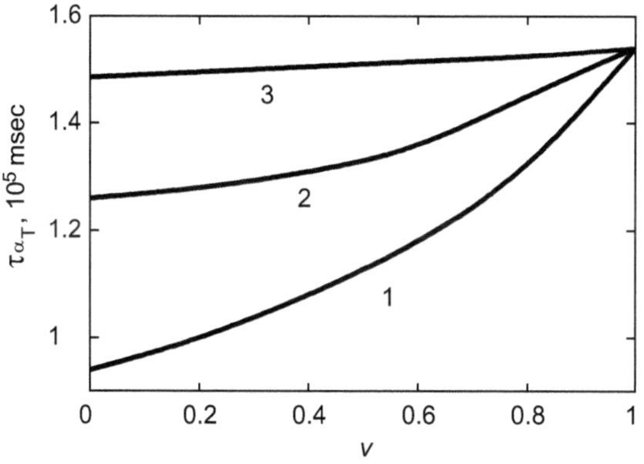

Figure 1.27 Concentration dependence of relaxation time of ME voltage coefficient: (1) $k = 1$, (2) $k = 0.5$, (3) $k = 0.1$.

Thus, in layered ferrite–piezoelectric composites, a strong relaxation of the ME susceptibility is observed, whereas, for the ME voltage coefficient, a strong inverse effect occurs. The relaxation time and relaxation frequency of the ME susceptibility can be changed over a wide range of values, by varying the volume fractions of the composite's component phases, and also by selection of the initial component phases as different piezoelectric and magnetostrictive phases can have different values of property tensors.

1.1.8.2 Bulk composites

As an example, consider the cubic model of bulk ferrite–piezoelectric composites [62] in which one phase has connectivity in all three directions and in which a second isolated phase does not have connectivity in any direction. Following known classifications [44], this specified composite has a connectivity of the type 3-0. For the piezoelectric and magnetostrictive phases, the equations for the strain, dielectric displacement, and the magnetic induction were presented earlier. In addition, in view of a finite electrical conductivity, the permittivities of the component phases are given by Eq. 1.55.

The general formulas for defining the effective parameters of the composite can then be obtained by averaging the expressions for components of the strains, and the electric and magnetic inductions following Eqs. 1.23–1.26. Then assume that the sample has the form of a disk that is oriented in the plane OX_1X_2, where the piezoelectric phase is polarized along the axis OX_3, an electric field with a frequency ω is applied along the same axis, and a magnetic bias and an alternate magnetic field are applied along axis OX_1.

Static and high-frequency ME susceptibilities, in addition to the relaxation time, can be obtained from solutions to Eq. 1.57 by applying boundary conditions for strains, stresses, and electric and magnetic fields. Then, suppose that the symmetry of the piezoelectric phase is ∞m, and that the magnetic phase possesses cubic symmetry. An analytical expression can then be obtained, but due to inconvenience of its form, solution of equations can only be fulfilled numerically [69]. As an example system to apply this numerical solution, we shall consider a composite consisting of poled PZT ferroelectric ceramics and nickel ferro-spinel, which has the following values of component phase material parameters:

$^{p}s_{11} = 15.3 \times 10^{-12}\ \mathrm{m^2/N}$, $^{p}s_{12} = -5 \times 10^{-12}\ \mathrm{m^2/N}$,
$^{p}s_{13} = -7.22 \times 10^{-12}\ \mathrm{m^2/N}$, $^{p}s_{33} = 17.3 \times 10^{-12}\ \mathrm{m^2/N}$,
$^{m}s_{11} = 15.3 \times 10^{-12}\ \mathrm{m^2/N}$, $^{m}s_{12} = -5 \times 10^{-12}\ \mathrm{m^2/N}$,
$^{m}q_{33} = -1880 \times 10^{-12}\ \mathrm{m/A}$, $^{m}q_{31} = 556 \times 10^{-12}\ \mathrm{m/A}$,
$^{p}d_{31} = -\ 50 \times 10^{-12}\ \mathrm{m/V}$, $^{p}d_{33} = -100 \times 10^{-12}\ \mathrm{m/V}$,
$^{m}\mu_{33}/\mu_0 = 2$, $^{p}\varepsilon/\varepsilon_0 = 1000$, $^{m}\varepsilon/\varepsilon_0 = 10$,
$^{m}\gamma = 10^{-5}\ (\mathrm{Ohm{\cdot}m})^{-1}$, $^{p}\gamma = 10^{13}\ (\mathrm{Ohm{\cdot}m})^{-1}$.

In Figs. 1.28 and 1.29, the frequency dependences of the effective piezoelectric modulus of this 0-3 composite is illustrated. With increase of the piezoelectric phase volume fraction, the relaxation strength increases and the relaxation frequency decreases. Thus, parameters of the piezoelectric relaxation can be varied by changing the relative volume fractions of the composite component phases.

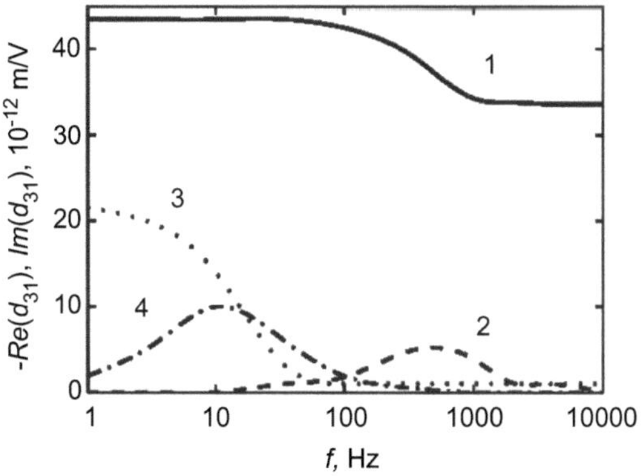

Figure 1.28 Frequency dependence of real (1, 3) and imaginary (2, 4) parts of the effective piezoelectric module d_{31}: (1, 2) $v = 0.9$, (3, 4) $v = 0.1$.

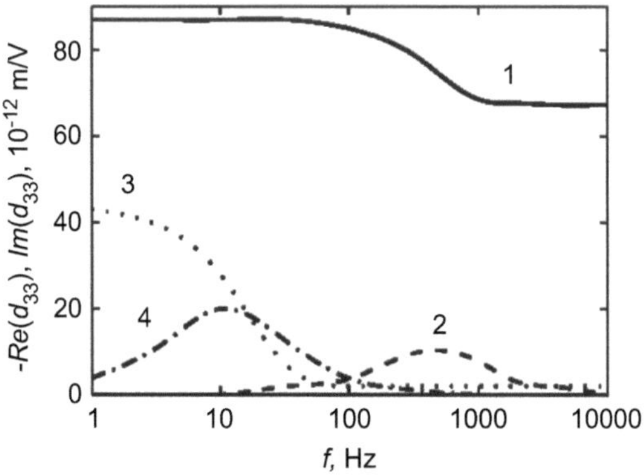

Figure 1.29 Frequency dependence of real (1, 3) and imaginary (2, 4) parts of the effective piezoelectric module d_{33}: (1, 2) $v = 0.9$, (3, 4) $v = 0.1$.

The relaxation of the effective dielectric permittivity is illustrated in Fig. 1.30. Under the conditions of ${}^{p}\gamma/{}^{m}\gamma << 1$, ${}^{p}\varepsilon/{}^{m}\varepsilon >> 1$, and $v << 1$, pronounced increases in the static permittivity can be seen, similar to that found earlier for layered composites. In this case, the relaxation frequency does not depend much on the volume fraction of the components, and is mainly defined by electric parameters. Interestingly, as shown in Fig. 1.30, there is a frequency for which the effective permittivity does not depend much on the piezoelectric phase volume fraction.

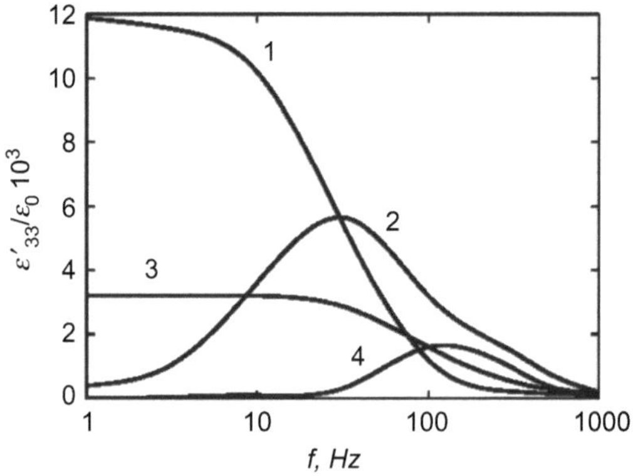

Figure 1.30 Frequency dependence of real (1, 3) and imaginary (2, 4) parts of effective permittivity: (1, 2) $v = 0.1$, (3, 4) $v = 0.9$.

The frequency dependence of the ME voltage coefficient is shown in Fig. 1.31. The real part of the ME voltage coefficient can be seen to increase with increasing frequency, that is, an inverse relaxation occurs. The relaxation spectrum is characterized by two regions of dispersion. The first covers the range of infra-low frequency. The relaxation time for this region depends weakly on the component phase volume fractions and is controlled by the discharge time of the piezoelectric phase capacitance through its own resistance. The second region of relaxation covers the frequency range over which the voltages on capacitance elements 2, 3, and 4 are equalized. In this region, the relaxation is defined by the charging time of the capacitance of element 2 through resistance of the ferrite phase component, and it varies as the relative volume fraction of the phases is changed.

Relaxation of the ME susceptibility is then shown in Fig. 1.32. The relaxation time of the ME susceptibility is defined by the charging time of the capacitance of element 2 through the resistance of the ferrite component. The relaxation strength decreases with an increase of the piezoelectric component volume fraction and for volumes with $v > 0.8$, it becomes negative. Thus, relaxation of the ME susceptibility depends on the volume fraction of component phases, and can be either direct or inverse. Interestingly, there is a specific value for the piezoelectric phase volume fraction for which there is no frequency dependence of the ME susceptibility. This specific value is defined by composite component parameters, as illustrated in Fig. 1.32. The static and high-frequency ME susceptibilities are then shown in Figs. 1.33 and 1.34. It can be seen that the static ME susceptibility has a peak for $v << 1$.

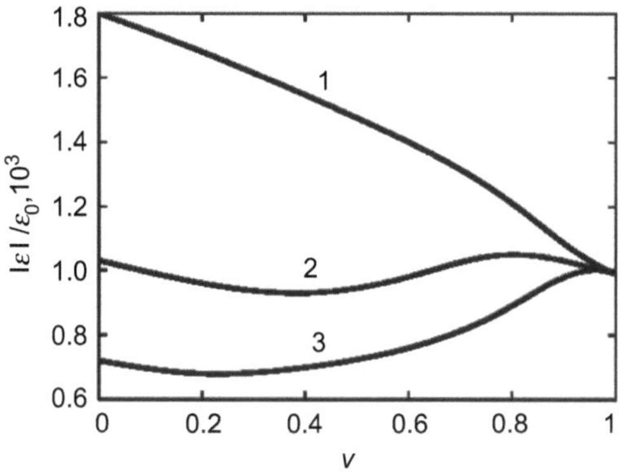

Figure 1.31 Concentration dependence of effective permittivity module: (1) f = 200 Hz, (2) f = 350 Hz, (3) f = 500 Hz.

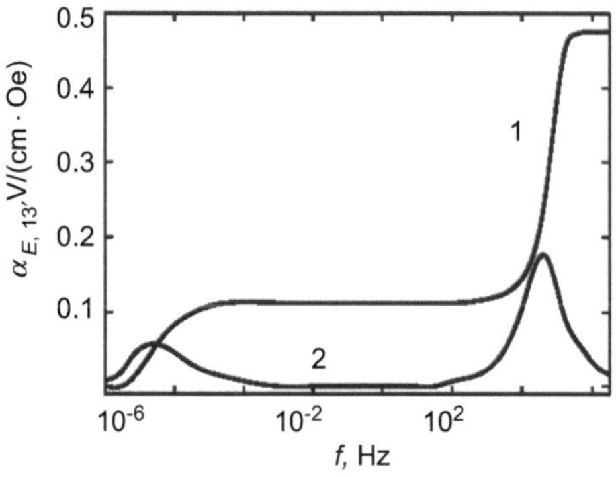

Figure 1.32 Frequency dependence of real (1) and imaginary (2) parts of ME voltage coefficient for $v = 0.6$.

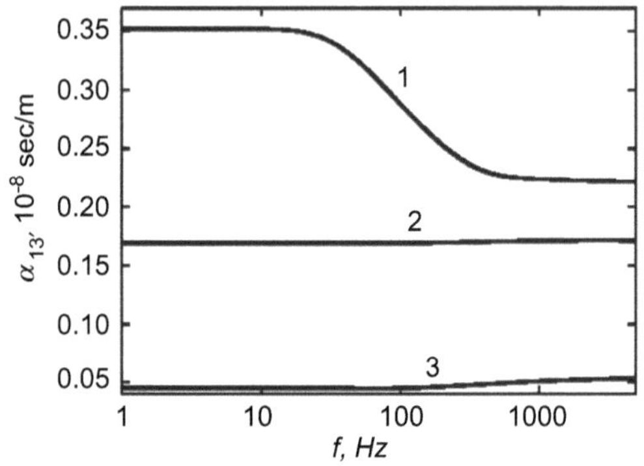

Figure 1.33 Frequency dependence of ME susceptibility real part: (1) $v = 0.5$, (2) $v = 0.8$, (3) $v = 0.95$.

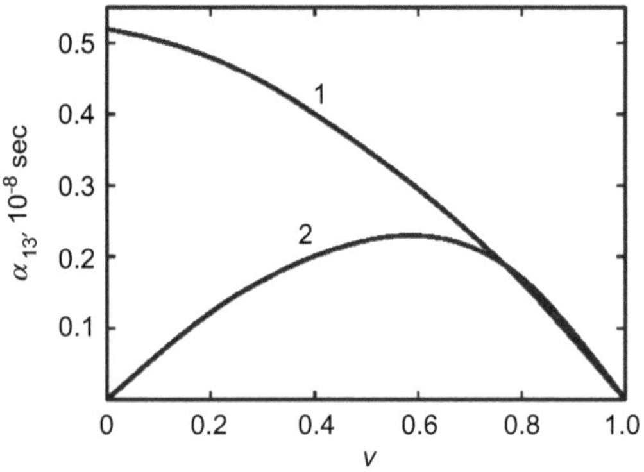

Figure 1.34 Dependence of static (1) and high-frequency (2) ME susceptibilities on volume fraction of the piezoelectric component.

In this chapter, a generalized theoretical model for low-frequency ME effects in layered composites was introduced. To describe the composite's physical properties, effective parameters were derived. This model allowed us to define on the basis of an exact solution the effective mechanical, electric, magnetic, and ME parameters of layered ferrite–piezoelectric composite. Expressions for the effective parameters (including ME susceptibility and ME voltage coefficient) were derived as functions of an interface coupling parameter, constituent phase material parameters, and relative volume fractions of phases. Longitudinal, transverse, and in-plane cases were all considered.

The approach predicted giant ME effects in ferrite–PZT ceramic composites. Predictions of the ME effect for various model ceramic composite systems were given including CFO–BTO, CFO–PZT, NFO–PZT, and lanthanum strontium manganite–PZT. It was shown that ME effect in ferrite–PZT systems is maximum for magnetic and electric fields applied along an in-plane orientation: however, this is yet to be experimentally verified. Using the model, a comparison of the ME parameters was then made between calculated values and experimental data. From this comparison, the importance of an interfacial coupling parameter between phases was inferred. This interphase interfacial connection parameter was shown to be weak for CFO–PZT and lanthanum strontium manganite–PZT composites but near ideal for NFO–PZT.

In addition, the generalized theory allows for modeling of the low-frequency ME effect in bulk composites. To describe these low-frequency composite properties, an effective medium method was used. Calculation of the effective permittivity and permeability, piezoelectric and piezomagnetic modulus, ME susceptibility, and ME voltage coefficient were performed as functions of volume fractions and component parameters. Composites with dimensional connectivities of types 3-0 and 0-3 were considered.

Larger ME coefficients were found for 3-0 composites with magnetic and/or electric fields applied along the longitudinal direction. For composites of CFO–PZT, values as high as 4 V/(cm · Oe) were predicted for the ME voltage coefficient. Fields applied along transverse orientations were found to have ME effects about 2–3.5× smaller than those for longitudinally applied fields. Furthermore, clamping was shown to significantly reduce the ME effect. Finally, in bulk ferrite–piezoelectric composites, large relaxation effects in the ME susceptibility were found with increasing frequency. The relaxation time and frequency of the ME susceptibility and voltage coefficient can be varied by changing the volume fraction of component phases or by changes in the composite component parameters.

1.2 Electromechanical Resonance Range

The magnetoelectric (ME) effect in composites is caused by mechanically coupled magnetostrictive and piezoelectric subsystems: it is present in neither subsystem separately. Under the magnetic field, which is due to the magnetostriction of the ferrite component, there are stresses that are elastically transmitted in the piezoelectric phase, resulting in polarization changes via piezoelectricity. Because the ME effect in composites is due to mechanically coupled piezoelectric and magnetostrictive subsystems, it sharply increases in the vicinity of the electromechanical resonance (EMR) frequency [71–75].

1.2.1 Longitudinal and Radial Modes

In this section, we present a theory for the resonance enhancement of magnetoelectric (ME) interactions at frequencies corresponding to electromechanical resonance (EMR). Frequency dependence of ME voltage coefficients are obtained using the simultaneous solution of electrostatic, magnetostatic, and elastodynamic equations. The ME voltage coefficients are estimated from known material parameters (piezoelectric coupling, magnetostriction, elastic constants, etc.) of composite components. It is shown that the resonance enhancement of ME interactions is observed at frequencies corresponding to EMR, and ME coupling in the EMR region exceeds the low-frequency value by more than an order of magnitude. It was found that the peak transverse ME coefficient at EMR is larger than the longitudinal one. The results of calculations obtained for a CFO–PZT composite are in good agreement with the experimental data.

Mechanical oscillations of a ME composite can be induced either by alternating magnetic or electric fields. If the length of the electromagnetic wave exceeds the spatial size of the composite by some orders of magnitude, then it is possible to neglect gradients of the electric and magnetic fields within the sample volume. Therefore, based on elastodynamics and electrostatics, the equations of medium motion are governed by

$$\bar{\rho}\frac{\partial u_{\mathrm{i}}}{\partial t^2} = V\frac{\partial^{\mathrm{p}}T_{\mathrm{ij}}}{\partial x_{\mathrm{i}}} + (1-V)\frac{\partial^{\mathrm{m}}T_{\mathrm{ij}}}{\partial x_{\mathrm{j}}} \quad , \tag{1.59}$$

where u_i is the displacement vector component,

$\bar{\rho} = V^{\mathrm{p}}\rho + (1-V)^{\mathrm{m}}\rho$ is the average mass density,

V is the ferroelectric volume fraction, and

${}^{\mathrm{p}}\rho$ and ${}^{\mathrm{m}}\rho$, ${}^{\mathrm{p}}T_{ij}$ and ${}^{\mathrm{m}}T_{ij}$ are the density and stress tensor components of ferroelectric and ferromagnetic phase, correspondingly.

Simultaneous solution of elasticity equations and Eq. 1.59 by use of appropriate boundary conditions allows one to find the ME voltage coefficient. Since the solution depends on the composite shape and orientation of applied electric and magnetic fields, we consider some of the most general cases in this chapter.

First, let us consider a composite that has the form of a narrow plate which has a length L, as shown in Fig. 1.35 [74, 77, 78].

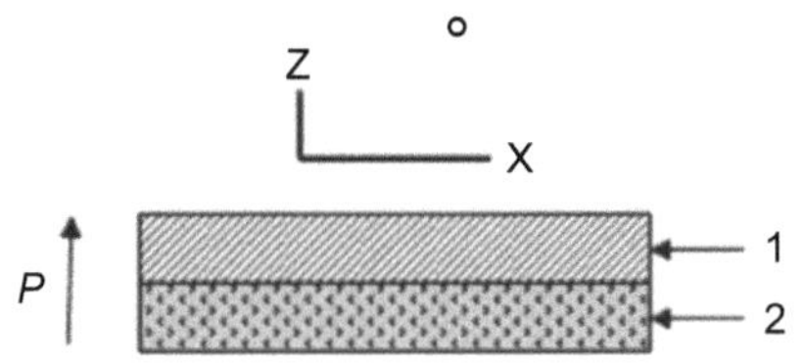

Figure 1.35 Schematic of bilayer of piezoelectric (1) and piezomagnetic (2) phases. The indicator specifies a direction of polarization.

Tensorial expressions for the strain in the ferromagnetic layers ${}^{\mathrm{m}}S_i$ and the strain ${}^{\mathrm{p}}S_i$ and electrical displacement D in the ferroelectric layers have the form for the bias field directed perpendicular to the sample plane (along z-axis):

$$ {}^{\mathrm{p}}S_1 = {}^{\mathrm{p}}s_{11}{}^{\mathrm{p}}T_1 + {}^{\mathrm{p}}s_{12}{}^{\mathrm{p}}T_2 + {}^{\mathrm{p}}d_{31}E_3, \tag{1.60} $$

$$ {}^{\mathrm{p}}S_2 = {}^{\mathrm{p}}s_{12}{}^{\mathrm{p}}T_1 + {}^{\mathrm{p}}s_{11}{}^{\mathrm{p}}T_2 + {}^{\mathrm{p}}d_{31}E_3, $$

$$ {}^{\mathrm{m}}S_1 = {}^{\mathrm{m}}s_{11}{}^{\mathrm{m}}T_1 + {}^{\mathrm{m}}s_{12}{}^{\mathrm{m}}T_2 + {}^{\mathrm{m}}q_{31}H_3, \tag{1.61} $$

$$ {}^{\mathrm{m}}S_2 = {}^{\mathrm{m}}s_{12}{}^{\mathrm{m}}T_1 + {}^{\mathrm{m}}s_{11}{}^{\mathrm{m}}T_2 + {}^{\mathrm{m}}q_{31}H_3, $$

$$ {}^{\mathrm{p}}D_3 = {}^{\mathrm{p}}\varepsilon_{33}E_3 + {}^{\mathrm{p}}d_{31}\,({}^{\mathrm{p}}T_1 + {}^{\mathrm{p}}T_2), \tag{1.62} $$

where ${}^{\mathrm{p}}T_i$ is the stress and ${}^{\mathrm{p}}s_{ii}$ is the compliance of the ferroelectric at constant electric field, ${}^{\mathrm{m}}T_i$ is the stress and ${}^{\mathrm{m}}s_{ii}$ is the compliance of the ferromagnetic at constant magnetic field, ${}^{\mathrm{p}}\varepsilon_{33}$ is the relevant component of the electrical permittivity, ${}^{\mathrm{p}}d_{31}$ is the piezoelectric coefficient of the ferroelectric, ${}^{\mathrm{m}}q_{ij}$ is the piezomagnetic coefficient of ferromagnetic, and E_3 and H_3 are ac electric and magnetic fields.

Close to EMR, we can assume ${}^{m}T_1 >> {}^{m}T_2$ and ${}^{p}T_1 >> {}^{p}T_2$ (axis 1 is directed along the plate length) such that ${}^{m}T_2$ and ${}^{p}T_2$ may be ignored.

Expressions for the stress components ${}^{p}T_1$ and ${}^{m}T_1$ can be found from Eqs. 1.60 and 1.61. Substituting these expressions into Eq. 1.59 yields a differential equation for u_x. Assuming harmonic motion along x, we get the solution of this equation in the form:

$$u_x = \omega A\cos(\omega x) + B\sin(\omega x), \tag{1.63}$$

$$k = \omega\sqrt{\bar{\rho}\left[\frac{V}{{}^{p}s_{11}} + \frac{1-V}{{}^{m}s_{11}}\right]^{-1}}$$

where ω is angular frequency. The integration constants A and B can be found from the boundary conditions. Assuming that the sample surfaces at $x = 0$ and $x = L$ are free from external stresses, we have the following boundary conditions

$$V{}^{p}T_1 + (1 - V){}^{m}T_1 = 0, \text{ at } x = 0 \text{ and } x = L\,. \tag{1.64}$$

For determining the ME voltage coefficient, we use the open circuit condition:

$$\int_0^L {}^{p}D_3 dx = 0 \qquad 1.65)$$

Substituting Eq. 1.61 into Eq. 1.65 and taking into account Eqs. 1.62 and 1.63, we can derive

$$\alpha_{E,33} = \frac{2{}^{p}d_{31}{}^{m}q_{31}{}^{p}s_{11}V(1-V)\tan(kL/2)}{s_1({}^{p}d_{31}^2 - {}^{p}s_{11}{}^{p}\varepsilon_{33})kL - 2{}^{p}d_{31}^2V{}^{m}s_{11}\tan(kL/2)} \tag{1.66}$$

where $s_1 = V{}^{m}s_{11} + (1 - V){}^{p}s_{11}$.

For transverse field orientation, magnetic induction B has the only component B_1 which should obey the condition $\partial B_1/\partial x = 0$ since B is divergence free. For this case, Eq. 1.53 should be written in the more convenient form:

$${}^{m}S_1 = {}^{m}s^{B}_{11}{}^{m}T_1 + {}^{m}g_{11}B_1, \tag{1.67}$$

where ${}^{m}s^{B}_{11}$ is compliance at constant magnetic induction and ${}^{m}g_{11}$ is piezomagnetic coefficient, ${}^{m}g_{11} = \partial{}^{m}S_1/\partial B_1$.

In a similar manner as the calculation above, the transverse ME voltage coefficient can shown to be

$$\alpha_{E,31} = \frac{2\,^{p}d_{31}\,^{m}g_{11}\mu_{\text{eff}}\,^{p}s_{11}V(1-V)\tan(kL/2)}{s_2(^{p}d_{31}^2 - {}^{p}s_{11}\,^{p}\varepsilon_{33})kL - 2\,^{p}d_{31}^2 V\,^{m}s_{11}^{B}\tan(kL/2)}, \quad (1.68)$$

where $s_2 = V^m s^B{}_{11} + (1 - V)^p s_{11}$. Effective permeability μ_{eff} can be found from constitutive equation

$$H_1 = -{}^{m}g_{11}\,^{m}T_1 + B_1/^{m}\mu_{11}, \quad (1.69)$$

where $^{m}\mu_{11}$ is permeability of magnetic phase.

Expressing $^{m}T_1$ from Eq. 1.67 and substituting it into Eq. 1.65 enables finding μ_{eff} after integrating over the sample length.

$$\mu_{\text{eff}} = \frac{s_2\,^{m}s_{11}^{Bm}\mu_{11}kL}{(^{m}s_{11}^{B} + {}^{m}g_{11}^2\,^{m}\mu_{11})kLs_2 + 2^{m}g_{11}^2\,^{m}\mu_{11}\,^{p}s_{11}(1-V)\tan(kL/2)}. \quad (1.70)$$

As one can see from Eqs. 1.52 and 1.54, the value of the ME coefficient under applied fields is directly proportional to the product of piezoelectric d_{31} and piezomagnetic q_{31} or g_{11} modulus. Bear in mind, that in reality, there are always loss factors that must be included, even in "perfect" materials if for no other reason than losses associated with electrical contacts. Said loss factors define the width of the resonant line, limiting the peak value of the ME coefficient. The width of the resonant peak can be varied through attenuation coefficients. Such coefficients are present in k and ω, as they are both complex parameters. We shall use a complex frequency $\omega(1 + i/Q)$ with Q to be determined experimentally.

The roots of the denominator in Eqs. 1.68 and 1.70 define the maxima in the frequency dependence of the ME voltage coefficient. In Figs. 1.36 and 1.37, the frequency dependence of the transverse and longitudinal ME voltage coefficients is shown for the bilayer of nickel ferrite (NFO) and lead zirconate titanate (PZT). In these figures, the resonance peaks caused by oscillations along the x-axes can be seen. The maximum value of the ME coefficient [5400 mV/(cm · Oe)] is observed for the transverse field orientation, whereas the value at frequency of 100 Hz is 144 mV/(cm · Oe). Thus, the resonant value of the ME coefficient exceeds the low-frequency value by a factor of about 30. For the longitudinal field orientation, the magnitude of the ME effect is smaller by one order of magnitude. This is explained by the fact that, for the longitudinal field orientation, the ME effect is significantly affected by the demagnetizing fields.

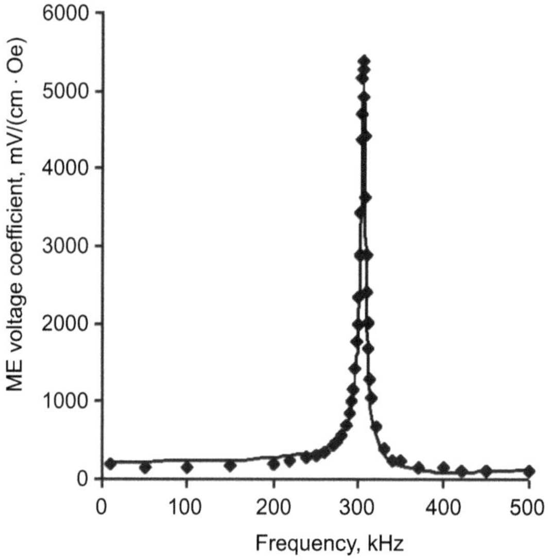

Figure 1.36 Frequency dependence of transverse ME voltage coefficient for the bilayer of NFO and PZT of 7.3 mm length. $Q = 250$ and PZT volume fraction is 0.6. There is a good agreement between calculation (solid line) and data (points).

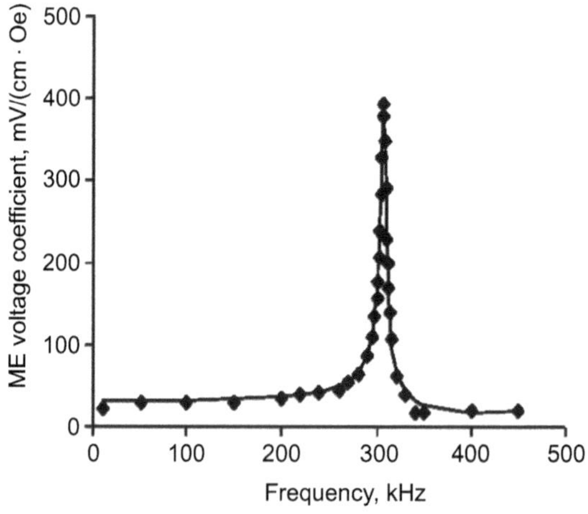

Figure 1.37 Frequency dependence of longitudinal ME voltage coefficient for the bilayer of NFO and PZT of 7.3 mm length. $Q = 105$ and PZT volume fraction is 0.6. There is a good agreement between calculation (solid line) and data (points).

1.2.2 Disc-Shaped Bilayer

Let us consider now a ferrite–piezoelectric disc-shaped composite of radius R and thickness d, which has thin metal electrodes deposited on bottom and top surfaces as shown in Fig. 1.38 [77–84].

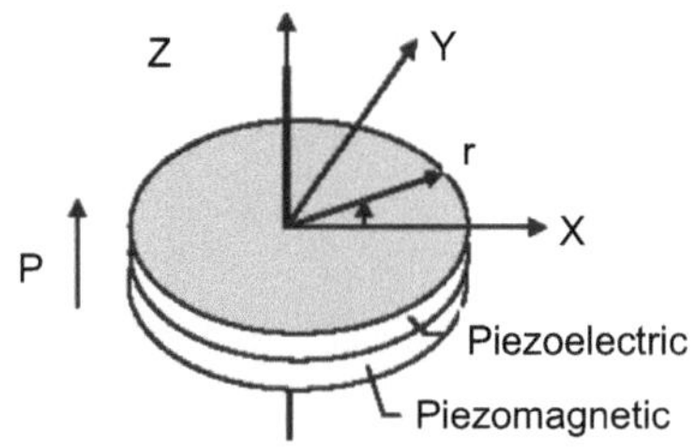

Figure 1.38 Disc-shaped bilayer of piezoelectric and piezomagnetic phases.

Assume that the sample is poled normal to the plane of the electrodes, along the z-axis. Both dc and ac magnetic fields can be directed either along the normal or in the plane of the contacts, which distinguishes the cases of longitudinal and transverse field orientations, respectively. Due to magnetostriction, application of an ac magnetic field will excite both thickness and radial oscillations. In what follows we consider the low-frequency radial oscillations. Radial oscillations will have a notably lower frequency than thickness mode ones, simply because of the geometrical size limitations of the thickness mode.

The disc is supposed to be thin (i.e., $d << R$). Since the surfaces are free from external forces, the normal components of the stress tensors will be zero. For a thin disc, it is reasonable to assume that the component of the stress tensor T_3 is zero not only on the surfaces, but also in the volume of the disc. In addition, since the top and bottom bases of the disc are equipotential surfaces, only the z component of the electric field vector is nonzero. Then, the constitutive equations are determined by Eqs. 1.60–1.62 for longitudinal field orientation.

It is convenient to take advantage of the symmetry of a disc by use of a cylindrical coordinate system (z, r, and θ). In this case, the strain and stress tensor components need to be transformed in by the following relations:

$$^iS_1 = {}^iS_{rr}\cos^2\theta - 2\,{}^iS_{r\theta}\sin\theta\cos\theta + {}^iS_{\theta\theta}\sin^2\theta, \tag{1.71}$$

$$^iS_2 = {^iS_{rr}} \sin^2\theta + 2{^iS_{r\theta}} \sin\theta \cos\theta + {^iS_{\theta\theta}} \cos^2 \theta, \tag{1.72}$$

$$^iT_{rr} = {^iT_1} \cos^2\theta + 2{^iT_6} \sin\theta \cos\theta + {^iT_2} \sin^2\theta, \tag{1.73}$$

$$^iT_{\theta\theta} = {^iT_1} \sin^2\theta - 2{^iT_6} \sin\theta \cos\theta + {^iT_2} \cos^2\theta, \tag{1.74}$$

$$^iT_{r\theta} = \sin\theta \cos\theta\, [{^iT_2} - {^iT_1}] + [\cos^2\theta - \sin^2\theta]{^iT_6}, \tag{1.75}$$

where i = p or m. The components of the strain tensor in cylindrical coordinates are defined via the vector displacement u, as follows:

$$^iS_{rr} = \partial u_r/\partial r \tag{1.76}$$

$$^iS_{\theta\theta} = u_r/r \tag{1.77}$$

$$^iS_{r\theta} = (1/r)\, \partial u_r/\partial\theta \tag{1.78}$$

The other necessary equation is the equation of medium motion transformed into cylindrical coordinates:

$$-\bar{\rho}\omega u_{\mathrm{r}} = V\left(\frac{\partial^{\mathrm{p}}T_{\mathrm{rr}}}{\partial r} + \frac{\partial^{\mathrm{p}}T_{\mathrm{r}\theta}}{r\partial\theta} + \frac{^{\mathrm{p}}T_{\mathrm{rr}} - {^{\mathrm{p}}T_{\theta\vartheta}}}{r}\right) +$$
$$(1-V)\left(\frac{\partial^{\mathrm{m}}T_{\mathrm{r}\theta}}{\partial r} + \frac{\partial^{\mathrm{m}}T_{\mathrm{r}\theta}}{r\partial\theta} + \frac{^{\mathrm{m}}T_{\mathrm{rr}} - {^{\mathrm{m}}T_{\theta\vartheta}}}{r}\right) \quad . \tag{1.79}$$

Solving the elasticity equations for stress components and substituting the found expressions into Eq. 1.79, we can obtain the equation for the radial displacements. However, the form of this equation depends on the electric and magnetic fields, orientations. Next, we consider the ME coupling for longitudinal and transverse fields, orientations separately.

In this case, the longitudinal orientation of dc and ac magnetic fields coincides with the direction of polarization. Taking into account the axial symmetry of disc-shaped sample, one can get the following expressions for nonvanishing strain components:

$$^{\mathrm{p}}S_{\mathrm{rr}} = {^{\mathrm{p}}s_{11}}{^{\mathrm{p}}T_{\mathrm{rr}}} + {^{\mathrm{p}}s_{12}}{^{\mathrm{p}}T_{\theta\theta}} + {^{\mathrm{p}}d_{31}}E_3,$$
$$^{\mathrm{m}}S_{\mathrm{rr}} = {^{\mathrm{m}}s_{11}}{^{\mathrm{m}}T_{\mathrm{rr}}} + {^{\mathrm{m}}s_{12}}{^{\mathrm{m}}T_{\theta\theta}} + {^{\mathrm{m}}q_{31}}H_3,$$
$$^{\mathrm{p}}S_{\theta\theta} = {^{\mathrm{p}}s_{12}}{^{\mathrm{p}}T_{\mathrm{rr}}} + {^{\mathrm{p}}s_{11}}{^{\mathrm{p}}T_{\theta\theta}} + {^{\mathrm{p}}d_{31}}E_3,$$
$$^{\mathrm{m}}S_{\theta\theta} = {^{\mathrm{m}}s_{12}}{^{\mathrm{m}}T_{\mathrm{rr}}} + {^{\mathrm{m}}s_{11}}{^{\mathrm{m}}T_{\theta\theta}} + {^{\mathrm{m}}q_{31}}H_3. \tag{1.80}$$

We get the equation of motion in the form:

$$-\omega^2\bar{\rho}u_r = \left[\frac{V}{{}^{\mathrm{p}}s_{11}(1-{}^{\mathrm{p}}v^2)} + \frac{1-V}{{}^{\mathrm{m}}s_{11}(1-{}^{\mathrm{m}}v^2)}\right]\left(\frac{\partial^2 u_r}{\partial r^2} + \frac{\partial u_r}{r\partial r} - \frac{\iota}{r}\right), \quad (1.81)$$

where ${}^{\mathrm{p}}v$ and ${}^{\mathrm{m}}v$ are Poisson's ratios for piezoelectric and magnetostrictive phases.

The solution of Eq. 1.81 is a linear combination of Bessel function of the first and second sort:

$$u_{\mathrm{r}} = c_1 J_1(kr) + c_2 Y_1(kr) \quad (1.82)$$

$$k = \omega\sqrt{\bar{\rho}\left[\frac{V}{{}^{\mathrm{p}}s_{11}(1-{}^{\mathrm{p}}v^2)} + \frac{1-V}{{}^{\mathrm{m}}s_{11}(1-{}^{\mathrm{m}}v^2)}\right]^{-1}}.$$

The constants of integration c_1 and c_2 should be found from boundary conditions: $u_r = 0$ at $r = 0$ and $T_{rr} = 0$ at $r = R$.

For obtaining the ME voltage coefficient, we use the open-circuit condition

$$\int_0^R r dr \int_0^{2\pi} d\theta D_3 = 0 \quad (1.83)$$

where electric induction is determined by the constitutive equation

$$D_3 = {}^{\mathrm{p}}d_{31}({}^{\mathrm{p}}T_{\mathrm{rr}}{}^{\mathrm{p}}T_{\theta\theta}) + {}^{\mathrm{p}}\varepsilon_{33}E_3. \quad (1.84)$$

Substituting Eq. 1.80 into Eqs. 1.82 and 1.83 and then into Eq. 1.74, we find the expressions for stress components. Once the stress components are determined, the ME voltage coefficient can be found from Eq. 1.83.

$$\alpha_{\mathrm{E,L}} = -\frac{2(1+v)(1-V)^{\mathrm{p}}s_{11}j_1(kR)^{\mathrm{p}}d_{31}{}^{\mathrm{m}}q_{31}}{(1-v)^{\mathrm{p}}s_{11}[aJ_0(kR)-(1-v)s_1J_1^*]^{p}\varepsilon_{33} + 2[aJ_0(kR)-s_3J_1^*]^{p}d_{31}^2}, \quad (1.85)$$

where $J_1^* = J_1(kR)$, $a = kRs_1$, $s_1 = V^{\mathrm{m}}s_{11} + (1 - V)^{\mathrm{p}}s_{11}$, and $s_3 = (1 - v)(1 - V)^{\mathrm{p}}s_{11} + 2V^{\mathrm{m}}s_{11}$. It should be noted that ${}^{\mathrm{m}}v$ is assumed to be equal to ${}^{\mathrm{p}}v = v$ for simplicity of Eq. 1.84.

In case of transverse fields' orientation, the in-plane dc and ac magnetic field vectors are perpendicular to the electric field vector field which is along the z-axis. Eqs. 1.80 should be replaced by

$$
\begin{aligned}
{}^{\mathrm{p}}S_{\mathrm{rr}} &= {}^{\mathrm{p}}s_{11}{}^{\mathrm{p}}T_{\mathrm{rr}}{}^{\mathrm{p}}s_{12}{}^{\mathrm{p}}T_{\theta\theta} + {}^{\mathrm{p}}d_{31}E_{3,}\\
{}^{\mathrm{m}}S_{\mathrm{rr}} &= {}^{\mathrm{m}}s_{11}^{\mathrm{B}}{}^{\mathrm{m}}T_{\mathrm{rr}} + {}^{\mathrm{m}}s_{12}^{\mathrm{B}}{}^{\mathrm{m}}T_{\theta\theta} + [{}^{\mathrm{m}}g_{11}\cos^2(\theta) + {}^{\mathrm{m}}g_{12}\sin^2(\theta)]B_1,\\
{}^{\mathrm{p}}S_{\theta\theta} &= {}^{\mathrm{p}}s_{12}{}^{\mathrm{p}}T_{\mathrm{rr}} + {}^{\mathrm{p}}s_{11}{}^{\mathrm{p}}T_{\theta\theta} + {}^{\mathrm{p}}d_{31}E_{3,}\\
{}^{\mathrm{m}}S_{\theta\theta} &= {}^{\mathrm{m}}s_{12}^{\mathrm{B}}{}^{\mathrm{m}}T_{\mathrm{rr}} + {}^{\mathrm{m}}s_{11}^{\mathrm{B}}{}^{\mathrm{m}}T_{\theta\theta} + [{}^{\mathrm{m}}g_{12}\cos^2(\theta) + {}^{\mathrm{m}}g_{11}\sin^2(\theta)]B_1,\\
{}^{\mathrm{p}}S_{\mathrm{r}\theta} &= ({}^{\mathrm{p}}s_{11} - {}^{\mathrm{p}}s_{12})^{\mathrm{p}}T_{\mathrm{r}\theta},\\
{}^{\mathrm{m}}S_{\mathrm{r}\theta} &= ({}^{\mathrm{m}}s_{11}^{\mathrm{B}} - {}^{\mathrm{m}}s_{12}^{\mathrm{B}})^{\mathrm{m}}T_{\mathrm{r}\theta} - \frac{1}{2}\sin(2\theta)({}^{\mathrm{m}}g_{11} - {}^{\mathrm{m}}g_{12})B_1
\end{aligned}
\tag{1.86}
$$

Solving Eqs. 1.86 for stress components and substituting these expressions into Eq. 1.79 results in the following form of equation of media motion:

$$
\begin{aligned}
-\omega^2\bar{\rho}u_r = &\left[\frac{V}{{}^{\mathrm{p}}s_{11}(1-{}^{\mathrm{p}}v^2)} + \frac{1-V}{{}^{\mathrm{m}}s_{11}(1-{}^{\mathrm{m}}v^2)}\right]\left(\frac{\partial^2 u_{\mathrm{r}}}{\partial r^2} + \frac{\partial u_{\mathrm{r}}}{r\partial r} - \frac{u_r}{r^2}\right) +\\
&+\left[\frac{V}{{}^{\mathrm{p}}s_{11}(1-{}^{\mathrm{p}}v)} + \frac{1-V}{{}^{\mathrm{m}}s_{11}(1-{}^{\mathrm{m}}v)}\right]\frac{\partial^2 u_{\mathrm{r}}}{r^2\partial\theta^2}
\end{aligned}
\tag{1.87}
$$

The solution of Eq. 1.79 corresponding to the radial mode is defined by Eq. 1.82 with

$$
k = \omega\sqrt{\bar{\rho}\left[\frac{V}{{}^{\mathrm{p}}s_{11}(1-{}^{\mathrm{p}}v^2)} + \frac{1-V}{{}^{\mathrm{m}}s_{11}^{\mathrm{B}}(1-{}^{\mathrm{m}}v^2)}\right]^{-1}}
$$

The constants of integration c_1 and c_2 should be found from boundary conditions: $u_r = 0$ at $r = 0$ and

$$
\int_0^{2\pi} T_{\mathrm{rr}}d\theta = 0 \quad \text{at} \quad r = R.
$$

The ME voltage coefficient can be found as

$$
\alpha_{\mathrm{E,T}} = \frac{(1+v)(1-V)^{\mathrm{p}}s_{11}J_1(kR)^{\mathrm{p}}d_{31}({}^{\mathrm{m}}g_{11} + {}^{\mathrm{m}}g_{12})\mu_{\mathrm{eff}}}{(1-v)^{\mathrm{p}}s_{11}[aJ_0(kR) - (1-v)s_1J_1^*]^{\mathrm{p}}\varepsilon_{33} + 2[aJ_0(kR) - s_3J_1^*]^{\mathrm{p}}d_{31}^2},
\tag{1.88}
$$

where $a = kRs_1$, $s_1 = V^{\mathrm{m}}s^{\mathrm{B}}{}_{11} + (1 - V)^{\mathrm{p}}s_{11}$, and $s_3 = (1 - v)(1 - V)^{\mathrm{p}}s_{11} + 2V^{\mathrm{m}}s^{\mathrm{B}}{}_{11}$. It should be noted that ${}^{\mathrm{m}}v$ is assumed to be equal to ${}^{\mathrm{p}}v$ for simplicity of Eq. 1.88.

Experimental investigations of the ME effect have been performed for bilayer of NFO and PZT. The samples were discs a the radius of R = 4.7 mm and were poled by dc electric field E = 4 kV/mm for 3 hours at 80°C. The bias field dependence of low-frequency ME effect was measured in the beginning. Measurements were done for both longitudinal and transverse orientations of electric and magnetic fields in the EMR region. Experimental data and corresponding theoretical estimates based on Eqs. 1.38 are given in Figs. 1.39 and 1.40 for longitudinal and transverse fields, respectively. Calculations were performed using the following values of the materials parameters:

for NFO, ${}^{m}s_{11} = 6.5 \times 10^{-12}$ m²/м, ${}^{m}s_{12} = -2.4 \times 10^{-12}$ m²/м,
${}^{m}q_{31} = 70 \times 10^{-12}$ м/A, ${}^{m}q_{11} = -430 \times 10^{-12}$ м/A,
${}^{m}q_{12} = 125 \times 10^{-12}$ м/A, ${}^{m}\varepsilon_{33}/\varepsilon_0 = 10$; and
for PZT ${}^{p}s_{11} = 15.3 \times 10^{-12}$ m²/м, ${}^{p}s_{12} = 5 \times 10^{-12}$ m²/м,
${}^{p}d_{31} = -175 \times 10^{-12}$ м/B, ${}^{p}\varepsilon_{33}/\varepsilon_0 = 1750$.

The attenuation parameter was empirically defined by the line-width of the EMR peak at its half-maximum point.

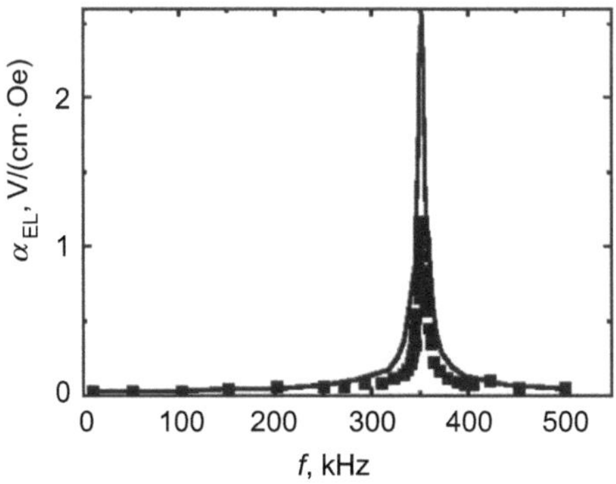

Figure 1.39 Frequency dependence of magnetoelectric voltage coefficient at longitudinal orientation of fields. (Solid line) theory, (points) experiment.

As can be seen in these figures, there is a relatively good agreement between theory and experimental data. The maximum value of the transverse ME coefficient for the disc-shaped composites was about 15 V/(cm · Oe) at resonance, whereas the low-frequency value was only about 0.16 V/(cm · Oe). The coefficient of attenuation was found to be smaller for transverse fields compared to longitudinal ones. This can be accounted for by the higher energy losses for longitudinal fields orientation due to eddy currents in the electrodes. In general, it was found that the transverse ME coefficient was larger than the longitudinal one. As mentioned earlier, this is also due to the fact that the demagnetization field appears at longitudinal orientations that reduces the piezomagnetic modulus.

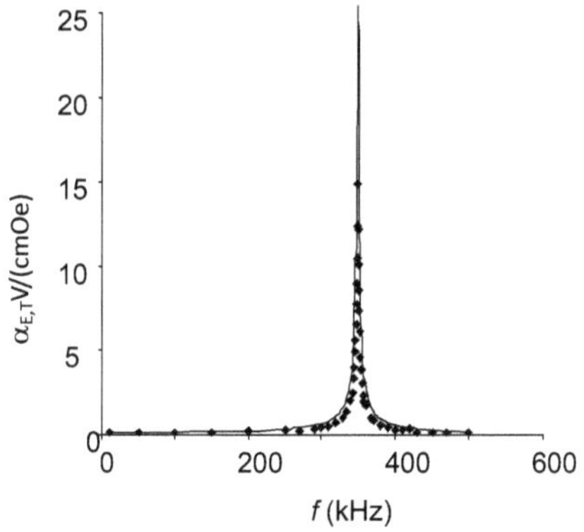

Figure 1.40 Frequency dependence of magnetoelectric voltage coefficient at transverse orientation of fields. (Solid line) theory, (points) experiment.

1.2.3 Bending Modes

A key drawback for ME effect at longitudinal modes is that the frequencies are quite high, on the order of hundreds of kHz, for nominal sample dimensions. The eddy current losses for the magnetostrictive phase can be quite high at such frequencies, in particular for transition metals and alloys and rare earth alloys such as Terfenol-D, resulting in an inefficient magnetoelectric energy conversion. In order to reduce the operating frequency, one must therefore increase the laminate size that is inconvenient for any applications. An alternative for getting a strong ME coupling is the resonance enhancement at bending modes of the composite. The frequency of the applied ac field is expected to be much lower compared to longitudinal acoustic modes. Recent investigations have shown a giant ME effect at bending modes in several layered structures [83–91]. In this section, we focus our attention on theoretical modeling of ME effects at bending modes.

An in-plane bias field is assumed to be applied to magnetostrictive component to avoid the demagnetizing field. The thickness of the plate is assumed to be small compared to remaining dimensions. Moreover, the plate width is assumed to be small compared to its length. In that case, we can consider only one component of strain and stress tensors in the EMR region. The equation of bending motion of bilayer has the form:

$$\nabla^2\nabla^2 w + \frac{\rho b}{D}\frac{\partial^2 w}{\partial \tau^2} = 0 \tag{1.89}$$

where $\nabla^2\nabla^2$ is biharmonic operator, w is the deflection (displacement in z direction), t and ρ are thickness and average density of sample, $b = {}^{p}t + {}^{m}t$, $\rho = ({}^{p}\rho{}^{p}t + {}^{m}\rho{}^{m}t)/b$, ${}^{p}\rho$, ${}^{m}\rho$ and ${}^{p}t$, ${}^{m}t$ are densities and thicknesses of piezoelectric and piezomagnetic, correspondingly, and D is cylindrical stiffness.

The boundary conditions for $x = 0$ and $x = L$ have to be used for finding the solution of above equation. Here L is length of bilayer. As an example, we consider the plate with free ends. At free end, the turning moment M_1 and transverse force V_1 equals zero: $M_1 = 0$ and $V_1 = 0$ at $x = 0$ and $x = L$, where

$$M_1 = \int_{A} zT_1 dz_1 \quad \text{and} \quad V_1 = \frac{\partial M_1}{\partial x}$$

A is the cross-sectional area of the sample normal to the x-axis. We are interested in the dynamic ME effect; for an ac magnetic field H applied to a biased sample, one measures the average induced electric field and calculates the ME voltage coefficient. Using the open circuit condition, the ME voltage coefficient can be found as:

$$\alpha_{E31} = \frac{E_3}{H_1} = -\frac{\int_{z_0-{}^{p}t}^{z_0} {}^{p}E_3 dz}{tH_1{}^{p}\varepsilon_{33}} \tag{1.90}$$

where E_3 and H_1 are the average electric field induced across the sample and applied magnetic field.

Solving the elasticity equations for stress components taking into account Eq. 1.80, and substituting the solutions into Eq. 1.80 enables one to obtain the explicit expression for ME voltage coefficient

$$\alpha_{E31} = \frac{A \cdot {}^{p}E_{31} \cdot {}^{p}Y \cdot {}^{m}q_{11}}{\Delta \cdot {}^{p}\varepsilon_{33}} F(kL), \tag{1.91}$$

$$A = \frac{{}^{m}Y \cdot {}^{m}t \cdot (2 \cdot z_0 + {}^{m}t) \cdot (2 \cdot z_0 - {}^{p}t)}{4 \cdot D \cdot ({}^{p}K_{31}^{2})}$$

where

$$k^4 = \omega^2 \rho t / D$$

D is cylindrical stiffness of sample and L is the sample length.

$$F(kL) = (4 \cdot r_2 \cdot r_3 - 2 \cdot r_2 \cdot r_3^2 + 2 \cdot r_2^2 \cdot r_4 - 2 \cdot r_2 \cdot r_4^2 + 4 \cdot r_1 \cdot r_4 - 2 \cdot r_4 \cdot r_1^2 - 2 \cdot r_4 - 2 \cdot r_2) \cdot a_1;$$

$$\Delta = (r_3^2 - r_2^2 - 2 \cdot r_1 \cdot r_3 + r_1^2 + r_4^2)kL + (4 \cdot r_2 \cdot r_3 - 2 \cdot r_2 \cdot r_3^2 + 2 \cdot r_2^2 \cdot r_4 - 2 \cdot r_2 \cdot r_4^2 + 4 \cdot r_1 \cdot r_4 - 2 \cdot r_4 \cdot r_1^2 - 2 \cdot r_4 - 2 \cdot r_2) \cdot a_1;$$

$r_1 = \cosh(kL)$, $r_2 = \sinh(kL)$, $r_3 = \cos(kL)$, $r_4 = \sin(kL)$,

${}^{p}Y$ and ${}^{m}Y$ are the modulus of elasticity of piezoelectric and piezomagnetic component.

$$a_1 = {}^{p}K_{31}^{2} \cdot {}^{p}Y^{E} \cdot ((z_0 - {}^{p}t)^3 - z_0^3) / [3 \cdot D \cdot (1 - {}^{p}K_{31}^{2})] + {}^{m}K_{31}^{2} \cdot {}^{m}Y \cdot ((z_0 + {}^{m}t)^3 z_0^3) / (3 \cdot D)$$

$$z_0 = \frac{1}{2} \frac{{}^{p}Y {}^{p}t^2 - {}^{m}Y {}^{m}t^2}{{}^{p}Y {}^{p}t + {}^{m}Y {}^{m}t}$$

The energy losses are taken into account by substituting ω for complex frequency $\omega' + i\omega''$ with $\omega''/\omega' = 10^{-3}$.

As an example, we apply Eq. 1.91 to the bilayer of permendur and PZT. Fig. 1.41 shows the frequency dependence of ME voltage coefficient at bending mode for free-standing bilayer with length 9.2 mm and thickness 0.7 mm for PZT volume fraction 0.67. Graph of $\alpha_{E,31}$ reveals a giant value $\alpha_{E\ 31}$ = 6.6 V/(cm · Oe) and the resonance peak lies in the infra-low frequency range. Fig. 1.42 reveals the theoretical and measured frequency dependencies of transverse ME voltage coefficients for a permendur–PZT bilayer that is free to bend at both ends.

According to our model, there is a strong dependence of resonance frequency on boundary conditions. The lowest resonance frequency is expected for the bilayer clamped at one end. One expects bending motion to occur at decreasing frequencies with increasing bilayer length or decreasing thickness.

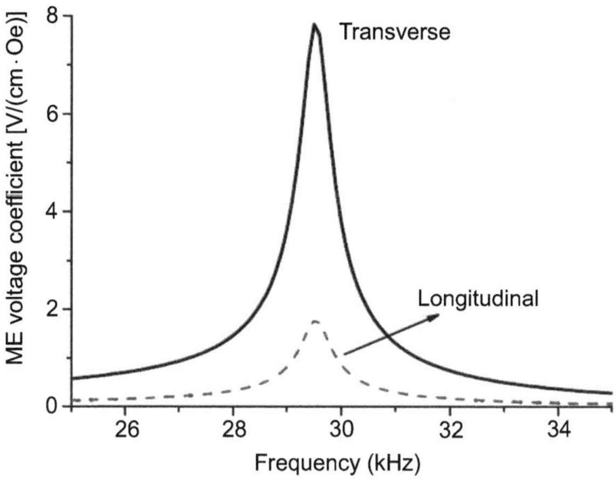

Figure 1.41 Frequency dependence of longitudinal and transverse ME voltage coefficients for a bilayer of permendur and PZT showing the resonance enhancement of ME interactions at the bending mode frequency. The bilayer is free to bend at both ends. The sample dimensions are L = 9.2 mm and total thickness t = 0.7 mm and the PZT volume fraction v = 0.6.

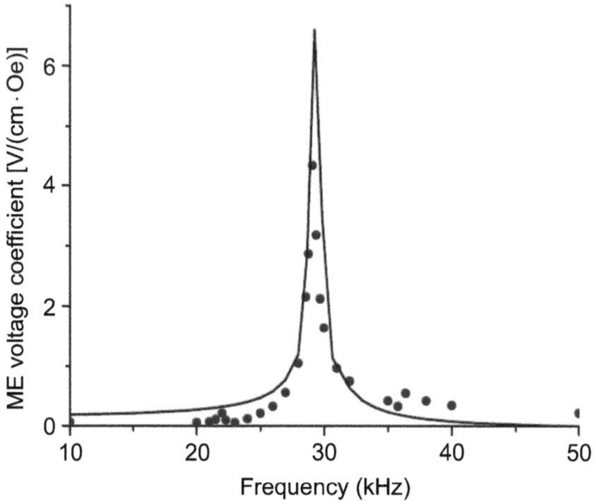

Figure 1.42 Theoretical (line) and measured (circles) frequency dependence of transverse ME voltage coefficients for a permendur–PZT bilayer that is free to bend at both ends and with v = 0.67.

1.2.4 Shear Vibrations

Mechanical oscillations of an ME composite can be induced by alternating magnetic and electric fields. If the length of the electromagnetic wave exceeds the spatial size of the composite by some orders of magnitude, then it is possible to neglect gradients of the electric and magnetic fields within the sample volume. Therefore, based on elastodynamics and electrostatics, the equation of shear medium motion of piezoelectric phase is governed by

$$\rho_{\mathrm{p}} = \frac{\partial^2 u_1}{\partial t^2} = \frac{\partial t_{\mathrm{p}5}}{\partial z}, \tag{1.92}$$

where u_i is the displacement vector component, and ρ_{p} is the mass density of piezoelectric. Similar equation can be written for magnetic component. Similar equations of media shear motion can be written for magnetostrictive phases and substrate. To solve these equations for displacement component, one should use the appropriate boundary conditions for stress and strain components on interfaces. Substituting the found solutions into the open circuit condition enables one to obtain the expression for ME voltage coefficient [91].

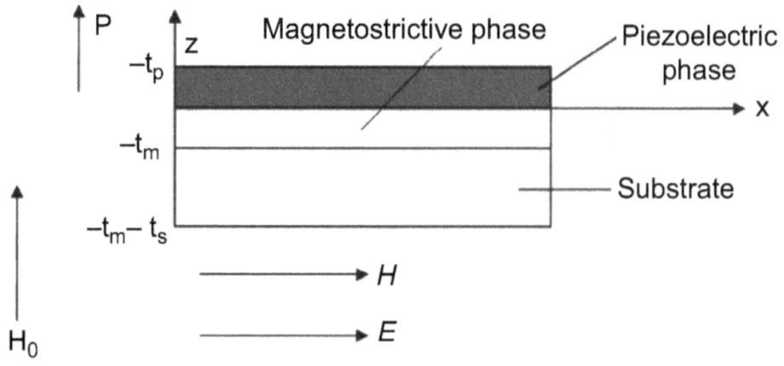

Figure 1.43 Bilayer of piezoelectric and ferrite in the *x–z* plane on the substrate. The poling electric field E_0 is along *z* direction and ac electric field *E* is along *x*. The bias magnetic field H_0 and the alternating magnetic field *H* are along *z* and *x*, respectively.

Expressions for the strain in the ferromagnetic layers $S_{\mathrm{m}i}$, the strain in the ferroelectric layers $S_{\mathrm{p}i}$, and strain in the substrate $S_{\mathrm{s}i}$ have the form:

$$\begin{cases} S_{\mathrm{p}5} = d_{15}E_1 + s_{\mathrm{p}44}T_{\mathrm{p}5}(z) \\ S_{5\mathrm{m}} = q_{15}H_1 + s_{\mathrm{m}44}T_{\mathrm{m}5}(z) \\ S_{\mathrm{s}5} = s_{\mathrm{s}44}T_{\mathrm{s}5}(z) \end{cases} \tag{1.93}$$

where $T_{\mathrm{p}i}$ is the stress and $s_{\mathrm{p}ii}$ is the compliance of the ferroelectric at constant electric field, $T_{\mathrm{m}i}$ is the stress and $s_{\mathrm{m}ii}$ is the compliance of the ferromagnetic at constant magnetic field, d_{ij} is the piezoelectric coefficient of the ferroelectric, q_{ij} is the piezomagnetic coefficient of ferromagnetic, and E_1 and H_1 are electric and magnetic felds, respectively.

Electric displacements D for the piezoelectric phases are given by

$$D_{\mathrm{p}}(z) = d_{15}T_{p5}(z) + \varepsilon_{33}E_1 \tag{1.94}$$

where ε_{33} is the relevant component of the electrical permittivity.

For determining the ME voltage coefficient, we use the open circuit condition:

$$\int_0^{t_{\mathrm{p}}} D_{\mathrm{p}}(z)\,dz = 0$$

Substituting Eq. 1.84 into Eq. 1.83 and solving the open circuit condition for E_1 enables finding the expression for the ME voltage coefficient $\alpha = E_1/H_1$ for the free-standing bilayer:

$$\alpha_{\mathrm{E}} = 2k_{\mathrm{p}}\frac{d_{15}}{\varepsilon_{33}}q_{15}[1-(1-Q_4)Q_2 - Q_4]\{[(2s_{\mathrm{m}44}Q_4 + s_{\mathrm{p}44}t_{\mathrm{p}}k_{\mathrm{m}}Q_3)k_{\mathrm{p}}Q_2$$
$$+ (k_{\mathrm{p}}^2t_{\mathrm{p}}s_{\mathrm{m}44}Q_4 - s_{\mathrm{p}44}k_{\mathrm{m}}Q_3)Q_1 - 2s_{\mathrm{m}44}k_{\mathrm{p}}Q_4]K_{\mathrm{p}}^2 - s_{\mathrm{p}44}k_{\mathrm{p}}t_{\mathrm{p}}k_{\mathrm{m}}Q_5$$
$$- k_{\mathrm{p}}^2t_{\mathrm{p}}s_{\mathrm{m}44}Q_6\}^{-1}, \tag{1.95}$$

where $k_{\mathrm{m}} = \omega\sqrt{\rho_{\mathrm{m}}s_{\mathrm{m}44}}$, $k_{\mathrm{p}} = \omega\sqrt{\rho_{\mathrm{p}}s_{\mathrm{p}44}}$, $Q_5 = Q_2Q_3$, $Q_6 = Q_1Q_4$,

$Q_1 = \sin(k_{\mathrm{p}}t_{\mathrm{p}})$, $Q_2 = \cos(k_{\mathrm{p}}t_{\mathrm{p}})$, $Q_3 = \sin(k_{\mathrm{m}}t_{\mathrm{m}})$, $Q_4 = \cos(k_{\mathrm{m}}t_{\mathrm{m}})$,

ω is the circular frequency.

As an example, Fig. 1.44 shows a resonance peak at shear modes in free-standing bilayer of lead zirconate titanate (PZT) and nickel ferrite $NiFe_2O_4$. The maximum value of the ME coefficient [20 V/(cm·Oe)] is observed at frequency of about 0.8 GHz. Placing the magnetostrictive–piezoelectric film on a substrate is expected to lead to two significant effects: a decrease in the EMR frequency due to clamping and a fine structure in the ME voltage versus frequency spectrum. Figure 1.45 shows a family of equally spaced peaks that are linear combinations of the modes corresponding to the substrate and bilayer structure. The frequency separation between two consecutive peaks is determined primarily by the substrate thickness. The envelope of the fine structure shows a maximum at 0.3 GHz. This maximum is due to EMR in ME bilayer. Peak value of ME voltage coefficient [2.2 V/(cm·Oe)] is a factor of about ten smaller than for free-standing bilayer.

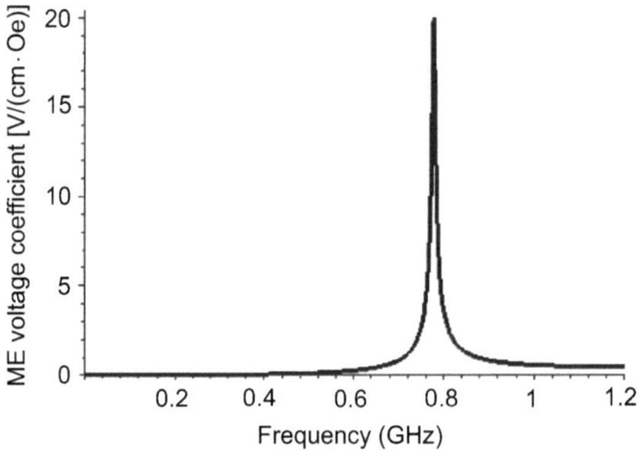

Figure 1.44 Frequency dependence of ME voltage coefficient for a free bilayer structure of PZT of thickness 300 nm and NFO of thickness 2 μm and without substrate.

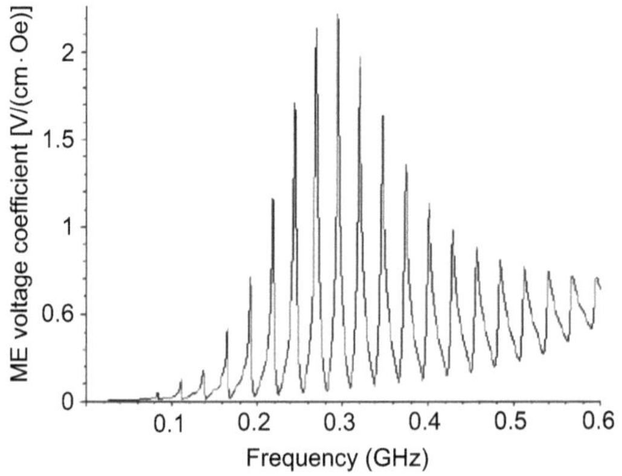

Figure 1.45 Frequency dependence of ME voltage coefficient for bilayer of 300 nm thick PZT and 2 μm-thick NFO on 100 μm-thick SrTiO3 substrate.

In this chapter, we have presented a theory for the resonance enhancement of ME interactions at frequencies corresponding to EMR. Frequency dependence for longitudinal and transverse ME voltage coefficients are obtained using the simultaneous solution of electrostatic, magnetostatic, and elastodynamic equations. The ME voltage coefficients are estimated from known material parameters (piezoelectric coupling, magnetostriction, elastic constants, etc.) of composite components.

It is shown that the ME coupling in the EMR region exceeds the low-frequency value by more than an order of magnitude. It was found that the peak transverse ME coefficient at EMR is larger than the longitudinal one. This is accounted for (i) by higher energy losses for longitudinal fields orientation due to eddy currents in the electrodes and (ii) by influence of demagnetization field that appears at longitudinal orientations that reduces the piezomagnetic modulus.

A theory has been described for ME interactions at shear modes in a magnetostrictive–piezoelectric film on a substrate in EMR region. For an NFO–PZT structure on $SrTiO_3$ substrate, the strength of ME interactions is weaker than for thick film bilayers due to the strong clamping effects of the substrate. For increasing substrate thickness in a bilayer, the ME coefficient drops rapidly and the EMR frequency decreases. For a magnetostrictive–piezoelectric film on a substrate, the ME voltage versus frequency profile shows a fine structure consisting of equally spaced peaks. The distance between the peaks is a function of the substrate thickness. The obtained phenomenon is of importance for the realization of multifunctional ME sensors/transducers operating at microwave frequencies.

The results of calculations obtained for a NFO spinel–PZT composite are in good agreement with the experimental data.

1.3 Ferromagnetic Resonance Range

The majority of prior work on magnetoelectric (ME) composites, both theoretical and experimental studies, has been devoted to the low-frequency range (10 Hz to 10 kHz). However, ME composites also offer important applications in the microwave range [30, 36, 49–55, 91–97]. In this frequency range, the ME effect reveals itself as a change in the magnetic permeability under an external electric field. Investigations of a ferromagnetic resonance (FMR) line shift by an applied electric field are easily performed for layered ferrite–piezoelectric structures [31]. In addition, layered composites are of interest for applications as electrically tunable microwave phase shifters, devices based on FMR, magnetically controlled electro-optical and/or piezoelectric devices, and electrically readable magnetic (ME) memories [30, 37, 38, 70, 92–100].

In this chapter, we discuss theoretical models of multilayer ME composites in the microwave range [101–104]. The model is a phenomenological theory, which is based on the magnetic susceptibility as function of strain, and this strain-dependence is defined both by piezoelectric and the magnetoelastic constants of piezoelectric and magnetostrictive phases, respectively.

1.3.1 Bilayer Structure

Let us again consider the simple model of a bilayer structure, consisting of a ferrite spinel with cubic (m3m) symmetry and poled lead zirconate titanate (PZT) with a symmetry of ∞m. The influence of an electric field on the piezoelectric PZT phase can be described as follows:

$$ {}^{p}T_{ij} = {}^{p}c_{ijkl}\,{}^{p}S_{kl} - e_{kij}E_k; \tag{1.96}$$

where E_k is a component of the electric field vector and ${}^{p}T_{ij}$, ${}^{p}S_{kl}$, e_{kij}, and ${}^{p}c_{ijk}$ are the components of the stress, strain, piezoelectric, and elastic stiffness tensors of the piezoelectric phase, respectively. The tensor coefficients of the elastic stiffness has the form:

$$ {}^{p}c = \begin{pmatrix} {}^{p}\tilde{n}_{11} & {}^{p}c_{12} & {}^{p}c_{13} & 0 & 0 & 0 \\ {}^{p}c_{12} & {}^{p}c_{11} & {}^{p}c_{13} & 0 & 0 & 0 \\ {}^{p}c_{13} & {}^{p}c_{13} & {}^{p}c_{11} & 0 & 0 & 0 \\ 0 & 0 & 0 & {}^{p}c_{44} & 0 & 0 \\ 0 & 0 & 0 & 0 & {}^{p}c_{44} & 0 \\ 0 & 0 & 0 & 0 & 0 & \frac{1}{2}\left({}^{p}c_{11} - {}^{p}c_{12}\right) \end{pmatrix} \tag{1.97}$$

and that of the piezoelectric coefficients is

$$e = \begin{pmatrix} 0 & 0 & 0 & 0 & e_{15} & 0 \\ 0 & 0 & 0 & e_{15} & 0 & 0 \\ e_{31} & e_{31} & e_{33} & 0 & 0 & 0 \end{pmatrix} \tag{1.98}$$

Substituting Eqs. 1.97 and 1.98 into Eq. 1.96, and by applying an electric field along the axis of polarization (i.e., $E_3 = E$, $E_1 = E_2 = 0$), we obtain the following stresses in the piezoelectric phase:

$$\begin{aligned} {}^{\mathrm{p}}T_{33} &= -e_{33}E + {}^{\mathrm{p}}c_{13}({}^{\mathrm{p}}S_{11} + {}^{\mathrm{p}}S_{22}) + {}^{\mathrm{p}}c_{33}{}^{\mathrm{p}}S_{33}, \\ {}^{\mathrm{p}}T_{11} &= -e_{31}E + {}^{\mathrm{p}}c_{11}{}^{\mathrm{p}}S_{11} + {}^{\mathrm{p}}c_{12}{}^{\mathrm{p}}S_{22} + {}^{\mathrm{p}}c_{13}{}^{\mathrm{p}}S_{33} = 0, \\ {}^{\mathrm{p}}T_{22} &= -e_{31}E + {}^{\mathrm{p}}c_{12}{}^{\mathrm{p}}S_{11} + {}^{\mathrm{p}}c_{11}{}^{\mathrm{p}}S_{22} + {}^{\mathrm{p}}c_{13}{}^{\mathrm{p}}S_{33} = 0. \end{aligned} \tag{1.99}$$

Now, assume that the axis of polarization in the piezoelectric phase coincides with the crystallographic [111] axis of the magnetostrictive phase. The stresses in the magnetostrictive phase considering it are then given by:

$$\begin{cases} {}^{\mathrm{m}}T_1 = \left(\dfrac{{}^{\mathrm{m}}c_{11} + {}^{\mathrm{m}}c_{12}}{2} + {}^{\mathrm{m}}c_{44}\right){}^{\mathrm{m}}S_1 + \dfrac{{}^{\mathrm{m}}c_{11} + 5{}^{\mathrm{m}}c_{12} - 2{}^{\mathrm{m}}c_{44}}{6}\,{}^{\mathrm{m}}S_2 \\ \qquad + \dfrac{{}^{\mathrm{m}}c_{11} + 2{}^{\mathrm{m}}c_{12} - 2{}^{\mathrm{m}}c_{44}}{3}\,{}^{\mathrm{m}}S_3 + \dfrac{{}^{\mathrm{m}}c_{11} - {}^{\mathrm{m}}c_{12} - 2{}^{\mathrm{m}}c_{44}}{3\sqrt{2}}\,{}^{\mathrm{m}}S_4, \\ {}^{\mathrm{m}}T_2 = \dfrac{{}^{\mathrm{m}}c_{11} + 5{}^{\mathrm{m}}c_{12} - 2{}^{\mathrm{m}}c_{44}}{6}\,{}^{\mathrm{m}}S_1 + \left(\dfrac{{}^{\mathrm{m}}c_{11} + {}^{\mathrm{m}}c_{12}}{2} + {}^{\mathrm{m}}c_{44}\right){}^{\mathrm{m}}S_2 \\ \qquad + \dfrac{{}^{\mathrm{m}}c_{11} + 2{}^{\mathrm{m}}c_{12} + 4{}^{\mathrm{m}}c_{44}}{3}\,{}^{\mathrm{m}}S_3, \\ {}^{\mathrm{m}}T_3 = \dfrac{{}^{\mathrm{m}}c_{11} + 2{}^{\mathrm{m}}c_{12} - 2{}^{\mathrm{m}}c_{44}}{3}\,{}^{\mathrm{m}}S_1 + \dfrac{{}^{\mathrm{m}}c_{11} + 2{}^{\mathrm{m}}c_{12} - 2{}^{\mathrm{m}}c_{44}}{3}\,{}^{\mathrm{m}}S_2 \\ \qquad + \dfrac{{}^{\mathrm{m}}c_{11} + 2{}^{\mathrm{m}}c_{12} - 2{}^{\mathrm{m}}c_{44}}{3}\,{}^{\mathrm{m}}S_3 - \dfrac{{}^{\mathrm{m}}c_{11} - {}^{\mathrm{m}}c_{12} - 2{}^{\mathrm{m}}c_{44}}{3\sqrt{2}}\,{}^{\mathrm{m}}S_4, \\ {}^{\mathrm{m}}T_4 = \dfrac{{}^{\mathrm{m}}c_{11} + 2{}^{\mathrm{m}}c_{12} - 2{}^{\mathrm{m}}c_{44}}{3}\,{}^{\mathrm{m}}S_1 - \dfrac{{}^{\mathrm{m}}c_{11} - 2{}^{\mathrm{m}}c_{12} - 2{}^{\mathrm{m}}c_{44}}{3\sqrt{2}}\,{}^{\mathrm{m}}S_2 \\ \qquad + \dfrac{{}^{\mathrm{m}}c_{11} - {}^{\mathrm{m}}c_{12} + {}^{\mathrm{m}}c_{44}}{3}\,{}^{\mathrm{m}}S_4 \end{cases} \tag{1.100}$$

To calculate the ME effect of bilayer structures in the FMR range, we should use the following procedure: (i) ${}^{\mathrm{m}}S_3$ is defined as a function of the stress ${}^{\mathrm{m}}T_3$; (ii) ${}^{\mathrm{p}}T_3$ is defined as a function of the strain ${}^{\mathrm{p}}S_3$; and (iii) use the known expressions for the dependence of the resonant magnetic field on stress to determine the FMR line shift. Thus, the task is reduced to the solutions of electrostatic equations under specific boundary conditions.

Let us then consider that the bilayer structure is mechanically clamped along the 3 axis of the ferrite and piezoelectric subsystems. In this case, the boundary conditions, without taking into account forces of friction, have the form:

$$\begin{aligned} {}^{\mathrm{p}}T_3 &= {}^{\mathrm{m}}T_3, \\ {}^{\mathrm{p}}S_3 &= -\frac{{}^{\mathrm{m}}v}{{}^{\mathrm{p}}v}\cdot mS_3, \end{aligned} \tag{1.101}$$

where ${}^{\mathrm{m}}v$ and ${}^{\mathrm{p}}v$ are the volume fractions of the magnetostrictive and piezoelectric phases, respectively. From Eqs. 1.99, 1.100, and 1.101, we can find that the stress along the 3 axis is given by:

$$ {}^{\mathrm{m}}T_3 = \frac{E_3\left(\dfrac{2\,{}^{\mathrm{p}}c_{13}e_{31}}{{}^{\mathrm{p}}c_{11}+{}^{\mathrm{p}}c_{12}} - e_{33}\right)}{1+\dfrac{{}^{\mathrm{m}}v}{{}^{\mathrm{p}}v}\left(\dfrac{1}{3\,{}^{\mathrm{m}}c_{44}}+\dfrac{1}{3({}^{\mathrm{m}}c_{11}+2\,{}^{\mathrm{m}}c_{12})}\right)\left({}^{\mathrm{p}}c_{33}-2\dfrac{{}^{\mathrm{p}}c_{13}^2}{{}^{\mathrm{p}}c_{11}+{}^{\mathrm{p}}c_{12}}\right)} \tag{1.102}$$

It is known that an applied electric voltage results in a change of the resonant magnetic field [102]. We limit ourselves to the case when both the magnetic and electric fields are oriented along the polarization axis of the piezoelectric phases, which also corresponds to the [111] axis of the magnetostrictive one. In this case, the shift of the resonant magnetic field is given by:

$$\delta H_{\mathrm{E}} = \frac{3\lambda_{111}\,{}^{\mathrm{m}}T_3}{M_0} = AE_3 \tag{1.103}$$

where M_0 is the saturation magnetization of the magnetostrictive layer and the ME constant is defined by:

$$A = \frac{3\lambda_{111}}{M_0} \frac{\left(\frac{2^{\mathrm{p}}c_{13}e_{31}}{{}^{\mathrm{p}}c_{11} + {}^{\mathrm{p}}c_{12}} - e_{33} \right)}{\left[1 + \frac{{}^{\mathrm{m}}v}{{}^{\mathrm{p}}v} \left(\frac{1}{3^{\mathrm{m}}c_{44}} + \frac{1}{3({}^{\mathrm{m}}c_{11} + 2^{\mathrm{m}}c_{12})} \right) \left({}^{\mathrm{p}}c_{33} - 2\frac{{}^{\mathrm{p}}c_{13}^2}{({}^{\mathrm{p}}c_{11} + {}^{\mathrm{p}}c_{12})} \right) \right]} \tag{1.104}$$

From Eqs. 1.103 and 1.104, it is possible to estimate the shift of the resonant field for a bilayer magnetostrictive–piezoelectric composite. The theoretical values for the ME constant $A = \delta \mathrm{H_E}/E_3$ were determined for composites of both nickel ferrite (NFO)–PZT and yttrium iron garnet (YIG)–PZT. For these calculations, the following materials parameters of two component phases were used:

YIG ${}^{\mathrm{p}}c_{11} = 12.6 \times 10^{10}$ N/m^2, ${}^{\mathrm{p}}c_{12} = 7.95 \times 10^{10}$ N/m^2, ${}^{\mathrm{p}}c_{13} = 8.4 \times 10^{10}$ N/m^2, ${}^{\mathrm{p}}c_{33} = 11.7 \times 10^{10}$ N/m^2, $e_{31} = -6.5$ Sim/m^2, and

$e_{33} = 23.3$ Sim/m^2

$NiFe_2O_4$ ${}^{\mathrm{m}}c_{11} = 22 \times 10^{10}$ N/m^2, ${}^{\mathrm{m}}c_{12} = 10.9 \times 10^{10}$ N/m^2,

${}^{\mathrm{m}}c_{44} = 8.12 \times 10^{10}$ N/m^2, $\lambda_{111} = -21.6 \times 10^{10}$, and $4\pi M_0 = 3200$ Gs.

In Fig. 1.46, the dependence of the ME constant A on the relative volume fraction of the component phases in the bilayer structure is shown. In a composite of NFO and PZT, the ME effect has been reported to be stronger than that of YIG–PZT. This is because NFO has a much higher magnetostriction than YIG. Another notable feature in Fig. 1.46 is the fact that theoretically the field shift should increase with increasing volume fraction of the piezoelectric phase. However, it should be borne in mind that the magnetic resonance line will become too weak if the volume fraction of the magnetostrictive phase is too low. This is uniquely different from results at lower frequencies, where the largest ME effect is found for an approximate volume fraction ratio of 50:50. Assuming that ${}^{\mathrm{m}}v/{}^{\mathrm{p}}v = 1$, we can estimate that the resonance line shift under applied electric field is given by the relation $\delta\mathrm{H} = \mathrm{AE}$, where $A \approx 2$ Oe·cm/kV for $NiFe_2O_4$–PZT and A ≈ 0.45 Oe·cm/kV for YIG–PZT.

The values for the calculated resonance line shift are in good agreement with prior experimental data: theoretical line shift is 0.45 Oe·cm/kV, which is comparable to the measured one of 0.2–0.56 Oe·cm/kV depending on relative volume fraction of magnetostrictive and piezoelectric phases. Thus, we can see that it is necessary to use a piezoelectric component with large piezoelectric coefficients, and a magnetostrictive one with small saturation magnetization and high magnetostriction.

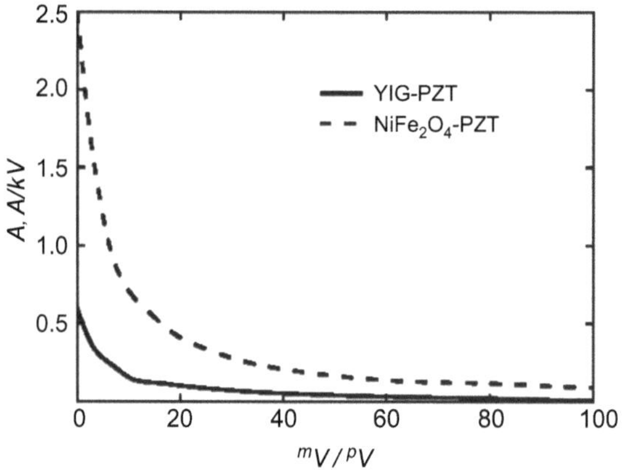

Figure 1.46 Dependence of magnetoelectric constant A on volume fraction of magnetostrictive and piezoelectric phases ${}^{m}v/{}^{p}v$. The results are for bilayer composites of nickel ferrite–PZT and YIG–PZT.

1.3.2 Basic Theory: Macroscopic Homogeneous Model

Now, let us assume that the composite is a macroscopically homogeneous material, that is, the ME composite can be considered as an effective media with an ME effect that is not observed separately in each component phase. In this case, the influence of an external electric field E on the magnetic resonance spectrum can be described by the term [102, 103–107]:

$$W = \int_{V} (W_0 + \Delta W_{\mathrm{ME}})\, d^3x \tag{1.105}$$

where W_0 is thermodynamic potential at $E = 0$, and

$$\Delta W_{\mathrm{ME}} = B_{ikn} E_i M_k M_n + B_{ijkn} E_i M_k M_n \tag{1.106}$$

where B_{ijk} and b_{ijkn} are the linear ME coefficients which are third-rank axial tensors. The potential ΔW_{ME} can be found by considering the elastic, magnetostrictive, piezoelectric, and electrostrictive contributions at specified boundary conditions.

First, let us obtain the expression for the magnetic susceptibility tensor of the ferrite phase in the presence of ME interactions. In this case, the composite will be influenced by an applied electric field, and also by the constant and variable magnetic fields that are necessary to observe magnetic resonance. Solving the linearized equation for the magnetization vector rotation, without taking into account losses and taking into account the effect of demagnetization factors, the following expression can be obtained [104] for the magnetic susceptibility tensor:

$$\chi^{\mathrm{m}} = \begin{bmatrix} \chi_1 & \chi_5 - i\chi_{\mathrm{a}} & 0 \\ \chi_5 - i\chi_{\mathrm{a}} & \chi_2 & 0 \\ 0 & 0 & 0 \end{bmatrix} \tag{1.107}$$

where

$$\chi_1 = D^{-1}\gamma^2 M_0[H_{03'} + M_0 \sum_i (N^i_{1'1'} - N^i_{3'3'})]$$

$$\chi_2 = D^{-1}\gamma^2 M_0[H_{03'} + M_0 \sum_i (N^i_{2'2'} - N^i_{3'3'})]$$

$$\chi_3 = -D^{-1}\gamma^2 M_0^2 \sum_i N^i_{1'2'};$$

$$\chi_a = D^{-1}\gamma M_0 \omega;$$

$$D = \omega_0^2 - \omega^2;$$

$$\omega_0^2 = \gamma^2[H_{03'} + \sum_i (N^i_{1'1'} - N^i_{3'3'})M_0][H_{03'} + \sum_i (N^i_{2'2'} - N^i_{3'3'})M_0] - (\sum_i N^i_{1'2'} M_0)^2,$$

and where γ is the magnetomechanical coefficient, ω is circular frequency, N_{kl} are the effective demagnetization factors describing the effective fields of the magnetic anisotropy, and 1′, 2′, and 3′ is a coordinates system for which the axis 3′ is directed along that of the spontaneous magnetization.

The following additional term ($i = E$) must be included into the magnetic anisotropy to account for the ME constants B_{ij}

$$N^{i'}_{k'n'} = 2B_{ikn}E_{oi}\beta_{k'k}\beta_{n'n} \tag{1.108}$$

where β is the direction cosines matrix of axes (1′, 2′, 3′) in respect to the axes of the crystallographic system (1, 2, 3). We must also account for losses in the equation for the magnetization vector rotation, which takes the form of $i\omega\alpha(\boldsymbol{M}_0 \times \boldsymbol{m})/M_0$, where α is the loss parameter and m *is* the ac magnetization. This results in a complex value for the magnetic susceptibility, given as $\chi = \chi' + i\chi''$ with

$$\chi' = \chi_0 \frac{\omega_0^2(\omega_0^2 - \omega^2 + 2\alpha^2\omega^2)}{(\omega_0^2 - \omega^2)^2 + 4\alpha^2\omega_0^2\omega^2},$$

$$\chi'' = \chi_0 \frac{\alpha\omega\omega_0(\omega_0^2 + \omega^2)}{(\omega_0^2 - \omega^2)^2 + 4\alpha^2\omega_0^2\omega^2}, \chi_0 = \gamma\frac{M_0}{\omega_0}.$$

As an example, we consider a special case when the magnetostrictive component is magnetized in a plane (110) under the corner θ to the cubic axis [001]. Theoretical calculations are essentially simplified by choice of magnetization direction coinciding to direction of stress ${}^{m}T_{33}$. Then the additional term of energy can be presented as

$$\begin{aligned}\Delta W_{\mathrm{ME}} &= \frac{3}{8M^2}[\lambda_{111} - \lambda_{100} + (\lambda_{100} - \lambda_{111}]\cos 2\theta)M_1^{2\mathrm{m}}T_{33} \\ &+ \frac{9}{32M^2}[\lambda_{111} - \lambda_{100} + (\lambda_{100} - \lambda_{111})\cos 4\theta]M_2^{2\mathrm{m}}T_{33} \\ &+ \frac{3}{8M^2}\left((\lambda_{100} - \lambda_{111})\sin 2\theta + \frac{3}{2}(\lambda_{100} - \lambda_{111})\sin 4\theta\right)M_2M_3{}^{\mathrm{m}}T_{33} \\ &+ \frac{3}{8M^2}[\left(\frac{-9\lambda_{111} - 7\lambda_{100}}{4}\right) + (\lambda_{111} - \lambda_{100})\cos 2\theta \\ &+ \frac{3}{4}(\lambda_{111} - \lambda_{100})\cos 4\theta]M_3^{2m}T_{33}.\end{aligned} \quad (1.109)$$

For the special case considered above when $\boldsymbol{M}_0//[111]$, we can obtain

$$\begin{aligned}\Delta W_{\mathrm{ME}} &= \left(\frac{\lambda_{111} - \lambda_{100}}{2}\right)M_1^{2\mathrm{m}}T_{33} + \left(\frac{\lambda_{111} - \lambda_{100}}{2}\right)M_2^{2\mathrm{m}}T_{33} \\ &+ \left(\frac{-\lambda_{111} - \lambda_{100}}{2}\right)M_3^{2\mathrm{m}}T_{33}.\end{aligned} \quad (1.110)$$

Using Eq. 1.110, it is then easy to show that the resonance line is shifted under the action of an applied electric field and using a linear approximation for $N_{\mathrm{kl}}^{\mathrm{E}}$, we have

$$\delta H_{\mathrm{E}} = -\frac{M_0}{Q_1}\left[Q_2(N_{11}^{\mathrm{E}} - N_{33}^{\mathrm{E}}) + Q_3(N_{22}^{\mathrm{E}} - N_{33}^{\mathrm{E}}) + Q_4N_{12}^{\mathrm{E}}\right], \quad (1.111)$$

where

$$Q_1 = 2H_3 + M_0 \sum_{i \neq \mathrm{E}} \left[(N_{11}^i - N_{33}^i) + (N_{22}^i - N_{33}^i) \right];$$

$$Q_2 = \left[H_3 + M_0 \sum_{i \neq \mathrm{E}} (N_{22}^i - N_{33}^i) \right];$$

$$Q_3 = \left[H_3 + M_0 \sum_{i \neq \mathrm{E}} (N_{11}^i - N_{33}^i) \right];$$

$$Q_4 = 2M_0 \sum_{i \neq \mathrm{E}} N_{12}^i.$$

Equation 1.111 allows us to define the ME constants of the composite, and hence, to interpret data concerning the ME effect.

1.3.3 Uniaxial Structure

Let us apply these equations to the consideration of an ME material with a uniaxial structure having 3m symmetry. In this case, expression for the thermodynamic potential can be written as:

$$\begin{aligned}\Delta W_{\mathrm{ME}} &= E_1[B_{11}(M_1^2 - M_2^2) - 2B_{22}M_1M_2 + 2B_{14}M_2M_3 + 2B_{15}M_1M_3] + \\ &+ E_1[B_{22}(M_2^2 - M_1^2) - 2B_{11}M_1M_2 + 2B_{15}M_2M_3 - 2B_{14}M_1M_3] + E_3(B_{33} - B_{31}) \\ &M_3^2 + E_1^2[(b_{11} - b_{12})M_1^2 + (b_{13} - b_{12})M_3^2 + 2b_{14}M_2M_3 - 2b_{25}M_1M_3 + \\ &+ 2b_{16}M_1M_2] + E_2^2[(b_{11} - b_{12})M_2^2 + (b_{13} - b_{12})M_3^2 - 2b_{14}M_2M_3 + \\ &+ 2b_{25}M_1M_3 - 2b_{16}M_1M_2] + E_3^2(b_{33} - b_{31})M_3^2 + 2E_2E_3(2b_{41}M_1^2 + b_{41}M_3^2 + \\ &+ 2b_{44}M_2M_3 + 2B_{45}M_1M_3 + 2B_{52}M_1M_2) + 2E_1E_3(-2b_{52}M_1^2 - b_{52}M_3^2 - \\ &- 2b_{45}M_2M_3 + 2b_{44}M_1M_3 + 2B_{41}M_1M_2) + 2E_1E_2(-2b_{16}M_1^2 - b_{16}M_3^2 + \\ &+ 2b_{25}M_2M_3 + 2b_{14}M_1M_3 + 2b_{66}M_1M_2),\end{aligned} \quad (1.112)$$

where the usual designations for two indexes is used: $B_{ijk} = B_{i\lambda}$ where $\lambda = 1, 2, 3, 4, 5, 6$ corresponds to $j = k = 1, j = k = 2, j = k = 3, j = 2$ and $k = 3, j = 1$ and $k = 3, j = 1$ and $k = 2$. Using these two index systems, we can then use either tensor $(3 \times 3 \times 3)$ or matrix (6×6) forms for the third-rank linear ME coefficient.

Using effective demagnetization factors, the effective magnetic field can be calculated as [103]:

$$\bar{H}_{\mathrm{E}} = -\partial W_{\mathrm{ME}} / \partial \bar{M} = -N^{\mathrm{E}} \bar{M}; \quad (1.113)$$

Equation 1.106 can be written in a system of coordinates (1′, 2′, 3′) for which the axis 3′ coincides with the direction of spontaneous magnetization M_0. The components of $H_{k'}^{\mathrm{E}}$ can then be given as:

$$H_{k'}^{\mathrm{E}} = \beta_{k'k\,k}^{\mathrm{E}}; \tag{1.114}$$

where the direction cosine matrix β is assumed to be equal to

$$\hat{\beta} = \begin{pmatrix} 1 & 0 & 0 \\ 0 & \cos\Theta & -\sin\Theta \\ 0 & \sin\Theta & \cos\Theta \end{pmatrix}. \tag{1.115}$$

Accordingly, we also have that

$$\mathrm{M}_K = \beta_{K'K}\mathrm{M}_{K'}. \tag{1.116}$$

By substituting Eq. 1.113 into Eq. 1.114 and by considering Eqs. 1.105 and 1.106, we can find that

$$N_{11}^{\mathrm{E}} - N_{33}^{\mathrm{E}} = 4(B_{11}E_1 - B_{22}E_2) + 2(b_{11} - b_{12})(E_1^2 - E_2^2)$$
$$+ 8(b_{41}E_2E_3 - b_{16}E_1E_2 - b_{52}E_1E_3) + 2g_2\cos^2\Theta + 2g_3\sin 2\Theta;$$
$$N_{22}^{\mathrm{E}} - N_{33}^{\mathrm{E}} = 2g_2\cos 2\Theta + 4g_3\sin 2\Theta;$$
$$N_{12}^{\mathrm{E}} = [-2(B_{11}E_2 + B_{22}E_1) + 2b_{16}(E_1^2 - E_2^2) + 4(b_{41}E_1E_3 + b_{52}E_2E_3$$
$$+ b_{66}E_1E_2)]\cos\Theta + [2(B_{14}E_2 - B_{15}E_1) + b_{25}(E_1^2 - E_2^2) - 4(b_{14}E_1E_2$$
$$+ b_{44}E_1E_3 + b_{45}E_2E_3)]\sin\Theta, \tag{1.117}$$

where

$$g_2 = -B_{11}E_1 + B_{22}E_2 + (B_{31} - B_{33})E_3 + (b_{12} - b_{13})E_1^2 + (b_{11} - b_{13})E_2^2$$
$$+ (b_{31} - b_{33})E_3^2 + 2(b_{16}E_1E_2 - b_{41}E_2E_3 + b_{52}E_1E_3); \tag{1.118}$$
$$g_3 = -B_{14}E_1 - b_{15}E_2 + b_{14}(E_2^2 - E_1^2) + 2(b_{25}E_1E_2 + b_{44}E_2E_3 - b_{45}E_1E_3);$$
$$B_{66} = (b_{11} - b_{12})/2.$$

Please note that indexes in the right-hand side of (1.117) correspond to those of the crystallographic system of coordinates. We then consider the case when the electric field is directed along the axis of symmetry, that is, $E_1 = E_2 = 0$, and $E_3 = E$. For this specific orientation, we obtain

$$N_{11}^{\mathrm{E}} - N_{33}^{\mathrm{E}} = 2[(B_{31} - B_{33})E + (b_{31} - b_{33})E^2]\cos^2\Theta;$$
$$N_{22}^{\mathrm{E}} - N_{33}^{\mathrm{E}} = 2[(B_{31} - B_{33})E + (b_{31} - b_{33})E^2]\cos^2\Theta; \tag{1.119}$$
$$N_{12}^{\mathrm{E}} = 0.$$

Taking into account the magnetic cubic anisotropy and the anisotropy form for the ferrite layer oriented along the crystallographic (111) plane (i.e., E // H // [111]), it is possible to find that the resonance frequency depends on magnetic and electric fields, as given by:

$$\omega/\gamma = H_3 + M_0[4/3 \cdot K_1/M_0^2 - 4\pi + 2(B_{31} - B_{33})E + 2(b_{31} - b_{33})E^2]; \tag{1.120}$$

where M_0 is now the effective saturation magnetization of the multilayer composite as an effective magnetic media. If we wish to only consider how the resonance frequency shifts under an applied electric field due to magnetoelectric interactions, we neglect all other contributions to Eq. 1.112 other than those provided via the linear ME coefficient B_{ijk}, which can be given as:

$$\delta H_{\mathrm{E}} = -2M_0(B_{33} - B_{31})E. \tag{1.121}$$

From Eqs. 1.104 and 1.120, we can estimate the value of the ME coefficient in the microwave range [103]: $2M_0(B_{31} - B_{33}) = 0.1$ A/kV for YIG–PZT composite, $2M_0(B_{31} - B_{33}) = 0.6$ A/kV for a lithium ferrite (LFO)–PZT composite, and $2M_0(B_{31} - B_{33}) = 1.4$ A/kV for a NFO–PZT composite. Similar analyses can be performed to determine the ME constants for other orientations of E and H fields.

Using the above values of the ME constant in the microwave range, it is now possible to define the changes in the magnetic susceptibility under an external electric field by Eq. 1.120. Results of such calculations of the magnetic susceptibility for bilayer composites are given in Figs. 1.47 and 1.48. In Fig. 1.47, the dependence of the imaginary part of the magnetic susceptibility on dc bias is shown for a bilayer disc composite, where both constant magnetic and electric fields are perpendicular to the plane of the sample. For $E = 0$, a typical resonant line is observed. External constant electric field simply results in a shift of the resonant magnetic field, where the magnitude of the shift is proportional to the ME constants, which in turn are defined by the piezoelectric and magnetoelastic constants. Prior investigations have shown shifts of 330 Oe for NFO–PZT and 22 Oe for YIG–PZT. The large ME effect in NFO is due to its much larger magnetostriction, in comparison to YIG. Clearly, the theoretical model presented above is an important tool for predicting ME interaction in the microwave frequency range. It allows us to define the ME constants and their dependence on crystallography, material systems, and boundary parameters.

Figures 1.47 and 1.48 illustrate the dependence of the resonant (9.2 GHz) magnetic susceptibility on an external electric field. The width of the resonant line in terms of electric field is inversely proportional to the value of $2M_0\,(B_{31} - B_{31})$. From Eq. 1.119, it then follows that a narrow resonant line is characteristic of a composite with strong ME effects. Accordingly, a bilayer composite of ferrite nickel–PZT has a narrower resonant line, relative to a similar composite structure with YIG–PZT.

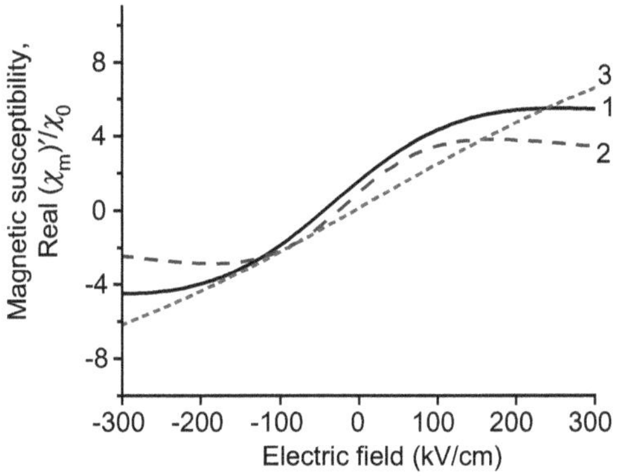

Figure 1.47 Dependence of the real part of magnetic susceptibility for layered structure at frequency 9.3 GHz on the electric field: (1) lithium ferrite–PZT, (2) nickel ferrite–PZT, (3) YIG–PZT.

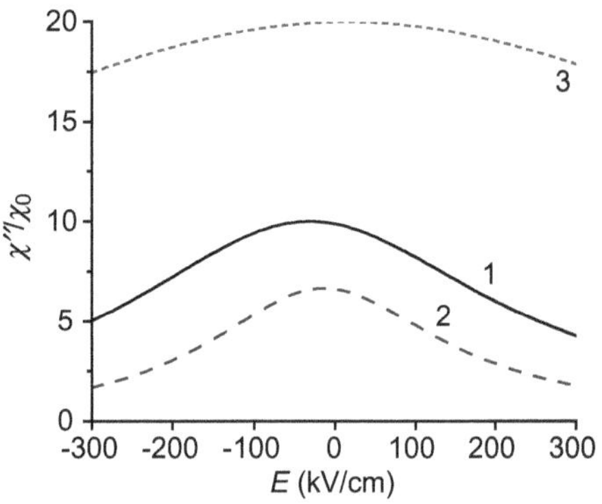

Figure 1.48 Dependence of the imaginary part of magnetic susceptibility for layered structure at frequency 9.3 GHz on the electric field: (1) lithium ferrite–PZT, (2) nickel ferrite–PZT, (3) YIG–PZT.

The magnetic resonance data shown above was taken under sweeping magnetic and electric fields of possible interests for applications in solid state electronics. Electric field tunability of the magnetic subsystem allows for simplified methods for frequency tuning of microwave signals. Considering that the dielectric breakdown strength of insulating thin layers that are well prepared can reach values of up to 300 kV/cm, it is possible to conceive an ME resonator based on a simple bilayer composite structure of nikel ferrite–PZT that can be tuned up to 925 MHz and up to 60 MHz for a similar structure of YIG–PZT.

1.3.4 Layered Composite with Single-Crystal Components

ME interaction in ferromagnetic–ferroelectric heterostructures is mediated by elastic strain. Under an external electric field *E*, the piezoelectric component phase strains. This strain is then transferred to the ferrite component phase, resulting in a shift of its resonant magnetic field. Modeling of the microwave ME effect in layered or bulk composites has revealed two important types of ME interaction [107, 108], which are as follows: (i) between microwave magnetic and microwave electric fields and (ii) between microwave magnetic and constant electric fields.

There are practical difficulties in observing the bias dependence of the magnetic susceptibility. Measurements on samples of bulk composites of 90%YIG–10%PZT have shown only weak ME interactions, due to a low volume fraction of PZT; but, when the composite was loaded with higher volume fractions of PZT, an FMR line broadening masked the effect of an external electric field. Such line broadening, however, can easily be eliminated in layered structure by using single crystal films of YIG.

Next, let us discuss prior experimental results of microwave ME interactions in bilayer structures of YIG and lead magnesium niobate–lead titanate (PMN–PT) single crystals, and compare data to theoretical predictions. Single crystal heterostructures are preferable for experimental measurements because: (i) a smaller FMR linewidth facilitates exact definition of the resonant magnetic field shift, and accordingly the ME constants; and (ii) theoretically, the ME constant of single crystals should be larger relative to a corresponding polycrystalline sample [109]. Bilayer structures (4×4 mm^2) were made using epitaxial YIG films of thickness 1–110 μm grown on (111) gadolinium gallium garnet (GGG) substrates that were 0.5 mm thick.

Measurements of the ME effect were then performed using an EPR spectrometer at a frequency 9.3 GHz. The input power was 0.1 mW, which corresponds to a magnetic field of about 130 A/m. The ME composite was placed outside of the resonator to eliminate potential overload of the resonator at the resonant absorption peak. Measurements were made under an electric field ($0 < E < 8$ kV/cm) applied perpendicular to the plane of composite for the following directions of magnetization: (i) H'', [011], and (ii) H'', (111).

For $E = 0$, the FMR linewidth was 0.5–4 Oe, depending on film thickness. Under external electric fields, the resonant magnetic field decreased with increasing E. The dependence of resonant magnetic field shift on external dc electric field is shown in Fig. 1.49: data are given for H'', E'', and for different thickness of YIG. A linear dependence of the resonant magnetic field shift on dc electric field can clearly be seen in this figure. The ME constant $A = dH_E/E$ can be estimated from data to be 5.4 A/kV for the composite with a YIG thickness of 4.9 μm. Increase of the YIG thickness to 58 μm resulted in a reduction of the ME constant to 4.4 A/kV.

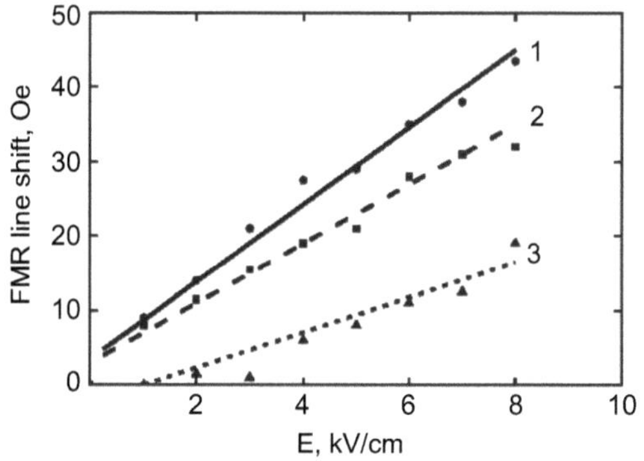

Figure 1.49 Magnetoelectric effect in two-layer structures YIG (111) on GGG and PMN–PT (100) at frequency 9.3 GHz. Static fields E and H are parallel to an axis YIG [111] and perpendicular to sample plane. Shift of resonant magnetic field is shown as function of E for different thickness of YIG film: (1) 4.9 μm, (2) 58 μm, (3) 110 μm.

Measurements of the resonant magnetic field shift were also performed for H'', (011) plane of YIG. Along this direction, the resonant magnetic field shift is not as large as that when H'', (111). Accordingly, the values of dH_E and A were smaller for H'', (011). Prior investigations have been performed on magnetostrictive YIG and piezoelectric PMN–xPT bilayer composites as a function of the relative volume fractions. Figure 1.50 shows that the microwave ME effect depends near linearly on the relative volume fraction. The shift in the resonant magnetic field induced by an external electric field can be seen to decrease with increasing volume fraction of YIG, consistent with the predictions of the last section.

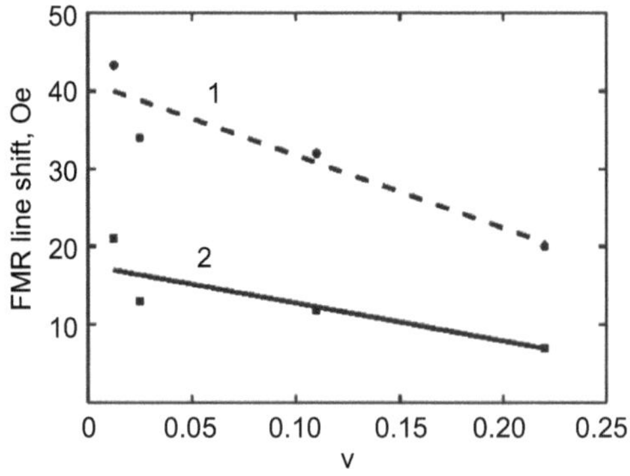

Figure 1.50 Measured shift of resonant magnetic field in electric field E = 8 kV/sm as function of volume fractions of YIG and PMN–PT: (1) H II [111], (2) H II [011].

For comparisons of the experimental data with theoretical predictions, we should consider the additional term for the thermodynamic potential given by Eq. 1.112, which describes influence of external constant electric field E on the magnetic resonance spectrum. Let us then consider a bilayer composite of YIG and PMN–PT. First, we should calculate the deformation induced in PMN–PT by application of E. The generalized Hooke's law in this case, including piezoelectric constitutive relations, can be given as:

$$
\begin{aligned}
{}^{p}S_3 &= {}^{p}d_{33}E + {}^{p}s_{13}({}^{p}T_1 + {}^{p}T_2) + {}^{p}s_{33}{}^{p}T_3, \\
{}^{p}S_1 &= {}^{p}d_{31}E + {}^{p}s_{11}{}^{p}T_1 + {}^{p}s_{12}{}^{p}T_2 + {}^{p}s_{13}{}^{p}T_3, \qquad (1.122) \\
{}^{p}S_2 &= {}^{p}d_{31}E + {}^{p}s_{12}{}^{p}T_1 + {}^{p}s_{11}{}^{p}T_2 + {}^{p}s_{13}{}^{p}T_3 .
\end{aligned}
$$

Now, assume that the direction of the spontaneous polarization in the piezoelectric phase coincides with the [111] axis of the ferrite phase. In this case, the compliance tensor of the ferrite phase is

$$^{m}s_{ijkl} = {}^{m}s_{i'j'k'l'}\,\beta_{ii'}\,\beta_{jj'}\,\beta_{kk'}\,\beta_{kk'}; \tag{1.123}$$

where (1, 2, 3) is a system of coordinates in which axis 3 is directed along direction of the spontaneous magnetization, and β is the direction of cosine matrix. For the ferrite phase we can then determine that

$$^{m}S_3 = {}^{m}s_{13}({}^{m}T_1 + {}^{m}T_2) + {}^{m}s_{33}{}^{m}T_3, \tag{1.124}$$

$$^{m}S_1 = {}^{m}s_{11}{}^{m}T_1 + {}^{m}s_{12}{}^{m}T_2 + {}^{m}s_{13}{}^{m}T_3, \tag{1.125}$$

$$^{m}S_2 = {}^{m}s_{12}{}^{m}T_1 + {}^{m}s_{11}{}^{m}T_2 + {}^{m}s_{13}{}^{m}T_3\,. \tag{1.126}$$

Next, let us use the usual boundary conditions for a mechanically free composite:

$$^{p}S_1 = {}^{m}S_1,\ {}^{p}S_2 = {}^{m}S_2,\ {}^{p}T_3 = {}^{m}T_3 = 0, \tag{1.127}$$

$$^{p}T_1 = -{}^{m}v/{}^{p}v\ {}^{m}T_1\,,\ {}^{p}T_2 = -{}^{m}v/{}^{p}v\ {}^{m}T_2; \tag{1.128}$$

where ^{m}v and ^{p}v are the volume fractions of the ferrite and piezoelectric phases, respectively. Solutions of Eqs. 1.122 and 1.123 by taking into account Eq. 1.124 give values for the stress on the magnetostrictive phase:

$$^{m}T_1 = {}^{m}T_2 = -{}^{p}d_{31}\,E\,{}^{p}v/[{}^{p}v({}^{m}s_{11} - {}^{m}s_{12}) + {}^{m}v({}^{p}s_{11} - {}^{p}s_{12})] \tag{1.125}$$

The shift of the resonant magnetic field (which depends on the volume fractions of magnetic and piezoelectric phases) can then be calculated from Eq. 1.104 by taking into account Eq. 1.118 for the following two cases in the (111) plane of YIG: (i) H II [111] and (ii) H II [011].

Figure 1.51 illustrates the dependence of the ME constant A (which is numerically equal to the shift of the resonant line under E = 1 kV/cm) for a bilayer composite of YIG + GGG and PMN–PT. The thicknesses of the YIG + GGG and PMN–PT layers were 0.5 mm and 0.1 mm, respectively. Calculations were performed using the following material parameters:

$^{p}d_{31} = -600 \times 10^{-12}$ m/V, $^{p}d_{33} = 1500 \times 10^{-12}$ m/V,

$^{p}s_{11} = 23 \times 10^{-12}$ m^2/N, $^{p}s_{12} = -8.3 \times 10^{-12}$ m^2/N,

$\lambda_{100} = -1.4 \times 10^{-6}$, $\lambda_{111} = -2.4 \times 10^{-6}$,

$4\pi M_0 = 1750$ Gs, $H_a = -42$ Oe,

$^{m}s_{11} = 4.8 \times 10^{-12}\,m^2/N$, and $\quad ^{m}s_{12} = -1.4 \times 10^{-12}\,m^2/N$.

Figure 1.51 shows (i) that the ME constant decreases with increasing YIG volume fraction; and (ii) that the ME interaction is larger for magnetic fields perpendicular to composite plane than for those parallel to it. In this figure, good qualitative and quantitative agreement can be seen between theory and data.

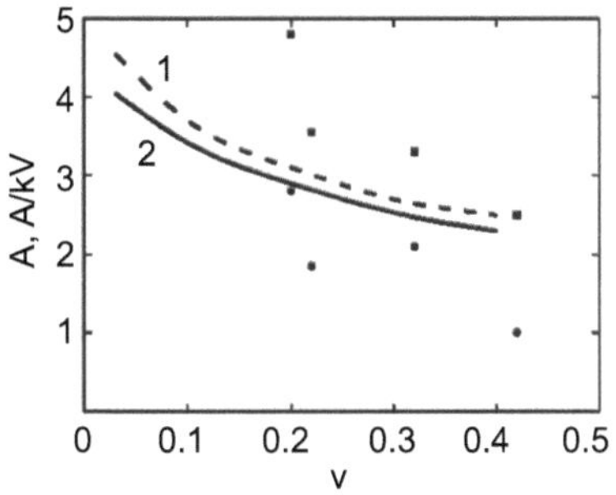

Figure 1.51 Comparison of calculation (line) and data of ME constant for magnetic field laying in sample plane (circles) and perpendicular to sample plane (squares). Values of ME constant are presented as function of (YIG + GGG) and PMN–PT volume fractions.

From the above results, we can conclude to obtain an optimum ME effect: (i) that the volume fraction of the piezoelectric phase should be high; (ii) that it is necessary to use a piezoelectric component phase with a large piezoelectric coefficient; and (iii) that it is necessary to use a magnetostrictive component phase with a small saturation magnetization and a high magnetostriction.

1.3.5 Electric Field–Induced Broadening of Magnetic Resonance Line

Here we discuss the microwave ME interactions in bilayers based on PZT and single crystal YIG with graded anisotropy induced by external electric field. The YIG plate is assumed to be bonded to PZT layer. Electric field applied to piezoelectric layer gives rise to flexural deformations due to the asymmetry of structure. Thickness-dependent strains are transferred into the ferrite phase and result in graded magnetic anisotropy. In addition to the advantage of low linewidths for accurate determination of ME coupling, our theory provides evidence for electric control of magnetic resonance line-width. As opposed to the prediction of variation in ferromagnetic resonance absorption due to the applied electric field to graded ferroelectric [33], we suggest using the ferrite–piezoelectric bilayer with single-sided support. In this case, the applied electric field gives rise to flexural deformations that result in the thickness-dependent induced anisotropy and thus in the shift and broadening of FMR line. This can be useful for microwave applications. Estimates have been obtained for magnetic susceptibility taking into account ME coupling in terms of component material parameters.

For calculating the electric field-induced anisotropy, we consider a bilayer of ferrite and piezoelectric. A bias field H_0 is applied to the ferrite component perpendicular to its plane. This field is supposed to be high enough to drive the ferrite to a saturated state. To find the magnetic susceptibility, we solve the linearized equation of motion of magnetization by using the effective demagnetization factor method. The magnetic susceptibility is known to be expressed by Eq. 1.107. Following types of anisotropy should be taken into account: form anisotropy, crystallographic anisotropy, and electric field-induced anisotropy.

We assume the poling axis of the piezoelectric phase to coincide with [111] axis of the magnetostrictive phase.

The layers are considered perfectly bonded together and the ferrite layer restricts piezoelectric layer's contraction by applying the external electric field. Since the forces are not applied centrally, they produce bending moments on each layer. To take into account the flexural deformations, the longitudinal strains of each layer should be linear functions of the vertical coordinate z_i:

$$^{m}S_1 = {}^{m}S_{10} + z_m/R; \quad (1.129)$$

$$^{p}S_1 = {}^{p}S_{10} + z_p/R; \quad (1.130)$$

where $^{m}S_{10}$ and $^{p}S_{10}$ are the centroidal strains along x-axis at $z_m = 0$ and $z_p = 0$ and R_1 is the radius of curvature, z_m and z_p are counted off from the middle plane of layers.

From geometric consideration, one can see that $^{m}S_{10} - {}^{p}S_{10} = h_m/R$ where $h_m = ({}^{m}t + {}^{p}t)/2$ with ^{m}t and ^{p}t being the thicknesses of piezomagnetic and piezoelectric layers.

Equilibrium implies that the sum of axial forces in the layers must be zero and the sum of the moments must equal to the sum of moment couples created by the shearing forces acting centrally on each layer:

$$F_{m1} + F_{p1} = 0, \tag{1.131}$$

$$F_{m1}h_m = M_{m1} + M_{p1}, \tag{1.132}$$

where $F_{i1} = \int_{-^it/2}^{^it/2} {}^iT_1 dz_i$ and $M_{i1} = \int_{-^it/2}^{^it/2} z_i {}^iT_1 dz_i$ are the transverse force and bending moment per unit width.

For solving Eqs. 1.132 for ${}^mS_{10}$ and R, stress components should be expressed from elasticity equations. Note that ${}^mT_1$ depends on z due to flexural deformations. Eq. 1.132 results in magnetic susceptibility that is z-dependent, too.

Bending the sample have two implications: (i) it gives rise to a decrease in the stress of ferrite layer and thus in ME interaction compared to the case when neglecting the flexural strains, and (ii) the actual manuscript provides evidence that flexural deformations lead to the electric field-induced broadening of magnetic resonance line.

Next we apply the proposed model to an important case of magnetic field H along [111] axis of YIG. For this case, the geometrical demagnetization factors are $N^F_{11} = N^F_{22} = 0$, $N^F_{33} = 4\pi$. Analytical expression for average magnetic susceptibility is too tedious; it is more practical to solve these equations numerically and to present them graphically. The results are for the following material parameters.

PZT: piezoelectric coefficient ${}^pd_{31} = -175{\cdot}\times 10^{-12}$ m/V, compliance constant ${}^ps_{11} = 15{\cdot}\times 10^{-12}$ m^2/N.

YIG: magnetostrictions $\lambda_{100} = -1.4{\cdot}\times 10^{-6}$; $\lambda_{111} = -2.4{\cdot}\times 10^{-6}$;

$4\pi M_0 = 1750$ G; anisotropy field $H_a = -42$ Oe; and

compliance constant ${}^ms_{11} = 4.8{\cdot}\times 10^{-12}$ m^2/N.

The bias field dependence of the imaginary part of the microwave susceptibility at 9.3 GHz is estimated for (111) YIG/PZT (Fig. 1.52).

Fig. 1.52 clearly shows two important effects: the resonance curve is broadened and is down shifted only by approximately

4 Oe. Applied electric field of 30 kV/cm essentially reduces the peak absorption, but the power is coupled to the sample over a wider range of bias field. The FMR linewidth increases to 20 Oe.

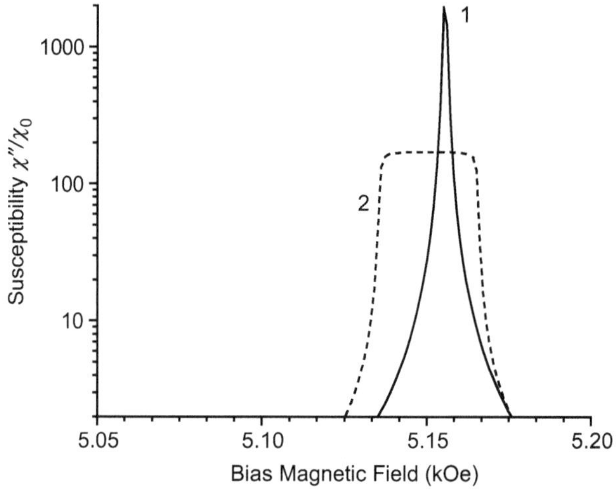

Figure 1.52 Theoretical magnetic field dependence of imaginary parts of magnetic susceptibility for the bilayer of (111) YIG and PZT at 9.3 GHz for $E = 0$ (1) and $E = 30$ kV/cm (2). The bias magnetic field and the electric fields are perpendicular to the sample plane and parallel to [111] axis of YIG. The YIG to PZT thickness ratio equals 1.2.

Figure 1.53 shows the magnetic field dependence of the FMR line broadening for the bilayer of PZT and YIG.

The results are for YIG and GGG layer thickness of 0.1 mm and 0.5 mm, correspondingly, at 9.3 GHz. For microwave measurements at 9.3 GHz, we used an electron spin resonance spectrometer. The incident power on the TE102 reflection cavity with Q of 2000 was about 0.2 mW and provided an rf magnetic field of 1.7 mOe. Weak rf field enabled one to avoid any sample heating. To reduce influence of high-permittivity piezoelectric layer on field distribution in the cavity, the laminate investigated (4 mm × 4 mm) was mounted outside the cavity holes for ME measurements. The cavity had holes at the center of the cavity length. The bias magnetic field and constant electric field were applied perpendicular to the sample plane and parallel to [111] axis of YIG. The power absorption versus H was observed for $E = 0 - 6$ kV/cm. The down-shift of FMR line was calculated as the difference between two bias fields corresponding to $E > 0$ and $E = 0$. Decrease in ME coupling strength due to clamping effect produced by the substrate is estimated similar to our previous calculations [33].

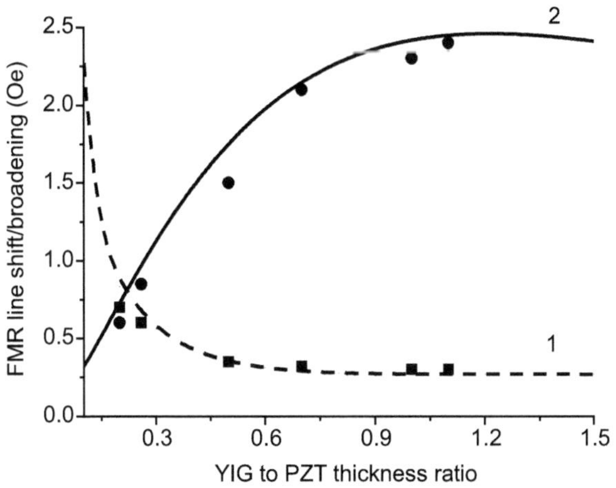

Figure 1.53 YIG to PZT thickness ratio dependence of FMR line shift (1) and broadening (2) for the bilayer of PZT and YIG at 9.3 GHZ for $E = 6$ kV/cm. Solid and dash lines correspond to theoretical estimates and circles and squarescorrespond to data on FMR line broadening and resonance line shift, respectively.

From Fig. 1.53, one can see that maximal FMR line shift is obtained for YIG layer thickness smaller as compared to PZT layer thickness that is in agreement with our previous model [106]. However, the maximal FMR line broadening is obtained for the YIG to PZT thickness ratio of 1.2. Tuning the electric field enables reaching the required FMR linewidth.

1.3.6 Resonance Line Shift by Electric Signal with Electromechanical Resonance Frequency

As we discussed in the last chapter, the electromechanical resonance (EMR) frequency range has a significant enhancement of the ME interaction. Here, we estimate the influence of an external electric field at the EMR frequency on the magnetic resonance spectrum for layered composite with a magnetostrictive ferrite component phase.

To determine the magnetic resonance line shift, we use Eq. 1.101. The calculation of the mechanical stress on the ferrite phase imposed by the piezoelectric on being driven in the EMR range can be obtained from the equation of motion in Eq. 1.73, which has a general solution of the form given in Eq. 1.74. The integration constants can be obtained by assuming that the composite is mechanically free. In this case, the stress in the ferrite component phase can be given as:

$$^{m}T_1 = \frac{[\sin(kx)tg(kL/2)+\cos(kx)]E_3{}^{p}d_{31}v}{^{p}s_{11}(1-v)+{}^{m}s_{11}v} \tag{1.133}$$

From Eq. 1.101, taking into account Eq. 1.126, it is possible to find out how the ME constant depends on the phase volume fractions. The results are shown in Fig. 1.54. In addition, the dependence of the magnetic resonance line shift on external electrical field is illustrated in Fig. 1.55.

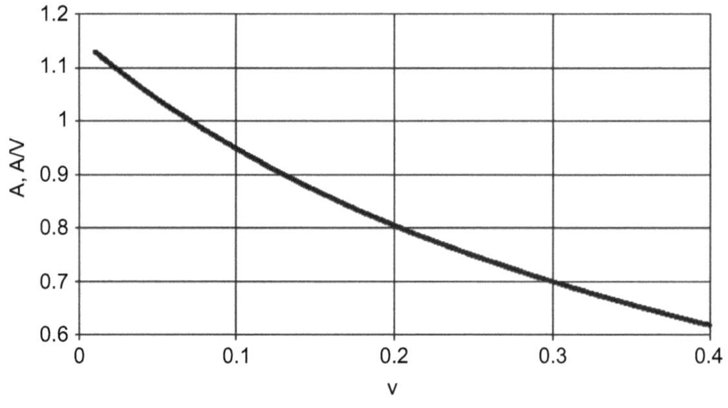

Figure 1.54 Calculated dependence of ME constant in the EMR range for magnetic field perpendicular to sample plane on (YIG + GGG) and PMN–PT volume fractions.

In these figures, it can be seen to achieve an equivalent resonant line shift under external electric field at the EMR frequency that a field of only 1/100th of that of a dc electric bias needs to be applied. These results show that the microwave ME properties should also be dramatically enhanced by driving the composite at its EMR frequency.

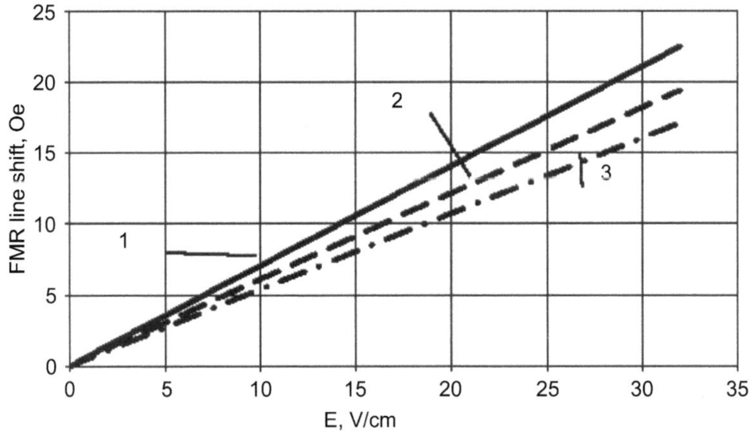

Figure 1.55 Calculated shift of resonant magnetic field in external electric field with EMR frequency for bias fields perpendicular to sample plan: (1) thickness of YIG film is 4.9 μm, (2) 58 μm, (3) 110 μm.

1.4 Magnetoacoustic Resonance Range

Now, let us consider the bilayer composite shown in Fig. 1.56 with ferrite and piezoelectric single crystal layers. The ferrite layer is supposed to be in a saturated single domain state [110–115]. Bias field is assumed to be applied perpendicular to the sample plane. This state has two important advantages. First, when domains are absent, acoustic losses are minimum. Second, the single-domain state under FMR provides the conditions necessary for achieving a large effective susceptibility.

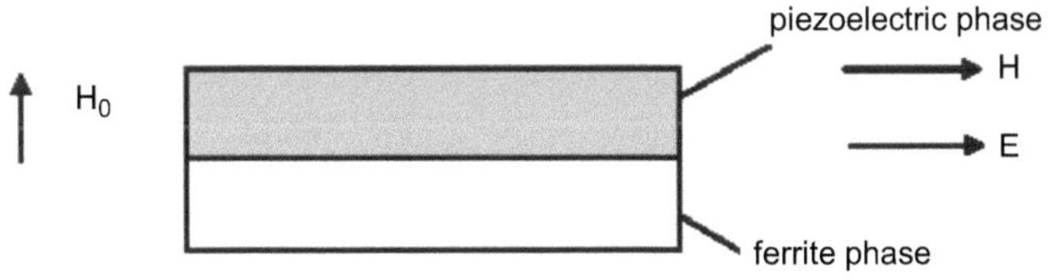

Figure 1.56 Two-layer structure on the basis of single crystal phases.

1.4.1 Direct Magnetoelectric Effect

The free energy density of a single crystal ferrite phase can be given as

$$^{m}W = W_{H} + W_{an} + W_{ma} + W_{ac}; \quad (1.134)$$

where $W_H = -M \cdot H$ is Zeeman's energy,

$W_{an} = K_1/M^4(M_1^2 M_2^2 + M_2^2 M_3^2 + M_3^2 M_1^2)$ is cubic anisotropy energy,

K_1 is a constant of the cubic anisotropy,

$W_{ma} = B_1/M^2(M_1^2\, {}^{m}S_1 + M_2^2\, {}^{m}S_2 + M_3^2\, {}^{m}S_3) + B_2/M^2(M_1 M_2\, {}^{m}S_6 + M_2 M_3\, {}^{m}S_4 + M_1 M_3\, {}^{m}S_5)$ is the magnetoelastic energy, B_1 and B_2 are magnetoelastic constants, and

$W_{ac} = ½\,{}^{m}c_{11}({}^{m}S_1^2 + {}^{m}S_2^2 + {}^{m}S_3^2) + ½\,{}^{m}c_{44}\,({}^{m}S_4^2 + {}^{m}S_5^2 + {}^{m}S_6^2) + {}^{m}c_{12}\,({}^{m}S_1\,{}^{m}S_2 + {}^{m}S_2\,{}^{m}S_3 + {}^{m}S_1\,{}^{m}S_3$ is the elastic energy. In Eq. 1.134, it is supposed that the material is uniformly magnetized. On the basis of a generalized Hooke's law, we can write the stresses of the piezoelectric component phase as:

$$^{p}T_4 = {}^{p}c_{44}\,{}^{p}S_4 - e_{p15}\,{}^{p}E_2,$$
$$^{p}T_5 = {}^{p}c_{44}\,{}^{p}S_5 - e_{p15}\,{}^{p}E_1, \quad (1.135)$$

where e_{p15} is a shear piezoelectric coefficient. The equations of motion for the ferrite and piezoelectric phases are then

$$\partial^2({}^{m}u_1)/\partial t^2 = \partial^2({}^{m}W)/(\partial x\, \partial^{m}S_1) + \partial^2({}^{m}W)/(\partial y \partial^{m}S_6) + \partial^2({}^{m}W)/(\partial z \partial^{m}S_5),$$
$$\partial^2({}^{p}u_1)/\partial t^2 = \partial({}^{m}T_1)/\partial x + \partial({}^{m}T_1)/\partial y + \partial({}^{m}T_1)/\partial z, \quad (1.136)$$
$$\partial^2({}^{p}u_2)/\partial t^2 = \partial({}^{m}T_2)/\partial x + \partial({}^{m}T_2)/\partial y + \partial({}^{m}T_2)/\partial z.$$

The equation of motion of the magnetization vector is given by:

$$\partial M/\partial t = -\gamma[M, H_{eff}], \quad (1.137)$$

where $H_{eff} = -\partial({}^{m}W)/\partial M$.

To simplify calculations, we assume waves with the circular polarization:

$$m^{+} = m_1 + im_2,$$
$$H^{+} = H_1 + iH_2, \quad (1.138)$$
$$E^{+} = E_1 + iE_2,$$
$$u^{+} = u_1 + iu_2,$$

where m is a variable magnetization and u is the displacement. By taking into account Eq. 1.113, Eqs. 1.111 and 1.112 take the form:

$$\omega\, m^+ = \gamma(H_0\, m^+ - 4\pi\, M_0\, m^+ - M_0\, H^+),$$

$$-\omega^2\, {}^{m}\rho\, {}^{m}u^+ = {}^{m}c_{44}\, \partial^2({}^{m}u^+)/\partial z^2, \qquad (1.139)$$

$$-\omega^2\, {}^{p}\rho\, {}^{p}u^+ = {}^{p}c_{44}\, \partial^2({}^{p}u^+)/\partial z^2.$$

The boundary conditions at the interface between layers and at the top/bottom planes of the composite are given by:

$${}^{m}u^+ = {}^{p}u^+ \text{ at } z = 0;$$

$${}^{m}c_{44}\, \partial({}^{m}u^+)/\partial z + B_2\, m^+/M_0 = 0 \text{ at } z = {}^{m}L;$$

$${}^{m}c_{44}\, \partial({}^{m}u^+)/\partial z + B_2\, m^+/M_0 = {}^{p}c_{44}\, \partial({}^{p}u^+)/\partial z - {}^{p}e_{15}\, {}^{p}E^+ \text{ at } z = 0; \quad (1.140)$$

$${}^{p}c_{44}\, \partial({}^{p}u^+)/\partial z - {}^{p}e_{15}\, {}^{p}E^+ = 0 \text{ at } z = -{}^{p}L;$$

where ${}^{m}L$ and ${}^{p}L$ are the thicknesses of the ferrite and piezoelectric layers, respectively. Finally, the electric field E induced in the piezoelectric component can be defined by the open electric circuit condition:

$$D^+ = \frac{1}{{}^{p}L} \int_{-{}^{p}L}^{0} {}^{p}D^+ dz = 0 \qquad (1.141)$$

where ${}^{p}D^+ = {}^{p}e_{15}\, {}^{p}S^+ + {}^{p}\varepsilon_{11}\, {}^{p}E^+$ is the dielectric displacement of the piezoelectric layer.

Substituting the solution of Eq. 1.139 into Eq. 1.141 and by taking into account Eq. 1.140, we can derive an expression for the ME voltage coefficient:

$$|E^+/H^+| = \gamma\, B_2\, {}^{p}c_{44}k_p\, {}^{p}e_{15}\, [1 - \cos(k_p\, {}^{p}L)]^2/\{(\omega - \gamma H_0 + 4\pi\gamma M_0)$$

$$\times [-\tfrac{1}{2}{}^{p}c_{44}\, {}^{p}\varepsilon_{33}\, k_p\, {}^{p}L \sin(2k_p\, {}^{p}L)\, ({}^{p}c_{44}k_p + {}^{m}c_{44}k_m) + [\cos(k_p\, {}^{p}L) - 1]$$

$$\times\, {}^{p}e_{15}{}^{2}[(\cos(k_p\, {}^{p}L)({}^{m}c_{44}k_m + 2\, {}^{p}c_{44}k_p) + {}^{m}c_{44}k_m]]\}, \qquad (1.142)$$

where

$$k_m = \omega\sqrt{\frac{{}^{m}\rho}{{}^{m}c_{44}}}, \qquad k_p = \omega\sqrt{\frac{{}^{p}\rho}{{}^{p}c_{44}}}$$

As follows from Eq. 1.142, there is a connection in the ferrite phase between the displacement and a homogeneous magnetization precession, through boundary conditions on the plate surfaces. Expression 1.142 shows that if the frequency of an applied magnetic field equals that of the magnetization precession ($\omega_0 = \gamma H_0 - 4\pi\gamma M_0$), then the value of the ME voltage coefficient will be increased significantly. This enhancement is due to a coupling between the strain induced by an applied magnetic field in the range of the magnetic resonance and a corresponding one in the piezoelectric phase induced by an applied electric field.

Figures 1.57 and 1.58 show the dependences of the ME voltage coefficient for a bilayer composite of single crystal YIG and PMN–PT, both of which were calculated by Eq. 1.117. Calculations were performed by introducing a complex frequency to account for magnetoacoustic loss. The real component of the frequency was taken as $\omega_r = 10^7$ rad/s.

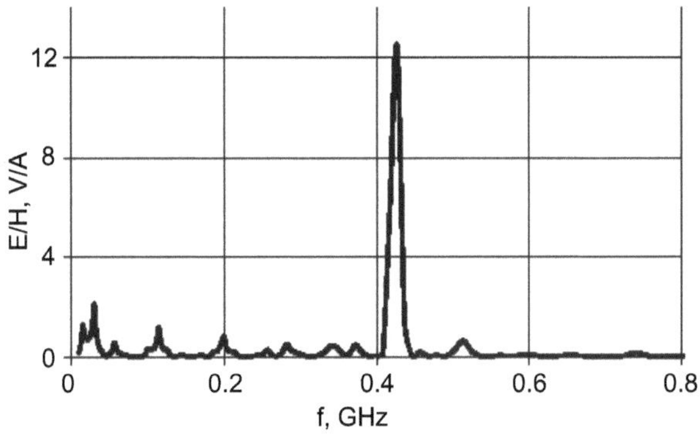

Figure 1.57 Frequency dependence of ME voltage coefficient for $L_1 = L_2 = 0.05$ mm, $H_0 = 1.9$ kOe.

For frequencies less than that of the homogeneous magnetization precession, the microwave ME voltage coefficient has a maximum when driven by an electric field in the EMR frequency range. Essential to this increase in the microwave ME voltage coefficient is that the EMR frequency of the sample is equal to the frequency of uniform magnetization precession in the ferrite phase. Even after including magnetoacoustic loss factors (as mentioned above), the microwave ME coefficient can reach giant values of up to 84 V/(cm · Oe).

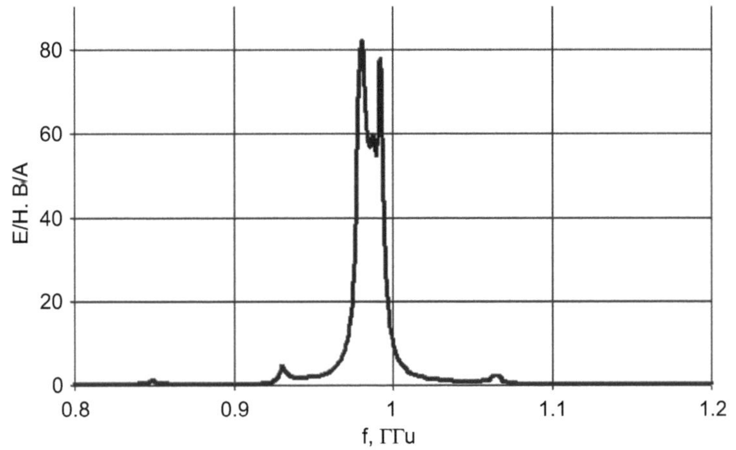

Figure 1.58 Frequency dependence of ME voltage coefficient for $L_1 = L_2 = 0.01$ mm, $H_0 = 2.1$ kOe.

1.4.2 Effects of Exchange Interactions on Magnetoacoustic Resonance

It is well known that the influence of exchange interactions on magnetoelastic waves is stronger with frequency. The exchange interaction determines the strength of the spin wave stiffness parameter and the frequency of the magnetoelastic modes. This section is concerned with the effect of magnetic exchange interactions on ME coupling at magnetoacoustic resonance (MAR) in a ferrite–piezoelectric nanobilayer [110].

The piezoelectric phase of thickness L_1 is supposed to be poled with an electric field E_0 along x direction and the in-plane microwave electric field E is along z. The bias magnetic field H_0 is along z and the alternating magnetic field h (along y) is assumed to be tangential to the bilayer plane. We first solve the equations of motion for magnetization for the ferrite and equations of mechanical displacements for the piezoelectric and ferrite.

The equations of media motion for the ferrite and piezoelectric phases and equation of motion of magnetization take the form, taking into account exchange interactions:

$$i\omega\tilde{m}_{\mathrm{x}} = -\gamma\left(H_0\tilde{m}_{\mathrm{y}} - H_{\mathrm{a}}a^2\frac{\partial^2\tilde{m}_{\mathrm{y}}}{\partial x^2}\right),$$

$$i\omega\tilde{m}_{\mathrm{y}} = \gamma\left((H_0 + 4\pi M_{\mathrm{s}})\tilde{m}_{\mathrm{x}} - H_{\mathrm{a}}a^2\frac{\partial^2\tilde{m}_{\mathrm{x}}}{\partial x^2} + B_2\right.$$

$$-\rho_{\mathrm{m}}\omega^2 U_{\mathrm{mz}} = c_{\mathrm{m44}}\frac{\partial^2 U_{\mathrm{mz}}}{\partial x^2} + \frac{B_2}{M_0}\frac{\partial\tilde{m}_{\mathrm{x}}}{\partial x},$$

$$D_z = \varepsilon_{\mathrm{p}}E_z + e_{\mathrm{p15}}\frac{\partial U_{\mathrm{pz}}}{\partial x},$$

$$-\rho_{\mathrm{p}}\omega^2 U_{\mathrm{pz}} = c_{\mathrm{p44}}\frac{\partial^2 U_{\mathrm{pz}}}{\partial x^2}, \tag{1.143}$$

where H_a is exchange field and a is lattice constant for YIG. Next we let $M_z = M_s$ and $h_z = 0$ since the ac magnetic field is directed along y-axis. Besides, for given composite orientation, we have $N_x = 4\pi$, $N_y = N_z = 0$.

Solving equations of motion of magnetization and media motion enables finding displacement of layers and induced voltage in piezoelectric layer at open circuit condition. For ME voltage coefficient, one obtains the following expression:

$$\alpha_{\mathrm{E}} = 4\pi i\omega\gamma e_{\mathrm{p15}}c_{\mathrm{p44}}k_{\mathrm{p}}k_{\mathrm{m}}B_2$$
$$\times(1-\cos k_{\mathrm{p}}L_1)(\cos k_{\mathrm{m}}L_2 - 1)/[\varepsilon_{\mathrm{p33}}c_{\mathrm{p44}}k_{\mathrm{p}}L_1(\rho_{\mathrm{m}}\omega^2\sin k_{\mathrm{m}}L_2\cos k_{\mathrm{p}}L_1$$
$$+c_{\mathrm{p44}}k_{\mathrm{p}}k_{\mathrm{m}}\sin k_{\mathrm{m}}L_2\cos k_{\mathrm{p}}L_2) + 4\pi e_{\mathrm{p15}}^2(\rho_{\mathrm{m}}\omega^2\sin k_{\mathrm{p}}L_1\cos k_{\mathrm{m}}L_2$$
$$+2c_{\mathrm{p44}}k_{\mathrm{p}}k_{\mathrm{m}}(1-\cos k_{\mathrm{p}}L_1)\cos k_{\mathrm{m}}L_2)(\omega^2 - \gamma^2(H_0 + 4\pi M_{\mathrm{s}})H_0)], \tag{1.144}$$

where L_1 and L_2 are thicknesses of piezoelectric and ferrite phases.

$$k_{\mathrm{m}} = \omega\sqrt{\rho_{\mathrm{m}}}\left[c_{\mathrm{m44}} + \frac{\gamma^2[B_2^2H_0 + 2M_{\mathrm{s}}a^2H_{\mathrm{a}}\omega^2\rho(H_0 + 2M_{\mathrm{s}})]}{M_{\mathrm{s}}[\omega^2 - \gamma^2H_0(H_0 + 4M_{\mathrm{s}})]}\right],$$

$$k_{\mathrm{p}} = \omega\sqrt{\frac{\rho_{\mathrm{p}}}{c_{\mathrm{p44}}}}$$

The theory developed here is applied to a YIG–PZT bilayer. The thicknesses of PZT and YIG layers are assumed to be 100 nm and 195 nm, respectively, so that the EMR thickness mode in PZT will have the same frequency as for FMR in YIG. Figure 1.59 shows the variation in α_E with f. The results are for bias fields H_0 = 1140 Oe for coincidence of FMR and fundamental EMR, and H_0 = 2850 Oe for the coincidence of FMR and higher harmonics of EMR. Consider the results for a bias field of 1140 Oe. A symmetric profile with a peak α_E of 65 V/(cm·Oe) is seen when the exchange parameter H_a = 0. When the effect of exchange is considered, one notices a sharpening of the resonance profile with a 30% increase in ME coeffcient and a down shift in the resonance frequency. One obtains the "exchange amplification" of ME effect. The appearance of a high-frequency shoulder is indicative of flow of rf energy to magnetoacoustic modes. Similar effects are expected for a frequency of about 10 GHz when the bias field is increased to 2850 Oe. When exchange interaction is considered, Figure 1.59 reveals a 60% increase in peak α_E and a broad shoulder due to magnetoacoustic excitations.

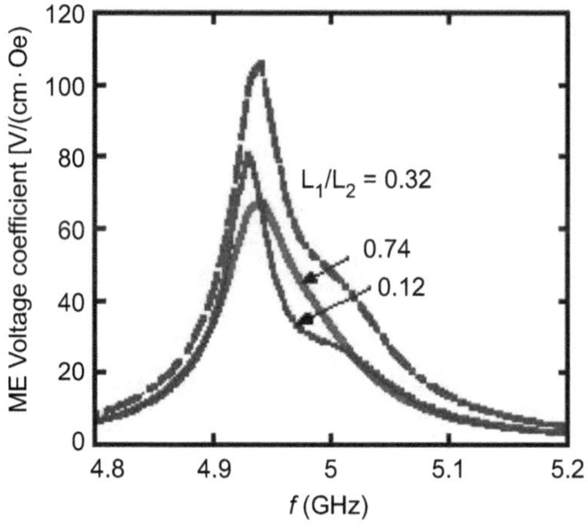

Figure 1.59 The ME voltage coefficient α_E *vs.* frequency profile for a bilayer of lead zirconate titanate (PZT) of thickness 100 nm and yttrium iron garnet (YIG) of thickness 195 nm. Estimates are for magnetic exchange b = 0 and $b \neq 0$ and dc magnetic fields of 1140 Oe and 2850 Oe.

Figure 1.60 shows the effect of PZT-to-YIG thickness on ME coupling. The results also correspond to volume dependence of α_E since the films are assumed to have equal surface areas. For a thickness ratio of 0.74, the figure shows a shallow and broad resonance. When the PZT thickness is reduced so that the thickness ratio is 0.32, one notices a sharp MAR and the shoulder due to spin wave modes is clearly resolved. With a further decrease in PZT thickness, there is reduction in peak α_E and an up shift in the spin-acoustic wave frequencies.

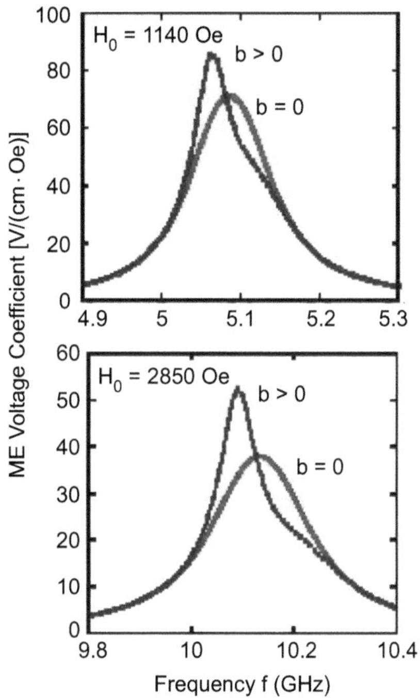

Figure 1.60 Results as in Fig. 1.59, but as a function of ratio of PZT (L_1) to YIG (L_2) thickness. The static magnetic field is 1100 Oe. The total thickness is half wavelength.

Results on thickness dependence of peak, α_{EE} obtained from profiles as in Fig. 1.60 are shown in Fig. 1.61. For small PZT thickness, α_{EE} remains small and shows an increase with increasing PZT thickness. Since EMR in PZT facilitates the maximum transfer of rf power to magnetic subsystem, increase in α_E with PZT thickness is expected. But for thickness ratios above 0.32, we see a drop in the ME coefficient. Thus maximum ME coupling is expected when the YIG layer is three times as thick as the PZT.

The theory discussed here and the estimates for YIG–PZT bilayers will be a roadmap for experiments on magnetoacoustic resonance and investigations on the effects of magnetic exchange interactions.

Such bilayers and ME phenomenon are potentially useful for devices such as resonators or phase shifters based on magnetoacoustic waves.

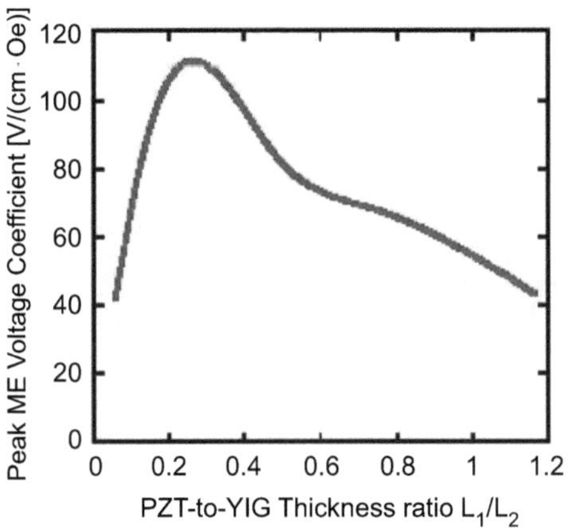

Figure 1.61 Variation of peak ME voltage coefficient α_E with the PZT to YIG thickness ratio. The bias field H_0 = 1100 Oe, $L_1 + L_2$ = half wavelength.

1.4.3 Electric Field–Induced Magnetic Excitations

This section is concerned with consideration of magnetic excitations in a ferrite–piezoelectric bilayer due to microwave electric field and ME interactions [108, 115]. The magnetic response is described in terms of ME susceptibility and a novel technique has been proposed for its determination for a yttrium iron garnet (YIG)–lead zirconate titanate (PZT) nanobilayer. It is assumed that the sample is positioned at the maximum of microwave electric field. An induced microwave magnetic field (parallel to the electric field) will result from ME interactions and, therefore, will lead to magnetic excitations in the bilayer. Such magnetic excitations originate from elastic modes in the piezoelectric component. These acoustic modes would in turn excite coupled magnetoelastic modes in the ferrite due to ME coupling. The excitations are standing waves along the thickness of the sample and the wavelength is determined by the thickness of PZT and YIG, and materials parameters. These coupled magnon–phonon modes will be in the microwave region of the electromagnetic spectrum for YIG.

Thus the focus here is high-frequency magnetic excitations, including ferromagnetic resonance (FMR) and ME susceptibility in a ferrite–ferroelectric bilayer. Traditional FMR at high powers in a ferrite will lead to nonlinear effects such as saturation of main resonance and subsidiary absorption. The idea here is to eliminate those effects by locating a bilayer at the position of maximum rf electric field.

We consider a ferrite–PZT bilayer as in Fig. 1.56 that is subjected to a bias field H_0 perpendicular its plane, along the z-axis. The piezoelectric phase is electrically polarized with a field E_0 parallel to z. The expression for the space-variant microwave magnetization m_k can be obtained by solving combined equations of medium motion for ferrite and piezoelectric phases and equation of motion of magnetization for the ferrite. Thus $\partial m/\partial z$ will appear in the equation for mechanical displacement for ferrite, and the term $\partial u/\partial z$ will be present in the equation of motion of magnetization. The magnetization m_k in terms of circularly polarized mechanical displacement ${}^{\mathrm{m}}u^{+}$ of the ferrite is obtained by solving the above equations and substituting them in:

$$m_{\mathrm{k}} = \frac{B_2\gamma \dfrac{\partial({}^{\mathrm{m}}u^{+})}{\partial z}}{\omega - \omega_{\mathrm{k}}} \tag{1.145}$$

The magnetic modes will have uniform magnetization in the plane of the film and a standing wave structure perpendicular to the film plane. Substituting the value for ${}^{\mathrm{m}}u^{+}$ into Eq. 1.145 yields:

$$m_{\mathrm{k}} = \frac{B_2\gamma e_{15}{}^{\mathrm{p}}E^{\mathrm{m}}k\sin[{}^{\mathrm{m}}k({}^{\mathrm{m}}L - z)][1-\cos({}^{\mathrm{p}}k^{\mathrm{p}}L)]}{{}^{\mathrm{p}}c_{44}{}^{\mathrm{p}}k\sin({}^{\mathrm{p}}k^{\mathrm{p}}L)\cos({}^{\mathrm{m}}k^{\mathrm{m}}L) + {}^{\mathrm{m}}c_{44}{}^{\mathrm{m}}k\sin({}^{\mathrm{m}}k^{\mathrm{m}}L)\cos({}^{\mathrm{p}}k^{\mathrm{p}}L)](\omega-\omega_{\mathrm{k}})} \tag{1.146}$$

where ${}^{\mathrm{m}}c^{+}{}_{44} = {}^{\mathrm{m}}c_{44} + \gamma(B_2{}^2 + \omega^2\, {}^{\mathrm{m}}\rho\, M_0H_{\mathrm{e}})/[M_0(\omega - \gamma H_0 - 4\pi\gamma M_0)$,

$${}^{\mathrm{m}}k = \omega\sqrt{\frac{{}^{\mathrm{m}}\rho}{{}^{\mathrm{m}}c_{44}^{+}}}, \qquad {}^{\mathrm{p}}k = \omega\sqrt{\frac{{}^{\mathrm{p}}}{{}^{\mathrm{p}}c}}.$$

Signal attenuation is taken into account by introducing a complex frequency and an imaginary component of $\omega'' = 10^{-3}\,\omega_{\mathrm{k}}$. This imaginary component corresponds to a Q-value of 1000 for resonance absorption in the ferrite.

Next we apply the theory to the specific case of YIG–PZT bilayer and calculate the ME susceptibility given by $\alpha = \mu_0 \partial m_k / \partial^p E$. The choice of YIG for the ferrite is because of low-losses at microwave frequency, a necessary condition for the observation of the enhancement in the ME coupling that is predicted by the theory. Figure 1.62(a) shows the susceptibility versus frequency *f*. We choose 100 nm YIG and 195 nm PZT so that the fundamental electromechanical resonance (EMR) will be around 6 GHz. A bias field of $H_0 = 2$ kOe is assumed so that it is smaller than the field H_r for the excitation of magnetic modes that include FMR. There are peaks in α at the fundamental EMR and higher order thickness modes. The susceptibility at the fundamental mode is an order of magnitude higher than the value at the higher harmonics.

Consider the results in Fig. 1.62(b) for a bias field H_0 corresponding to magnetic resonance in YIG. When H_0 is set equal to $H_r = \omega/\gamma + 4\pi M_0$, α is expected to show a dramatic increase in magnitude, as in Fig. 1.62(b and c), due to coincidence of resonance character for the mechanical displacement and magnetization. When the frequencies of magnon and phonon modes are matched, there is efficient transfer of energy between the electric and magnetic subsystems. In Fig. 1.62(b) for $H_0 = 3.86$ kOe, the fundamental acoustic mode coincides with uniform precession magnon mode that results in a sixty-fold increase in α. When H_0 is increased to 6 kOe so that uniform precession frequency coincides with the higher order EMR mode, one expects two orders of magnitude increase in α as in Fig. 1.62(c).

The *z*-dependence of the ME susceptibility is shown in Fig. 1.63. The results for 100 nm YIG and 195 nm PZT bilayer show the average value of the ME susceptibility at each *z*, with $z = 0$ representing the interface. As one moves along $+z$ away from the interface, the susceptibility decreases linearly to zero on the outer surface of YIG.

Next we consider measurements of ME susceptibility for the case of a PZT–YIG bilayer. One could use a resonant cavity with the sample located at the ac electric field maximum. The dc magnetic bias field is selected so that homogeneous precession frequency coincides with the fundamental EMR mode. Thus the microwave electric field will result in FMR in YIG and absorption of microwave power. The electric field-induced magnetization is equivalent to that induced by a microwave magnetic field:

$$H = m_k \frac{\gamma H_0 - 4\pi\gamma M_0 - \omega}{\gamma M_0} \tag{1.147}$$

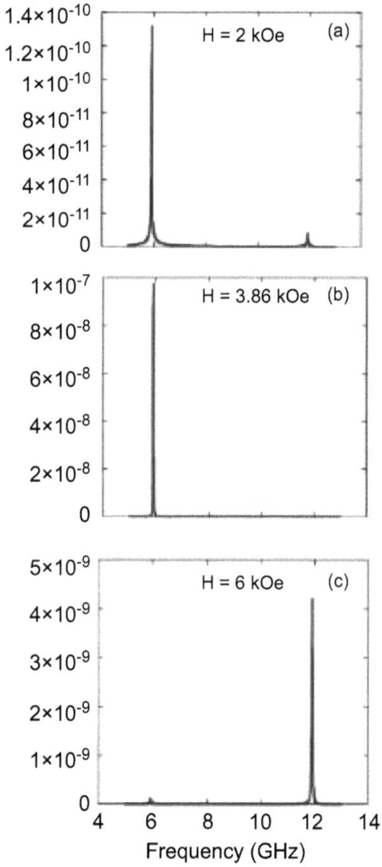

Figure 1.62 Theoretical frequency dependence of the ME susceptibility at the ferrite–piezoelectric interface ($z = 0$) for a bilayer of YIG and PZT subjected to dc and ac fields. The bilayer is assumed at the microwave electric field maximum. (a) Results are for $H_0 = 2$ kOe that is well below the magnetic resonance field for YIG. The peaks correspond to elastic modes in PZT. (b) ME susceptibility for $H_0 = 3.86$ kOe, corresponding to coincidence of fundamental elastic mode in PZT and magnetic resonance in YIG. (c) Similar data at $H_0 = 6$ kOe, at coincidence of higher order elastic mode and magnetic modes.

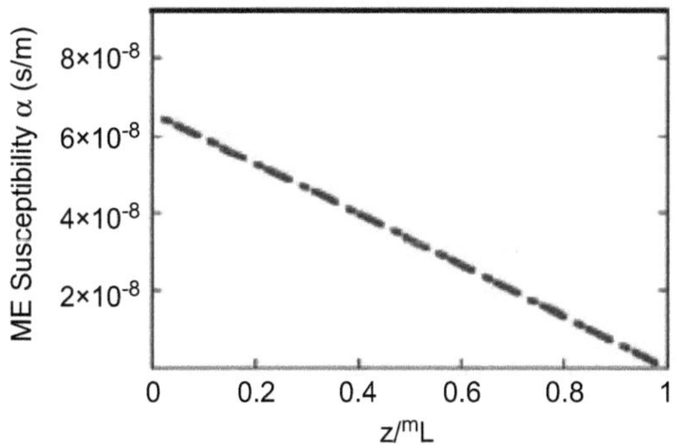

Figure 1.63 The average ME susceptibility *vs.* *z* for a YIG–PZT bilayer, with $z = 0$ is the interface and ^mL is the thickness of YIG. The susceptibility decreases from a maximum value to zero along the thickness of YIG.

Eq. 1.147 is obtained by solving the equation of motion of magnetization for a YIG plate being placed in an antinode of ac magnetic field. The absorbed power P is given by $P = k_1H^2$ with $k_1 = \pi M_0\omega V/\Delta H$, where V is volume of YIG and ΔH is half-width of resonance line. The ME susceptibility can be determined from data on power absorbed.

For using ME composites in microwave devices, the electronic noise is a crucial issue. Noise can be produced by several different effects. Thermal noise and shot noise are known to be inherent to all electronic devices. Flicker noise, also known as 1/f noise, can be produced by ferrite–piezoelectric composites. Using the equivalent circuit of composite in the form of capacitor C connected in parallel with the loss resistance R which produces noises, the following expressions for C and R can be obtained: $C = \varepsilon' A/d$ and $R = 1/(\omega C \tan \delta)$, where ε', A, d and δ are real part of permittivity, cross-sectional area, sample thickness, and dielectric loss angle, correspondingly. The dielectric loss angle was recently reported for the composite of nickel–zinc spinel ferrite and barium strontium titanate. The ferrite volume fraction dependence of loss angle tangent is shown in Fig. 1.64 for frequency of 1 and 10 GHz.

A theoretical model has been discussed for ME effects in a single-crystal ferrite–piezoelectric bilayer in the magnetoelastic resonance region. The theory predicts giant ME effect at MAR. The enhancement arises from interaction between elastic modes and the uniform precession spin-wave mode, resulting in magnetoelastic modes. The peak ME voltage coefficient appears at the coincidence of acoustic resonance and FMR frequencies. Estimates for nominal bilayer parameters for nickel ferrite–PZT and YIG–PZT predict MAR at 5–10 GHz and ME voltage coefficient on the order of 80–480 V/(cm · Oe) taking into account exchange interaction. In our calculations, we suppose that layer thickness is sufficiently large to neglect the influence of strain relaxation on average stresses in the structures that determine the ME voltage coefficient. This is valid for layer thickness of 30–50 nm or more. If necessary, the lattice mismatch effect can be taken into account by using the Landau–Ginsburg–Devonshire phenomenological thermodynamic theory.

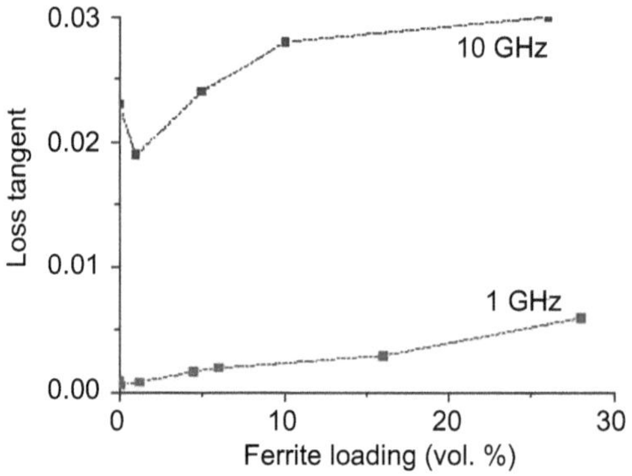

Figure 1.64 The ferrite volume fraction dependence of loss angle tangent for frequency of 1 and 10 GHz and the composite of nickel–zinc spinel ferrite and barium strontium titanate.

Obtained estimates will certainly be of interest for potential device applications based on multilayer composites with ME effects. Microwave devices based on ME effects have unique advantages over traditional ferrite and semiconductor analogues [116–117]. In general, the control can have micro and macro nature and be connected with the change of electric field strength, tension of mechanical field, magnetic field strength, temperature, and their combined effect and also with a change of medium activity conditions.

1.5 Nomograph Method for Predicting Magnetoelectric Coupling

We presented here a new rapid test method for ME laminates based on nomographs. This method can be used when rough estimates are relevant and helpful. To plot the ME parameters versus initial material parameters and component volume fractions, we modelled the magnetic field induced ME effect for several operational modes and laminate composites of most importance to application in electronic devices. In particular, we considered longitudinal and bending modes, laminates of homogeneous and compositionally stepped piezoelectrics and magnetostrictive materials. Using nomographs for a quick test of ME composites for applications where an approximate answer is appropriate and useful was reported recently.

1.5.1 Low-Frequency Magnetoelectric Coupling

Here we present a systematic study on low-frequency and resonance ME effects in magnetostrictive–piezoelectric bilayers and trilayers including structures of PZT bimorph and magnetostrictive layer or PZT and ferromagnetic layer with stepped q. The step in q is accomplished with the use of Ni with negative q and permendur that has positive q.

We consider more often used in practice the transverse fields' orientation that corresponds to E and δE being applied along the X_3 direction, and H and δH along the X_1 direction (in the sample plane).

For symmetric trilayer structures, using the 1-D approximations, the expression for transverse ME voltage coefficient takes on the form:

$$\alpha_{E,31} = \frac{V(1-V)x}{\varepsilon_0[{}^{m}s_{11}V + {}^{p}s_{11}(1-V)]} \tag{1.148}$$

where $x = {}^{m}q_{11}{}^{p}d_{31}\varepsilon_0/{}^{p}\varepsilon_{33}$, ${}^{p}s_{11}$, ${}^{m}s_{11}$, ${}^{p}d_{31}$, and ${}^{m}q_{11}$ are compliance and piezoelectric and piezomagnetic coupling coefficients for piezoelectric and piezomagnetic layers, respectively, ${}^{p}\varepsilon_{33}$ is the permittivity of piezoelectric layer. In Eq. 1.148, the electromechanical coupling factor is assumed to satisfy the condition:

$$^{p}K_{31}^{2} = \frac{^{p}d_{31}^{2}}{^{p}s_{11}\,^{p}\varepsilon_{33}} \ll 1.$$

For convenience we suggest using the nomograph method to estimate the ME voltage coefficients from given parameters of composite components.

Figs. 1.65 and 1.66 present the ME voltage coefficients as a function of piezoelectric volume fraction. As an example, estimates are obtained for transverse fields' orientation (in-plane ac and dc magnetic fields and out-of-plane ac electric field and poling direction).

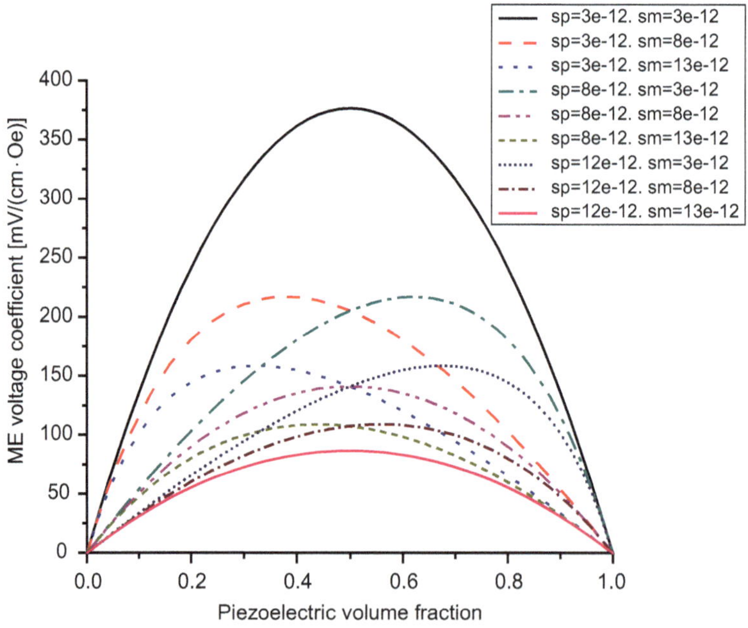

Figure 1.65 Piezoelectric volume fraction dependence of transverse ME voltage coefficient for symmetric layered structure of magnetostrictive and piezoelectric components with different compliances for $x = 0.5 \times 10^{-22}$ (in SI units). At axial mode, this graph is applicable to the quantity $\alpha_E \pi^2/(8Q_a)$ *vs.* piezoelectric volume fraction for both symmetric and asymmetric laminates.

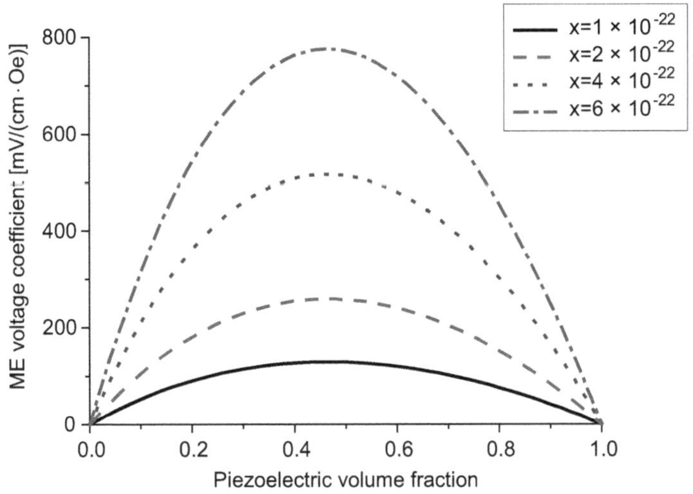

Figure 1.66 Piezoelectric volume fraction dependence of low-frequency transverse ME voltage coefficient for symmetric layered structure of magnetostrictive and piezoelectric components for ${}^{m}s_{11}$ = 20·× 10^{-12} m^2/N and ${}^{p}s_{11}$ = 16 × 10^{-12} m^2/N and for different *x*-values. At axial mode, this graph is applicable to the quantity $\alpha_E\pi^2/(8Q_a)$ *vs.* piezoelectric volume fraction for both symmetric and asymmetric laminates.

For the bilayer structure, the ME voltage coefficient should be calculated taking into account the flexural deformations. On the foregoing assumptions, our model enables deriving the explicit expression for ME voltage coefficient:

$$\frac{\delta E_3}{\delta E_1} = \frac{[1^{p}s_{11} + {}^{m}s_{11}r^3]^{m}q_{11}\dfrac{{}^{p}d_{31}}{{}^{p}\varepsilon_{33}}}{{}^{p}s_{11}[2r^{m}s_{11}(2+3r+2r^2)+{}^{p}s_{11}]+{}^{m}s_{11}^2 r^4}. \tag{1.149}$$

where $r = {}^{p}t/{}^{m}t$ with ${}^{p}t$ and ${}^{m}t$ denoting the thickness of piezoelectric and magnetostrictive layer, correspondingly.

Eq. 1.149 is written in a simplified form under assumption ${}^{p}K_{31}^2$ << 1 similarly to deriving Eq. 1.148. Piezoelectric volume fraction dependence of ME voltage coefficient is shown in Figs. 1.67 and 1.68 for different values of compliance and product of piezoelectric and piezomagnetic coefficient.

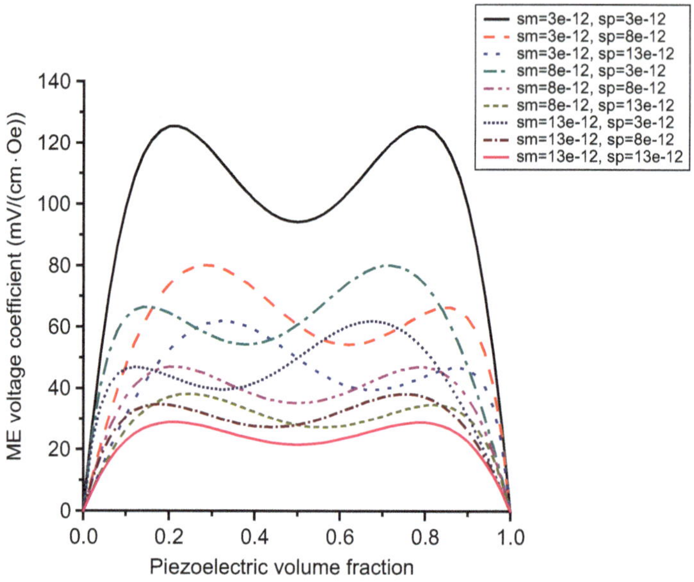

Figure 1.67 Piezoelectric volume fraction dependence of transverse ME voltage coefficient for bilayer of magnetostrictive and piezoelectric components with different compliances for $x = 0.5 \times 10^{-22}$ (in SI units). At axial mode, this graph is applicable to the quantity $\alpha_E\pi^2/(8Q_a)$ *vs.* piezoelectric volume fraction for both symmetric and asymmetric laminates.

Next we consider the low-frequency magnetoelectric coupling in trilayer of stepped piezoelectric layer and magnetostrictive component. Two PZT layers are assumed to be polarized in opposite directions to enable observing the ME coupling due to flexural deformations. The obtained ME coefficients for the structure based on piezoelectric bimorph is found two times higher compared to the traditional bilayer of PZT and Metglas.

To get the explicit expression for ME voltage coefficient, we consider a layered structure of piezoelectric bimorph and magnetostrictive phase which is assumed to consist of several Metglas layers. Two layers of piezoelectric bimorph are polarized in opposite directions. The sample is supposed to be a thin plate with thickness t that is small compared to width w. The sample length L is assumed to be much greater than width w. The poling direction is parallel to thickness direction Z. The applied dc bias and ac magnetic fields are parallel to x-axis (length direction). In this case, only the x-components of stress are nonzero. The magnetic field-produced strain of the magnetostrictive layer is restricted by piezoelectric bimorph. The asymmetry of the sample leads to the stress-produced bending moment with respect to y-axis because the mechanical forces are not applied centrally. Thus, flexural deformations appear. For taking into account the flexural deformations, we assume the longitudinal axial strains of each layer to be linear functions of the vertical coordinate.

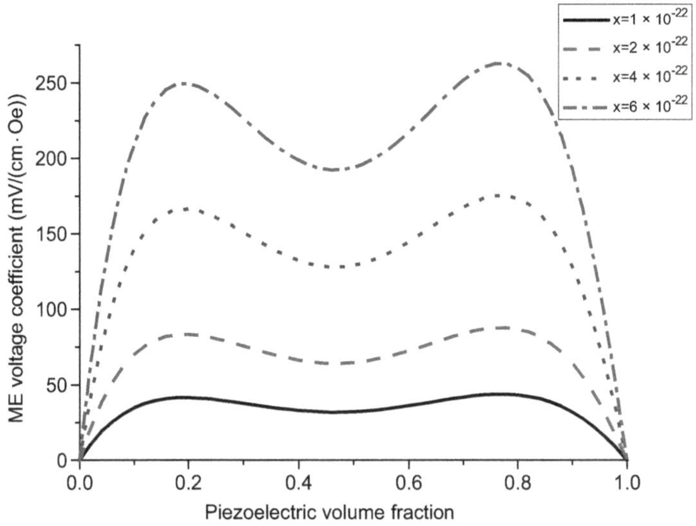

Figure 1.68 Piezoelectric volume fraction dependence of transverse ME voltage coefficient for bilayer of magnetostrictive and piezoelectric components for ${}^{m}s_{11} = 20 \times 10^{-12}$ m^2/N and ${}^{p}s_{11} = 16 \times 10^{-12}$ m^2/N at different x values in SI units.

The theoretical modeling is based on the well-known equations for the strain, electric displacement, and magnetic field of piezoelectric and magnetostrictive layers. Assuming the equal thickness of PZT layers, we can substitute the value of stress into open electric circuit condition and obtain the analytical expression for ME voltage coefficient:

$$\alpha_{E,31} = \frac{{}^{p}d_{31}}{tH_1{}^{p}\varepsilon_{33}} \left(\int_{-{}^{p}t_{1/2}}^{{}^{p}t_{1/2}} ({}^{p1}T_1 + {}^{p2}T_1)dz \right)$$

where H_1 is the applied ac magnetic field. An approximate expression for ME voltage coefficient in an explicit form can be derived assuming the electromechanical coupling coefficient squared to be small compared to unity.

$$\alpha_{E31} = \frac{3^{m}q_{11}{}^{m}s_{11}(1-v)v^{3p}d_{31}}{2\{(^{m}s_{11} - {}^{p}s_{11})v[(^{m}s_{11} - {}^{p}s_{11})v^{3} + 4^{p}s_{11}(v^{2}+1) - 6^{p}s_{11}v] + {}^{p}s_{11}^{2}\}^{p}\varepsilon_{33}}, \tag{1.150}$$

where piezoelectric volume fraction is defined as $v = (^{p1}t + {}^{p2}t)/t$.

Equation 1.150 shows that ME voltage coefficient is basically determined by piezoelectric coefficient of piezoelectric phases and piezomagnetic coefficient of magnetic phases and piezoelectric volume fraction. Thus, one can utilize the most appropriate piezoelectric and magnetic phases and optimal thickness of the components to obtain an increase in the ME coupling strength for the layered structures based on piezoelectric bimorph and magnetostrictive component. Figs 1.69 and 1.70 show the piezoelectric volume fraction dependence of ME voltage coefficient for layered structure of piezoelectric bimorph and Metglas for piezoelectric coefficient of 175 pm/V and 1500 pm/V, correspondingly. Estimated piezoelectric volume fraction dependence of ME voltage coefficient for layered structure of piezoelectric bimorph/Metglas is shown in Fig. 1.71 for different thickness ratio $r = {}^{p1}t/{}^{p2}t$ (^{p1}t and ^{p2}t are the thicknesses of upper and lower PZT layer).

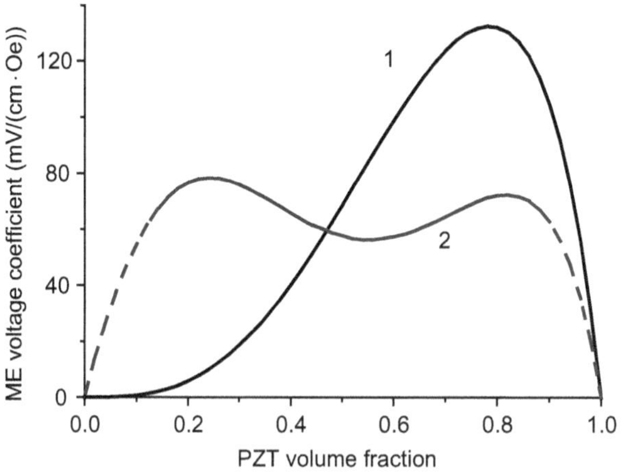

Figure 1.69 Piezoelectric volume fraction dependence of ME voltage coefficient for layered structure of piezoelectric bimorph and Metglas (1) and piezoelectric/Metglas bilayer (2) for piezoelectric coefficient of 175 pm/V.

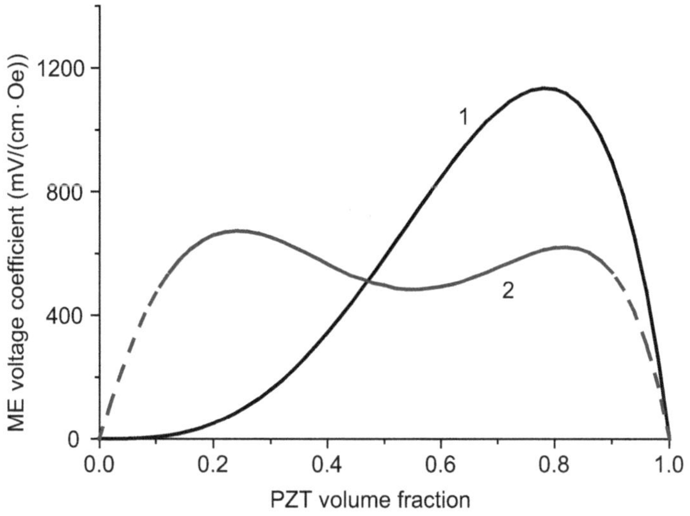

Figure 1.70 Piezoelectric volume fraction dependence of ME voltage coefficient for layered structure of piezoelectric bimorph and Metglas (1) and piezoelectric/Metglas bilayer (2) for piezoelectric coefficient of 1500 pm/V.

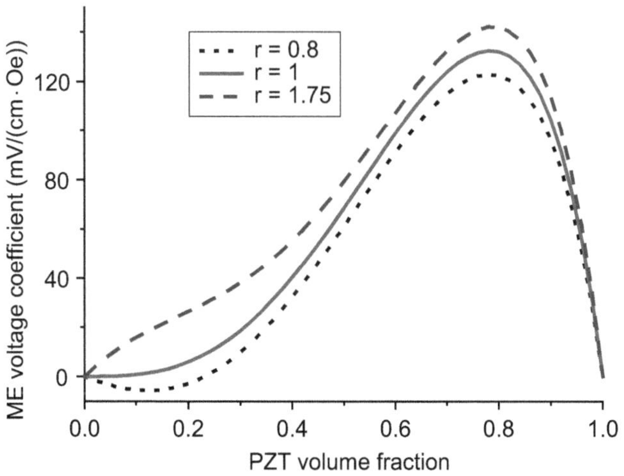

Figure 1.71 Estimated piezoelectric volume fraction dependence of ME voltage coefficient for layered structure of piezoelectric bimorph/Metglas at different thickness ratio $r = {}^{p1}t/{}^{p2}t$ (${}^{p1}t$ and ${}^{p2}t$ are the thicknesses of upper and lower PZT layer).

1.5.2 Magnetoelectric Coupling at Bending Mode

Next we consider ME coupling under small-amplitude flexural oscillations of a bilayer rigidly clamped at one end. The bilayer deflection should obey the equations of bending motion provided in our models in Ref. [11]. To solve these equations, we used the boundary conditions that the bilayer deflection and its derivative vanish at clamped end of the bilayer and rotational moment and transverse force vanish at free end. The resonance condition is $\cosh(kL)\cdot\cos(kL) = -1$ with k being the wave number.

The ME voltage coefficient at bending mode frequency can be estimated as:

$$\frac{\delta E_3}{\delta H_1} = \frac{0.0766 Q_{\mathrm{b}}{}^{\mathrm{m}}t({}^{\mathrm{m}}t + 2z_0)(2z_0 - {}^{\mathrm{p}}t){}^{\mathrm{p}}d_{31}{}^{\mathrm{m}}q_{11}}{D\,{}^{\mathrm{p}}s_{11}{}^{\mathrm{m}}s_{11}{}^{\mathrm{p}}\varepsilon_{33}} \tag{1.151}$$

where D and Q_{a} are cylindrical stiffness sample and the quality factor for the bending resonance. The peak ME voltage coefficient is dictated by Q value, piezoelectric and piezomagnetic coupling coefficients, elastic compliances, and volume fractions of initial components. Eq. 1.151 can be put in a more convenient form:

$$\alpha_{\mathrm{E},31} = \frac{0.0766\,{}^{\mathrm{m}}t({}^{\mathrm{m}}t + 2z_0)(2z_0 - {}^{\mathrm{p}}t)x_1}{D\,{}^{\mathrm{p}}s_{11}{}^{\mathrm{m}}s_{11}\varepsilon_0} \tag{1.152}$$

where $x_1 = {}^{\mathrm{p}}d_{31}{}^{\mathrm{m}}q_{11}Q_{\mathrm{b}}\varepsilon_0/{}^{\mathrm{p}}\varepsilon_{33}$. Piezoelectric volume fraction dependence of peak ME voltage coefficient at bending mode of magnetostrictive–piezoelectric bilayer is shown in Figs. 1.72 and 1.73 for different combinations of the piezoelectric and magnetostrictive layer compliances.

The bending resonance frequency is determined by expression:

$$f_{\mathrm{r}} = \frac{1.758}{\pi L^2}\sqrt{\frac{D}{{}^{\mathrm{p}}\rho^{\mathrm{p}}t + {}^{\mathrm{m}}\rho^{\mathrm{m}}t}}$$

and depends mainly on elastic compliances and volume fractions of initial components, and ratio $L/t^{1/2}$. Piezoelectric volume fraction dependence of bending resonance frequency of magnetostrictive–piezoelectric bilayer is presented in Figs. 1.74 and 1.75 for different combinations of the piezoelectric and magnetostrictive layer compliances.

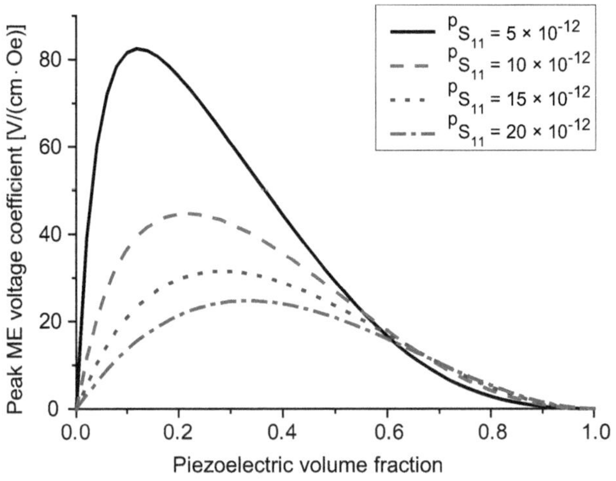

Figure 1.72 Piezoelectric volume fraction dependence of peak ME voltage coefficient at bending mode of magnetostrictive–piezoelectric bilayer for $^{m}s_{11}$ = 20×10^{-12} m^2/N and different values of $^{p}s_{11}$. Value of x_1 is assumed to be equal to 4×10^{-20} (in SI units).

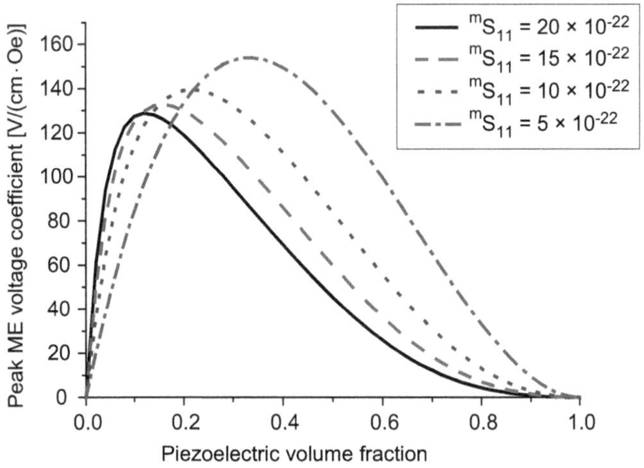

Figure 1.73 Piezoelectric volume fraction dependence of peak ME voltage coefficient at bending mode of magnetostrictive–piezoelectric bilayer for $^{p}s_{11}$ = 5×10^{-12} m^2/N and different values of $^{m}s_{11}$. Value of x_1 is assumed to be equal to 5×10^{-20} (in SI units).

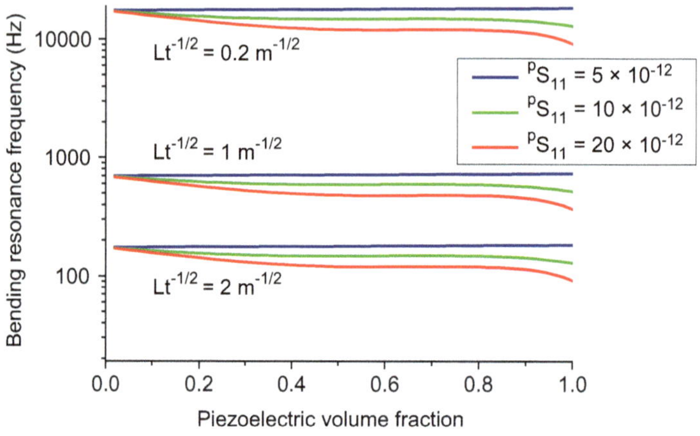

Figure 1.74 Piezoelectric volume fraction dependence of bending resonance frequency of magnetostrictive–piezoelectric bilayer for ${}^{m}s_{11} = 5 \times 10^{-12}$ m²/N and different values of ${}^{p}s_{11}$ and $Lt^{-1/2}$.

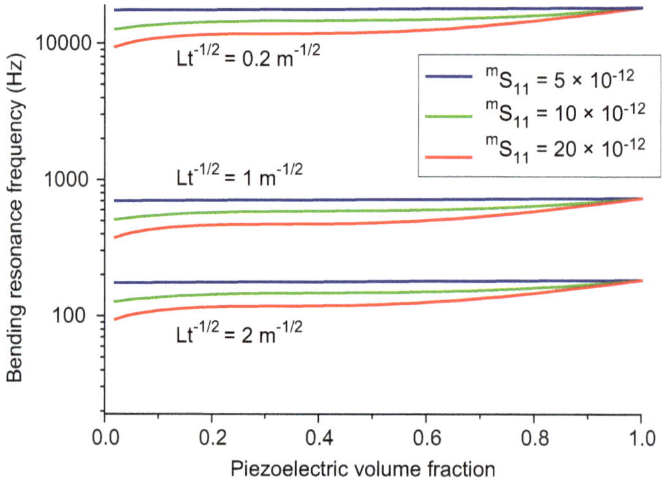

Figure 1.75 Piezoelectric volume fraction dependence of bending resonance frequency of magnetostrictive–piezoelectric bilayer for ${}^{p}s_{11} = 5 \times 10^{-12}$ m²/N and different values of ${}^{m}s_{11}$ and $Lt^{-1/2}$.

The model for enhanced ME interactions at electromechanical resonance in graded ME composites considers longitudinal acoustic modes in a bilayer of ferroelectric phase with a linearly graded piezoelectric coefficient and permittivity and a ferromagnetic phase with linear grading of piezomagnetic coefficient. Our focus here is modeling ME effects at bending modes in composite with stepped *d*.

Solving the equations of bending motion 1.89 and 1.151 enables finding the strains and stresses taking into consideration the boundary conditions at sample ends. The ME voltage coefficient can be found as

$$\alpha_{E31} = \frac{{}^{m}Y{}^{p}Y{}^{p}Y_{31}{}^{m}q_{11}t_{m}(2z_0 + t_m)}{4{}^{p}\varepsilon_{33}DkL(t_{p1} + t_{p2} + t_m)(1 + r_1 r_3)} \times [2z_0(t_{p2} - t_{p1}) - 2t_{p1}t_{p2} - t_{p2}^2 + t_{p1}^2](r_1 r_4 + r_2 r_3) \quad (1.153)$$

where $r_1 = \cos(kL)$; $r_2 = \sin(kL)$; $r_3 = \cosh(kL)$; $r_4 = \sinh(kL)$;

$$k^4 = \omega^2({}^{p}\rho{}^{p}t + {}^{m1}\rho{}^{m1}t + {}^{m2}\rho{}^{m2}t)D^{-1}$$

$$z_0 = \frac{{}^{p}Y{}^{p}t^2 - {}^{m1}Y{}^{m1}t^2 - {}^{m2}Y{}^{m2}t(2{}^{m1}t + {}^{m2}t)}{2({}^{p}Y{}^{p}t + {}^{m1}Y{}^{m1}t + {}^{m2}Y{}^{m2}t)}$$

D is cylindrical stiffness of the sample.

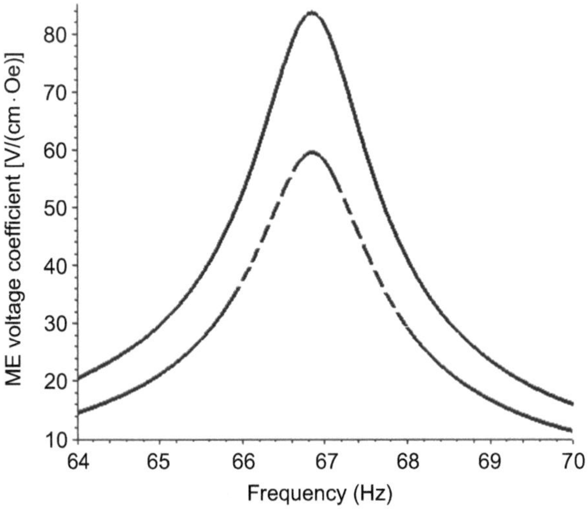

Figure 1.76 Frequency dependence of ME voltage coefficient for layered structure of piezoelectric bimorph/Metglas (solid line) and for bilayer of piezoelectric/Metglas (dashed line) for piezoelectric coefficient of 175 pm/V. Thickness of first and second piezoelectric and Metglas layers equals 0.1, 0.3, and 0.2 mm, correspondingly.

As expected, ME coupling at bending mode is determined by product of piezoelectric coefficient and piezomagnetic coefficients of original phases according to Eq. 1.153. A dramatic enhancement of ME interaction occurs in the EMR region. Neglecting the piezoelectric and piezomagnetic properties results in the following resonance condition: $\cos(kL)\cosh(kL) = -1$. This expression agrees with resonance condition for a bar free at one end and rigidly clamped at the other end. Frequency dependence of ME voltage coefficient for layered structure of piezoelectric bimorph/Metglas is shown in Figs. 1.76 and 1.77 for different values of piezoelectric coefficient.

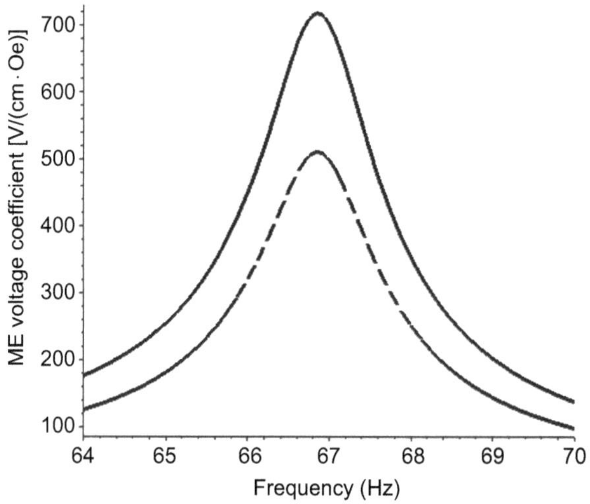

Figure 1.77 Frequency dependence of ME voltage coefficient for layered structure of piezoelectric bimorph/Metglas (solid line) and for bilayer of piezoelectric/Metglas (dashed line) for piezoelectric coefficient of 1500 pm/V. Thickness of first and second piezoelectric and Metglas layers equals 0.1, 0.3, and 0.2 mm, correspondingly.

1.5.3 Magnetoelectric Coupling at Axial Mode

Next we consider small-amplitude axial oscillations of the layered structures formed by magnetostrictive and piezoelectric phases. To solve the equation of media motion, we used the boundary conditions for a bilayer that is free at both ends. The fundamental EMR frequency is given by:

$$f = \frac{1}{2L}\sqrt{\frac{{}^{\mathrm{p}}s_{11} + r\,{}^{\mathrm{m}}s_{11}}{{}^{\mathrm{p}}s_{11}\,{}^{\mathrm{m}}s_{11}(r\,{}^{\mathrm{p}}\rho + {}^{\mathrm{m}}\rho)}}, \tag{1.154}$$

and the peak ME voltage coefficient at axial mode frequency is

$$\frac{\delta E_3}{\delta H_1} = \frac{8Q_{\mathrm{a}}}{\pi^2}\frac{r\,{}^{\mathrm{m}}q_{11}\dfrac{{}^{\mathrm{p}}d_{31}}{{}^{\mathrm{p}}\varepsilon_{33}}}{(r\,{}^{\mathrm{m}}s_{11} + {}^{\mathrm{p}}s_{11})(r+1)}, \tag{1.155}$$

or

$$\frac{\alpha_{\mathrm{E}}}{Q_{\mathrm{a}}} = \frac{8}{\pi^2}\frac{V(1-V)\,{}^{\mathrm{m}}q_{11}\dfrac{{}^{\mathrm{p}}d_{31}}{{}^{\mathrm{p}}\varepsilon_{33}}}{[V\,{}^{\mathrm{m}}s_{11} + (1-V)\,{}^{\mathrm{p}}s_{11}]}$$

where Q_{a} is the quality factor for the EMR resonance.

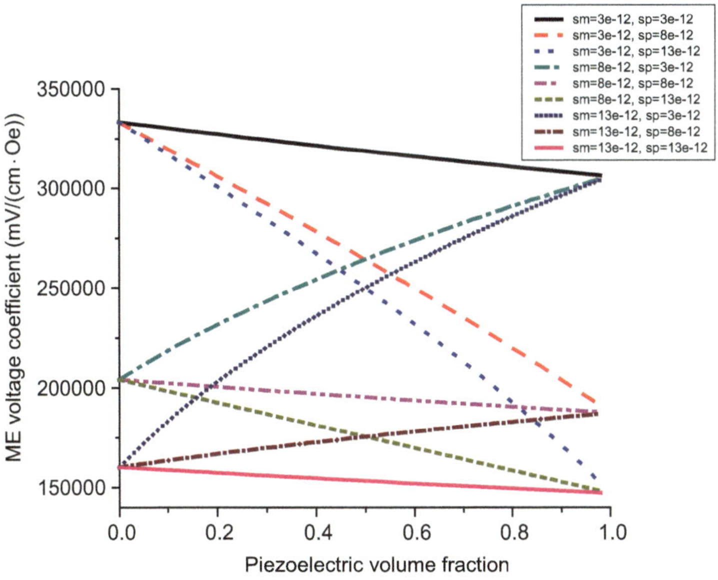

Figure 1.78 EMR frequency *vs.* piezoelectric volume fraction for longitudinal mode of 10 mm long magnetostrictive–piezoelectric layered structure.

It should be noted that Eqs. 1.154 and 1.155 for resonance frequency and ME voltage coefficient are valid for both bilayer and trilayer structures. It is easily seen from Eq. 1.153 that the piezoelectric volume fraction dependence of ME voltage coefficient divided by Q value is similar to that of low frequency ME coefficient. EMR frequency versus piezoelectric volume fraction is shown in Figs. 1.78 and 1.79.

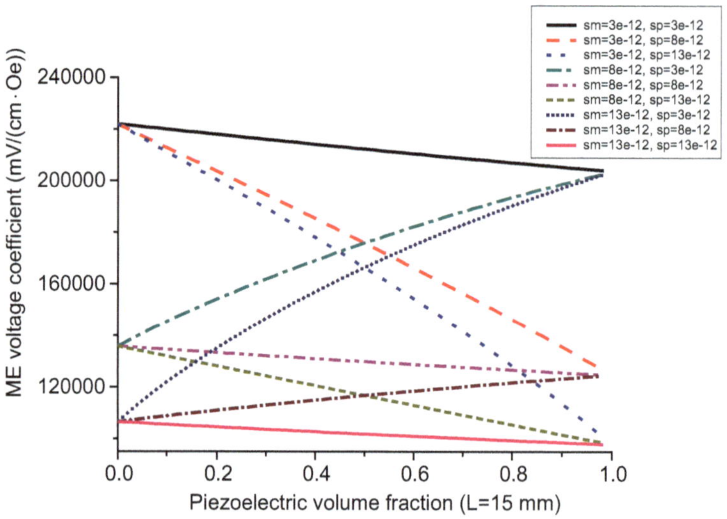

Figure 1.79 EMR frequency versus piezoelectric volume fraction for longitudinal mode of 15 mm long magnetostrictive–piezoelectric layered structure.

1.5.4 Magnetoelectric Coupling in FMR Region

The shift of magnetic resonance field can be expressed by Eq. 1.111 in the linear approximation in demagnetization factors due to electric field induced stress.

As an example, we consider a specific case of magnetic field H along [111] axis. The shift of FMR field versus ferrite volume fraction is shown in Figs. 1.80 and 1.81. Electric field dependence of FMR field shift is presented in Fig. 1.82.

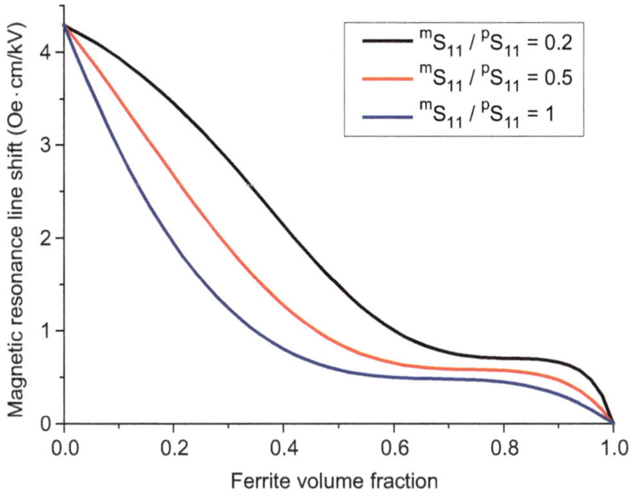

Figure 1.80 Ferrite volume fraction dependence of magnetic resonance line shift at E = 1 kV/cm for ferrite–piezoelectric bilayer for $\lambda_{111}/M_s = 0.16 \times 10^{-8}$ Oe^{-1}

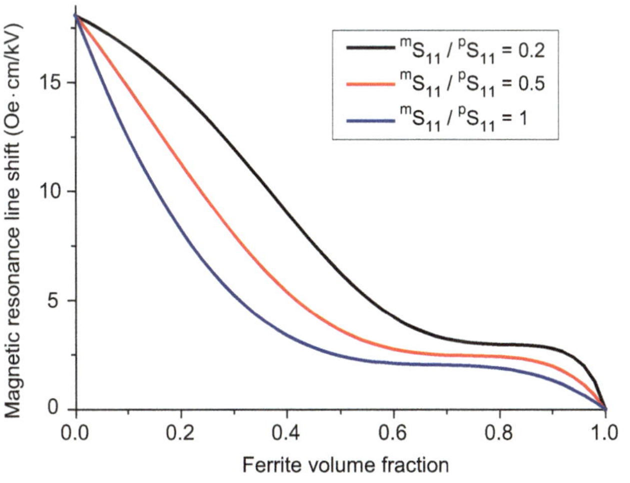

Figure 1.81 Ferrite volume fraction dependence of magnetic resonance line shift at E = 1 kV/cm for ferrite–piezoelectric bilayer for $\lambda_{111}/M_s = 0.68 \times 10^{-8}$ Oe^{-1}.

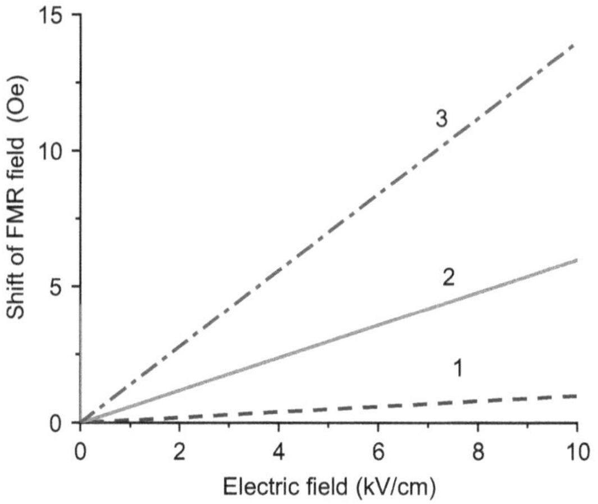

Figure 1.82 Estimated shift of FMR field *vs.* applied electric field at 9.3 GHz for the bilayers of YIG and PZT (1), NFO and PZT (2), and LFO and PZT (3) with equal thicknesses of magnetic and piezoelectric components.

A careful analysis of obtained monograms enables us to make the following conclusions. Figures 1.65 and 1.66 present the low-frequency ME voltage coefficients as a function of piezoelectric volume fraction for transverse fields' orientation for symmetric structures. It should be noted that ME voltage coefficients for longitudinal (out-of-plane electric and magnetic fields) and in-plane longitudinal fields orientations can be easily obtained by replacing the piezomagnetic coefficient q_{11} in Eq. 1.150 with q_{31} for longitudinal and replacing the piezoelectric coefficient d_{31} in Eq. 1.150 with d_{33} for in-plane longitudinal fields' orientation. The theory predicts the highest ME coupling for in-plane fields.

Piezoelectric volume fraction dependence of ME voltage coefficient for bilayers reveals a double maximum that is due to the fact that the strain produced by the magnetic component consists of two components: longitudinal and flexural (Figs. 1.67 and 1.68). In the absence of flexural strain the maximum ME coefficient occurs for certain value of volume fraction. Since the flexural strain is of opposite sign relative to longitudinal strain and reaches its maximum value approximately for the same volume fraction, the two types of strains combine to produce suppression of $\alpha_{E,31}$ in the middle and a double maximum in the ME coefficient piezoelectric volume dependence of ME voltage coefficient as in Figs. 1.67 and 1.68.

ME coefficient in structures of piezoelectric bimorph/Metglas in Figs. 1.69 and 1.70 is found to be approximately 2 times larger in the low-frequency region and 1.5 times larger in the bending mode region than in the traditional piesoelectric/Metglas bilayer. The observed enhancement in ME coupling strength is related to equal signs of induced voltage in both PZT layers with opposite poling directions due to the flexural deformations. The flexural deformations distribution provides increase in induced voltage at specified PZT volume fraction.

The resonance enhancement of ME coupling in the EMR region is obvious from Figs. 1.72 and 1.73 and the ME coefficient at resonance is more than two orders of magnitude higher compared to low-frequency values. The coefficient $\alpha_{E,13}$ increases with increasing piezoelectric volume fraction, attains a peak value and then decreases with increasing V. Similarly to ME coupling in the low-frequency region, the theory predicts the highest ME coupling for in-plane fields. The piezoelectric volume fraction dependence of EMR frequency can be estimated from Figs. 1.74, 1.75, 1.78, and 1.79.

From the above results in Figs 1.80–1.82, we can conclude to obtain an optimum ME effect in FMR region: (i) that the volume fraction of the piezoelectric phase should be high; (ii) that it is necessary to use a piezoelectric component phase with a large piezoelectric coefficient; and (iii) that it is necessary to use a magnetostrictive component phase with a small saturation magnetization and a high magnetostriction [33].

To obtain the estimates of ME coefficients from nomographs referred to above, one should use the material parameters of composite components. The relevant parameters of several materials that are most often used in ME structures are given in Table 1.2.

As an example of ME structure, we consider the bilayer of Ni and PZT with piezoelectric volume fraction 0.5. Based on data in Table 1.2, we get

$$^{m}s_{11} = 20 \times 10^{-12}\ \mathrm{m^2/N},\ ^{p}s_{11} = 15.3 \times 10^{-12}\ \mathrm{m^2/N},$$

$$^{p}d_{31} = -175 \times 10^{-12}\ \mathrm{m/V},\ ^{m}q_{11} = -4140 \times 10^{-12}\ \mathrm{m/A},\ ^{p}\varepsilon_{33}/\varepsilon_0 = 1750.$$

Fig. 1.68 reveals the low-frequency ME voltage coefficient $\alpha_{E\,31} = 190$ mV/(cm · Oe). Then, Fig. 1.69 gives the peak ME voltage coefficient $\alpha_{E\,31} = 20$ V/(cm · Oe) at bending resonance frequency and Fig. 1.66 gives the peak ME voltage coefficient $\alpha_{E\,31} = 70$ V/(cm · Oe) at axial resonance frequency. Q-value is assumed to be equal to 100.

Table 1.2 Material parameters for piezoelectric and magnetostrictive materials used for fabrication of layered structures

Material	s_{11} (10^{-12} m^2/N)	q_{33} (10^{-12} m/A)	d_{31} (10^{-12} m/V)	d_{33} (10^{-12} m/V)	λ_{100} (10^{-6})	$\varepsilon_{33}/\varepsilon_0$
PZT	15.3	–	–175	400	–	1750
BTO	7.3	–	–78		–	1345
PMN–PT	23	–	–600	1500		5000
Langasite	8.8	–	$d_{14} = -3.65 \times 10^{-12}$ m/V	$d_{11} = 6.3 \times 10^{-12}$ m/V		50
Langatite	9.8	–	$d_{14} = -2.81 \times 10^{-12}$ m/V	$d_{11} = 7.4 \times 10^{-12}$ m/V		77
Quartz	12.8	–	$d_{14} = -0.67 \times 10^{-12}$ m/V	$d_{11} = 2.3 \times 10^{-12}$ m/V		4.68
YIG	6.5		–	–	1.4	10
NFO	6.5	–680	–	–	23	10
LFO	35				46	10
Ni	20	–4140				
Terfenol–D	33.3	15707	–	–		
Metglas	10	14000	–	–		

1.6 Conclusions

We discussed here the ME properties of ferrite–piezoelectric composites, to create new ME composites with enhanced ME couplings that would enable them for applications in functional electronics devices. To address this important scientific and technical goal, a generalization of various theoretical and experimental studies of ME composites has been given. One of the main tasks according to the formulated approach is a comparative analysis of ME composites that have different connectivity types. The relative simplicity of manufacturing multilayer composites with a 2-2 type connectivity having giant ME responses is an important benefit. In addition, composites with 3-0 and 0-3 connectivity types can also be made in considerable quantity by a minimum monitoring of the synthesis process.

Availability of theoretical models for composite properties is necessary to interpret experimental data and to restrict oneself among the multitude of composite configurations. Theoretical estimations of the ME voltage coefficient for series and parallel composite models, and also a cubic model for composites with a 3-0 connectivity type, are known. However, as already mentioned, ME voltage coefficient was computed as the ratio of an electric field induced in the piezoelectric phase to the magnetic field applied to the magnetic one: that is, $\alpha_E = {}^pE/{}^mH$. But, in reality, the internal fields in composite components can be significantly different from the external fields. In particular, formulas predicted a maximum ME voltage coefficient in a pure piezomagnetic phase (i.e., ${}^pv = 0$): which distinctly mismatches reality. In addition, generalized models for composite based on an effective medium method were presented. These offer determination of the effective composite parameters with phase connectivity types of 2-2, 3-1, 1-3, 3-0, and 0-3 that are based on exact solutions. The ultimate purpose of any theoretical work must be to predict the ME parameters—both susceptibility and voltage coefficients—as these are the most basic parameters of magnetoelectricity.

To describe the composite's physical properties, the exact solution of elastostatic and electrostatic equations were obtained. Expressions for the ME susceptibility and ME voltage coefficient were derived as functions of an interface coupling parameter, constituent phase material parameters, and relative volume fractions of phases. Longitudinal, transverse, and in-plane cases were all considered. For a bilayer that is an asymmetric structure, the influence of flexural deformations of sample on ME output was estimated.

The ME interactions in a functionally graded ferroelectric–ferromagnetic bilayer is discussed. The grading-related thickness dependence of piezomagnetic or/and piezoelectric coefficients leads to an additional flexural strain and the theory predicts an enhancement in the strength of low-frequency ME coupling compared to homogeneous compositions. In graded bilayers the ME voltage coefficient is expected to increase by 50%.

Studies on ME effects have been performed on magnetization-graded ferromagnetic–piezoelectric composites. The grading is accomplished with the use of nickel zinc ferrite and pure Ni. Upon grading the ferromagnetic layer, hysteresis and remanence are observed in q versus H and ME voltage coefficient versus H. The observed hysteresis and remnant ME voltage coefficient are attributed to a grading-related magnetization at the Ni–NZFO interface that results in the ME coupling for zero external magnetic bias field.

It is shown that the ME coupling in the EMR region exceeds the low-frequency value by more than an order of magnitude. It was found that the peak transverse ME coefficient at EMR is larger than the longitudinal one. This is accounted for (i) by higher energy losses for longitudinal fields, orientation due to eddy currents in the electrodes and (ii) by influence of demagnetization field that appears at longitudinal orientations which reduces the piezomagnetic modulus. Predictions of the ME effect for various model composite systems were given including CFO–PZT and lanthanum strontium manganite–PZT. It was shown that ME effect in ferrite–PZT systems is maximum for in-plane magnetic and electric fields. The theoretical estimates of ME parameters were compared with experimental data.

We presented here the theoretical analysis of high frequency ME effects for a simple two-layer structure and a detailed treatment for electric field–induced shift for ferromagnetic resonance field in multilayers. Our calculations predict a strong microwave ME effect in composites consisting of a magnetic phase that has a large magnetostriction and a large volume fraction. Bending the sample gives rise to a decrease in the stress of ferrite layer and thus in ME interaction at FMR compared to the case when neglecting the flexural strains. In addition, the actual manuscript provides evidence that flexural deformations lead to the electric field-induced broadening of magnetic resonance line.

The results obtained in this study are of importance for quantitative information on microwave ME effects and for potential device applications based on multilayer composites. Microwave devices based on ME effects have unique advantages over traditional ferrite and semiconductor analogs.

The specific focus is on ME coupling at MAR, that is, at the coincidence of electromechanical resonance in the piezoelectric phase and ferromagnetic resonance in a tangentially magnetized ferrite. When exchange is ignored, estimated ME coefficient versus frequency profile shows a giant magnetoelectric coefficient at MAR. Exchange interactions will facilitate flow of microwave energy to magnetoacoustic spin waves, leading to the presence of a secondary peak in α_E versus f profiles. Estimates of the ME coefficient with and without exchange as and as a function of layer thickness are discussed for PZT – YIG bilayers.

ME effects occur over a broad frequency bandwidth, extending from the quasi-static to millimeter ranges. This offers important opportunities in potential device applications. It makes possible new concepts in sensing, gyrators, microwave communications, phase shifters, just to name a few. It also complicates the understanding of magnetoelectricity, as there are significant changes in its spectra with frequency. There are strong enhancements in the ME coefficients near both the electromechanical and magnetic resonances. In addition, there is the important problem of studying dispersion in the ME parameters over a broad frequency range of $10^{-3} < f < 10^{10}$ Hz. Relaxation parameters depend on connectivity type, composite geometry and structure, and volume fraction of constituent phases.

We presented here a new quick test of ME composites using nomographs and showed its use in applications where an approximate answer is appropriate and useful. To draw the graphs for ME voltage coefficients, we derived approximate expressions in explicit form for magnetically induced ME effect for different operational modes and laminate composite configurations including symmetrical and asymmetrical structures.

Another possible use of nomographs is predicting the volume fractions of composite components and sample geometry for specific composition to provide the strongest ME coupling.

In addition, a nomogram may be used to verify an estimate obtained from another exact calculation method notwithstanding the fact that the nomogram's accuracy is limited by the precision of reproducing the physical parameters.

References

1. L. D. Landau, E. M. Lifshitz, *Statistical Physics*, 3rd ed., Pergamon, 1980, p. 562.
2. I. E. Dzyaloshinskii, On the magneto-electrical effect in antiferromagnets, *Sov. Phys. JETP*, **10**, 628 (1960).
3. D. N. Astrov, Magnetoelectric effect in chromium oxide, *Sov. Phys. JETP*, **13**, 729 (1961).
4. V. J. Folen, G. T. Rado, E. W. Stalder, Anisotropy of the magnetoelectric effect in Cr_2O_3, *Phys. Rev. Lett.*, **6**, 607 (1961).
5. E. Asher, The interaction between magnetization and polarization: phenomenological symmetry consideration, *J. Phys. Soc. Jpn.*, **28**, 7 (1969).
6. R. P. Santoro, R. E. Newnham, Survey of magnetoelectric materials, Technical Report AFML TR-66-327, *Air Force Mater. Lab.*, Ohio, 1966.
7. G. Aubert, A novel approach of the magnetoelectric effect in antiferromagnets, *J. Appl. Phys.*, **53**, 8125 (1982).
8. W. Opechovski, Magnetoelectric symmetry, In *Proceedings of Symposium on Magnetoelectric Interaction in Crystals, USA, 1973,* Eds. A. Freeman and H. Schmid, Gordon and Breach Science Publishers, New York, **47** (1975).
9. T. H. O'Dell, *The Electrodynamics of Magnetoelectric Media,* North-Holland Publishing Company, Amsterdam, 1970, p. 304.
10. Jr. W. F. Brown, *et al.*, Upper bound on the magnetoelectric susceptibility, *Phys. Rev.*, **168**, 574 (1968).
11. A. Freeman, H. Schmid, *Magnetoelectric Interaction Phenomena in Crystals*, London. N.-J. Paris: Gordon and Breach, 1975, p. 228.

12. M. I. Bichurin, V. M. Petrov, Influence of electric field on antiferromagnetic resonance spectrum in iron borate, *Phys. Solid State*, **29**, 2509 (1987).
13. G. A. Smolenskii, I. E. Chupis, Ferroelectromagnets, *Sov. Phys. Usp.* **25**, 475–493 (1982)
14. H. Schmid, A. Janner, H. Grimmer, *et al.*, Proceedings of the 2nd International Conference on Magnetoelectric Interaction Phenomena in Crystals (MEIPIC-2, Ascona) *Ferroelectrics*, **161–162**, 748 (1993).
15. M. I. Bichurin, Proceedings of the 3rd International Conference on Magnetoelectric Interaction Phenomena in Crystals (MEIPIC-3, Novgorod) *Ferroelectrics*, **204**, 356 (1997).
16. M. I. Bichurin, Proceedings of the Fourth Conference on Magnetoelectric Interaction Phenomena in Crystals (MEIPIC-4, Veliky Novgorod) *Ferroelectrics*, **279–280**, 386 (2002).
17. M. Fiebig, V. V. Eremenko, I. E. Chupis, Proceedings of the Fifth Conference on Magnetoelectric Interaction Phenomena in Crystals (MEIPIC-5, Sudak) Kluwer Academic Publishers, NATO Sciences Series, 2004, p. 334.
18. J.-P. Rivera, A short review of the magnetoelectric effect and related experimental techniques on single phase (multi-)ferroics, *Eur. Phys. J. B*, **71**, 299–313 (2009).
19. K. F. Wang, J.-M. Liu, Z. F. Ren, Multiferroicity: the coupling between magnetic and polarization orders, *Adv. Phys.*, **58**, 321 (2009).
20. J. Van Suchtelen, Product properties: a new application of composite materials, *Philips Res. Rep.*, **27**, 28 (1972).
21. J. Van Suchtelen, Nonstructural application of composite materials, *Ann. Chim. Fr.*, **5**, 139 (1980).
22. B. D. H. Tellegen, The gyrator, a new electric network element, *Philips Res. Rep.*, **3**, 81 (1948).
23. J. Van Den Boomgard, *et al.*, An *in situ* grown eutectic magnetoelectric composite materials: Part I, *J. Mater. Sci.*, **9**, 1705 (1974).
24. A. M. J. G. Van Run, *et al.*, An *in situ* grown eutectic magnetoelectric composite materials: Part II, *J. Mater. Sci.*, **9**, 1710 (1974).
25. J. Van Den Boomgard, A. M. J. G. Van Run, Poling of a ferroelectric medium by means of a built-in space charge field with special reference to sintered magnetoelectric composites, *Solid State Commun.*, **19**, 405 (1976).
26. I. Bunget, V. Raetchi, Magnetoelectric effect in the heterogeneous system NiZn ferrite—PZT ceramic, *Phys. Stat. Sol.*, **63**, 55 (1981).

27. I. Bunget, V. Raetchi, Dynamic magnetoelectric effect in the composite system of NiZn ferrite and PZT ceramics, *Rev. Roum. Phys.*, **27**, 401 (1982).

28. J. Zhai, Z. Xing, S. Dong, *et al.*, Magnetoelectric laminate composites: An overview, *J. Am. Ceram. Soc.*, **91**, 351–358 (2008).

29. S. Priya, R. Islam, S. Dong, D. Viehland, Recent advancements in magnetoelectric particulate and laminate composites, *J. Electroceram.*, **19**, 149–166 (2007).

30. M. I. Bichurin, Yu. N. Venevtsev, O. S. Didkovskaya, *et al.*, Magnetoelectric materials: technology features and application perspectives, In *Magnetoelectric Substances*, Nauka, Moscow, Russia, 1990, p. 118–132 (in Russian).

31. M. I. Bichurin, V. M. Petrov, Magnetic resonance in layered ferrite–ferrielectric structures, *Sov. Phys. JETP*, **58**, 2277 (1988).

32. M. I. Bichurin, O. S. Didkovskaya, V. M. Petrov, S. E. Sofronev, Resonant magnetoelectric effect in composite materials, *Izv. vuzov., ser. Physic.*, **1**, 121 (1985) (in Russian).

33. M. Bichurin, V. Petrov, *Modeling of Magnetoelectric Effects in Composites*, Springer, 2014.

34. M. Fiebig, Revival of the magnetoelectric effect, *J. Phys. D: Appl. Phys.*, **38**, R1 (2005).

35. M. I. Bichurin, D. Viehland, G. Srinivasan, Magnetoelectric interactions in ferromagnetic—piezoelectric layered structures: phenomena and devices, *J. Electroceram.*, **19**, 243 (2007).

36. C.-W. Nan, M. I. Bichurin, S. Dong, *et al.*, Multiferroic magnetoelectric composites: historical perspectives, status, and future directions, *J. Appl. Phys.*, **103**, 031101 (2008).

37. M. I. Bichurin, D. Viehland, *Magnetoelectricity in Composites*, Pan Stanford Publishing, 2012, p. 251.

38. G. Srinivasan. Magnetoelectric composites, *Annu. Rev. Mater. Res.*, **40**, 153 (2010).

39. C. Ederer, Nicola A. Spaldin, Magnetoelectrics: A new route to magnetic ferroelectrics, *Nat. Mater.*, **3**, 849–851 (2004).

40. W. Eerenstein, N. D. Mathur, J. F. Scott, Multiferroic and magnetoelectric materials, *Nature*, **442**, 759–765 (2006).

41. R. Ramesh, Nicola A. Spaldin, Multiferroics: progress and prospects in thin films, *Nat. Mater.*, **6**, 21–29 (2007).

42. S.-W. Cheong, M. Mostovoy, Multiferroics: a magnetic twist for ferroelectricity, *Nat. Mater.,* **6**, 13–20 (2007).

43. J. Van Den Boomgard, A. M. J. G. Van Run, J. Van Suchtelen, Magnetoelectricity in piezoelectric–magnetostrictive composites, *Ferroelectrics*, **10**, 295 (1976).

44. R. E. Newnham, D. P. Skinner, L. E. Cross, Connectivity and piezoelectric–pyroelectric composites, *Mater. Res. Bull.*, **13**, 525 (1978).

45. G. Harshe, J. O. Dougherty, R. E. Newnham, Theoretical modelling of multilayer magnetoelectric composites, *Int. J. Appl. Electromagn. Mater.*, **4**, 145 (1993).

46. G. Harshe, J. P. Dougherty, R. E. Newnham, Theoretical modelling of 3-0, 0-3 magnetoelectric composites, *Int. J. Appl. Electromagn. Mater.*, **4**, 161 (1993).

47. K. Mori, M. Wuttig, Magnetoelectric coupling in terfenol-D/polyvinylidene difluoride composites, *Appl. Phys. Lett.*, **81**, 100 (2002).

48. J. Ma, J. Hu, Z. Li, C.-W. Nan, Recent progress in multiferroic magnetoelectric composites: from bulk to thin films, *Advanced Materials,* **23**, 1062–1087 (2011).

49. M. I. Bichurin, V. M. Petrov, R. V. Petrov, F. I. Bukashev, A. Y. Smirnov, Electrodynamic analysis of strip line on magnetoelectric substrate, *Ferroelectrics,* **280**, 203–209 (2002).

50. M. I. Bichurin, V. M. Petrov, R. V. Petrov, *et al.*, Magnetoelectric sensor of magnetic field, *Ferroelectrics,* **280**, 199–202 (2002).

51. M. I. Bichurin, V. M. Petrov, R. V. Petrov, *et al.*, Magnetoelectric microwave devices, *Ferroelectrics,* **280**, 211–218 (2002).

52. A. S. Tatarenko, M. I. Bichurin, V. Gheevarughese, *et. al.,* Microwave magnetoelectric effects in ferrite–piezoelectric composites and dual electric and magnetic field tunable filters, *J. Electroceram*, **24,** 5 (2010).

53. J. Zhai, J. Li, D. Viehland, M. I. Bichurin, Large magnetoelectric susceptibility: The fundamental property of piezoelectric and magnetostrictive laminated composites, *J. Appl. Phys.*, **101**, 014102 (2007).

54. J. Lou, G. N. Pellegrini, M. Liu, *et al.*, Inequivalence of direct and converse magnetoelectric coupling at electromechanical resonance, *Appl. Phys. Lett.*, **100**, 102907 (2012).

55. M. I. Bichurin, V. M. Petrov, G. Srinivasan, Modelling of magnetoelectric effect in ferromagnetic/piezoelectric multilayer composites, *Ferroelectrics,* **280**, 165 (2002).

56. M. I. Bichurin., V. M. Petrov, G. Srinivasan, Theory of magnetoelectric effects in ferromagnetic ferroelectric layer composites, *J. Appl. Phys*, **92**, 7681 (2002).

57. M. I. Bichurin, V. M. Petrov, G. Srinivasan, Theory of low-frequency magnetoelectric coupling in magnetostrictive–piezoelectric bilayers, *Phys. Rev. B*, **68**, 054402 (2003).

58. M. I. Bichurin, V. M. Petrov, G. Srinivasan, Low-frequency magnetoelectric effects in ferrite–piezoelectric nanostructures, *J. Magn. Magn. Mater.*, **321**, 846–849 (2009).

59. V. Gheevarughese, U. Laletsin, V. M. Petrov, *et al.*, Low-frequency and resonance magnetoelectric effects in lead zirconate titanate and single-crystal nickel zinc ferrite bilayers, *J. Mater. Res.*, **22**, 2130–2135 (2007).

60. V. M. Petrov, G. Srinivasan, M. I. Bichurin, *et al.*, Theory of magnetoelectric effect for bending modes in magnetostrictive–piezoelectric bilayers, *J. Appl. Phys.*, **105**, 063911 (2009).

61. I. A. Osaretin, R. G. Rojas, Theoretical model for the magnetoelectric effect in magnetostrictive/piezoelectric composites, *Phys. Rev. B*, **82**, 174415 (2010).

62. V. M. Petrov, M. I. Bichurin, V. M. Laletin, *et al.*, Modeling of magnetoelectric effects in ferromagnetic/piezoelectric bulk composites, In *Magnetoelectric Interaction Phenomena in Crystals-NATO Science Series II*, Vol. 164, Eds. M. Fiebig, V. V. Eremenko, and I. E. Chupis, pp. 65–70, Kluwer Academic Publishers, London (2004).

63. V. M. Petrov, G. Srinivasan, U. Laletsin, *et al.*, Magnetoelectric effects in porous ferromagnetic–piezoelectric bulk composites: Experiment and theory, *Phys. Rev. B*, **75**, 174422 (2007).

64. V. M. Petrov, G. Srinivasan, Enhancement of magnetoelectric coupling in functionally graded ferroelectric and ferromagnetic bilayers, *Phys. Rev. B*, **78**, 184421 (2008).

65. S. K. Mandal, G. Sreenivasulu, V. M. Petrov, G. Srinivasan, Magnetization-graded multiferroic composite and magnetoelectric effects at zero bias, *Phys. Rev. B*, **84**, 014432 (2011).

66. C.-S. Park, D. Avirovik, M. I. Bichurin, *et al.*, Tunable magnetoelectric response of dimensionally gradient laminate composites, *Appl. Phys. Lett.*, **100**, 212901 (2012).

67. A. Novick, B. Berry, *Anelastic Relaxation in Crystalline Solids*, Academic Press, New York, 1972, p. 677.

68. V. M. Petrov, M. I. Bichurin, G. Srinivasan, Maxwell–Wagner relaxation in magnetoelectric composites, *Tech. Phys. Lett.*, **30**, 81 (2004).
69. V. M. Petrov, M. I. Bichurin, G. Srinivasan, *et al.*, Dispersion characteristics for low-frequency magnetoelectric coefficients in bulk ferrite-piezoelectric composites, *Solid State Commun.*, **142**, 515 (2007).
70. M. Bichurin, V. Petrov, S. Priya, A. Bhalla, Multiferroic magnetoelectric composites and their applications, *Adv. Cond. Matter Phys.*, 2012, 129794 (2012).
71. M. I. Bichurin, V. M. Petrov, S. V. Averkin, E. Liverts, Present status of theoretical modeling the magnetoelectric effect in magnetostrictive-piezoelectric nanostructures. Part I: Low frequency and electromechanical resonance ranges, *Journal of Applied Physics*, **107**, 053904 (2010).
72. M. I. Bichurin, V. M. Petrov, R. V. Petrov, Direct and inverse magnetoelectric effect in layered composites in electromechanical resonance range: A review, *J. Magn. Magn. Mater.*, **324**, 3548–3550 (2012).
73. M. I. Bichurin, V. M. Petrov, R. V. Petrov, S. Priya, Electromechanical resonance in magnetoelectric composites: direct and inverse effect. *Solid State Phenomena*, **189**, 129–143 (2012).
74. M. I. Bichurin, I. A. Kornev, V. M. Petrov, I. V. Lisnevskaya, Investigation of magnetoelectric interaction in composite, *Ferroelectrics*, **204**, 289–297 (1997).
75. N. Zhang, V. M. Petrov, T. Johnson, *et al.*, Enhancement of magnetoelectric coupling in a piezoelectric–magnetostrictive semiring structure, *J. Appl. Phys.*, **106**, 126101 (2009).
76. M. I. Bichurin, V. M. Petrov, G. A. Semenov, Magnetoelectric material for components of radioelectronic devices, Patent No. RU 2363074 C1 (submitted 11.03.2008, published 27.07.2009).
77. M. I. Bichurin, V. M. Petrov, S. V. Averkin, A. V. Filippov, Electromechanical resonance in magnetoelectric layered structures, *Phys. Solid State*, **52**, 2116 (2010).
78. M. I. Bichurin, D. A. Filippov, V. M. Petrov, *et al.*, Resonance magnetoelectric effects in layered magnetostrictive-piezoelectric composites. *Phys. Rev. B*, **68**, 132408 (1–4) (2003).
79. M. I. Bichurin, V. M. Petrov, Modeling of magnetoelectric interaction in magnetostrictive-piezoelectric composites, *Adv. Cond. Matter Phys.*, **2012**, 798310 (2012).

80. M. I. Bichurin, V. M. Petrov, Magnetoelectric effect in the electromechanical resonance range, In *Magnetoelectricity in Composites*, Eds. M. I. Bichurin and D. Viehland, Pan Stanford Publishing (2012), p. 91–104.

81. M. I. Bichurin, Magnetoelectric interaction in magnetically ordered materials (Review), In *Magnetoelectricity in Composites*, Eds. M. I. Bichurin and D. Viehland, Pan Stanford Publishing (2012), p. 1–24.

82. M. I. Bichurin, V. M. Petrov, Magnetoelectric effect in magnetostriction–piezoelectric multiferroics, *Low Temp. Phys.*, **36**, 544–549 (2010).

83. V. M. Petrov, G. Srinivasan, M. I. Bichurin, T. A. Galkina, Theory of magnetoelectric effect for bending modes in magnetostrictive–piezoelectric bilayers., *J. Appl. Phys.*, **105**, 063911 (2009).

84. V. M. Petrov, M. I. Bichurin, V. V. Zibtsev, *et al.*, Flexural deformation and bending mode of magnetoelectric nanobilayer, *J. Appl. Phys.*, **106**, 113901 (2009).

85. L. Y. Fetisov, N. S. Perov, Y. K. Fetisov, *et al.*, Resonance magnetoelectric interactions in an asymmetric ferromagnetic–ferroelectric layered structure, *J. Appl. Phys.*, **109**, 053908 (2011).

86. G. Sreenivasulu, S. K. Mandal, V. M. Petrov, *et al.*, Bending resonance in a magnetostrictive–piezoelectric bilayer and magnetoelectric interactions, *Integr. Ferroelectr.*, **126**, 87–93 (2011).

87. S. K. Mandal, G. Sreenivasulu, V. M. Petrov, G. Srinivasan, Flexural deformation in a compositionally stepped ferrite and magnetoelectric effects in a composite with piezoelectrics, *Appl. Phys. Lett.*, **96**, 192502 (2010).

88. G. Sreenivasulu, V. M. Petrov, L. Y. Fetisov, *et al.*, Magnetoelectric interactions in layered composites of piezoelectric quartz and magnetostrictive alloys, *Phys. Rev. B*, **86**, 214405 (2012).

89. V. M. Petrov, M. I. Bichurin, G. Srinivasan, *et al.*, Bending modes and magnetoelectric effects in asymmetric ferromagnetic–ferroelectric structure, *Solid State Phenom.*, **190**, 281–284 (2012).

90. V. M. Laletin, V. M. Petrov, Nonlinear magnetoelectric response of a bulk magnetostrictive–piezoelectric composite, *Solid State Commun.*, **151**, 1806–1809 (2011).

91. M. I. Bichurin, R. V. Petrov, V. M. Petrov Magnetoelectric effect at thickness shear mode in ferrite–piezoelectric bilayer, *Appl. Phys. Lett.*, **103**, 0929021 (2013).

92. S. Tatarenko, M. I. Bichurin, Microwave magnetoelectric devices, *Adv. Cond. Matter Phys.*, **2012**, 286562 (2012).

93. N. Li, M. Liu, Z. Zhou, *et al.*, Electrostatic tuning of ferromagnetic resonance and magnetoelectric interactions in ferrite–piezoelectric heterostructures grown by chemical vapor deposition, *Appl. Phys. Lett.*, **99**, 192502 (2011).

94. G. Srinivasan, A. S. Tatarenko, V. Mathe, M. I. Bichurin, Microwave and MM-wave magnetoelectric interactions in ferrite–ferroelectric bilayers, *Eur. Phys. J. B*, **71**, 371–375 (2009).

95. R. V. Petrov, R. Pandey, G. Srinivasan, M. I. Bichurin, A magnetic field controlled negative-index microwave lens, *Microw. Opt. Technol. Lett.*, **50**, 2804 - 2807 (2008).

96. S. Tatarenko, G. Srinivasan, M. I. Bichurin, Magnetoelectric microwave phase shifter, *Appl. Phys. Lett.*, **88**, 183507 (2006).

97. M. Bichurin, V. Petrov, A. Zakharov, *et al.*, Magnetoelectric interactions in lead-based and lead-free composites, *Materials*, **4**, 651–702 (2011).

98. S. Dong, J. Zhai, J. F. Li, *et al.*, Magnetoelectric gyration effect in $Tb_{1-x}Dy_xFe_{2-y}$/Pb(Zr,Ti)O_3 laminated composites at the electromechanical resonance, *Appl. Phys. Lett.*, **89**, 243512 (2006).

99. M. I. Bichurin, V. M. Petrov, Magnetoelectric effects in nanocomposites, In *Magnetoelectricity in Composites*, Eds. M. I. Bichurin and D. Viehland, Pan Stanford Publishing (2012), p. 91–104.

100. R. V. Petrov, G. Srinivasan, M. I. Bichurin, D. Viehland, Three-dimensional left-handed material lens, *Appl. Phys. Lett.*, **91**, 104103 (2007).

101. M. I. Bichurin, V. M. Petrov, Composite magnetoelectrics: their microwave properties, *Ferroelectrics*, **162**, 33–35 (1994).

102. M. I. Bichurin, V. M. Petrov, Influence of external electric field on magnetic resonance frequency in magnetic ferroelectrics, *Ferroelectrics*, **167**, 147–150 (1995).

103. M. I. Bichurin, I. A. Kornev, V. M. Petrov, *et al.*, Theory of magnetoelectric effects at microwave frequencies in a piezoelectric/magnetostrictive multilayer composite, *Phys. Rev. B*, **64**, 094409 (2001).

104. M. I. Bichurin, V. M. Petrov, Yu. V. Kiliba, G. Srinivasan, Magnetic and magnetoelectric susceptibilities of a ferroelectric/ferromagnetic composite at microwave frequencies, *Phys. Rev. B*, **66**, 134404 (2002).

105. V. M. Petrov, G. Srinivasan, T. A. Galkina, Microwave magnetoelectric effects in bilayers of single crystal ferrite and functionally graded piezoelectric, *J. Appl. Phys.*, **104**, 113910 (2008).

106. M. I. Bichurin, V. M. Petrov, T. A. Galkina, Microwave magnetoelectric effects in bilayer of ferrite and piezoelectric, *Eur. Phys. J. Appl. Phys.*, **45**, 30801 (2009).

107. S. Shastry, G. Srinivasan, M. I. Bichurin, *et al.*, Microwave magnetoelectric effects in single crystal bilayers of yttrium iron garnet and lead magnesium niobate–lead titanate, *Phys. Rev. B,* **70**, 064416 (2004).

108. M. I. Bichurin, V. M. Petrov, S. V. Averkin, E. Liverts, Present status of theoretical modeling the magnetoelectric effect in magnetostrictive–piezoelectric nanostructures. Part II: Magnetic and magnetoacoustic resonance ranges, *J. Appl. Phys.*, **107**, 053905 (2010).

109. M. I. Bichurin, V. M. Petrov, O. V. Ryabkov, *et al.*, Theory of magnetoelectric effects at magnetoacoustic resonance in single-crystal ferromagnetic–ferroelectric heterostructures, *Phys. Rev. B,* **72**, 060408 (2005).

110. O. V. Ryabkov, V. M. Petrov, M. I. Bichurin, G. Srinivasan, Magnetoacoustic resonance in tangentially magnetized ferrite–piezoelectric bilayers, *Tech. Phys. Lett.,* **32**, 1021–1023 (2006).

111. O. V. Ryabkov, S. V. Averkin, M. I. Bichurin, *et al.*, Effects of exchange interactions on magnetoacoustic resonance in layered nanocomposites of yttrium iron garnet and lead zirconate titanate, *J. Mater. Res.,* **22**, 2174–2178 (2007).

112. M. I. Bichurin, V. M. Petrov, S. V. Averkin, *et al.*, Modelling of magneto-acoustic resonance in ferrite–piezoelectric bilayers, *J. Phys. D: Appl. Phys.,* **42**, 215001 (2009).

113. V. M. Petrov, V. V. Zibtsev, G. Srinivasan, Magnetoacoustic resonance in ferrite–ferroelectric nanopillars, *Eur. Phys. J. B,* **71**, 367–370 (2009).

114. V. Petrov, G. Srinivasan, O. V. Ryabkov, *et al.*, Microwave magnetoelectric interactions in ferrite–piezoelectric nanobilayers: Theory of electric field induced magnetic excitations, *Solid State Commun.,* **144**, 50–53 (2007).

115. M. I. Bichurin, Magnetoelectrics in microwave range, *Ferroelectrics,* **161**, 53–57 (1994).

116. M. I. Bichurin, Resonance microwave magnetoelectric effect, In *Magnetoelectric Substances,* Nauka, Moscow, Russia, p. 53–67 (1990) (in Russian).

117. M. Bichurin, V. Petrov, R. Petrov, *et al.*, Nomograph method for predicting magnetoelectric coupling, *J. Magn. Magn. Mater.,* **412** , 1–6 (2016).

Chapter 2

Applications of Magnetoelectric Composites

2.1 ME Inductance

Layered magnetostrictive–piezoelectric structures with inhomogeneous components can be used in inductive elements with electric control [1]. The inductor core is implemented as a magnetostrictive–piezoelectric composite which enables varying the inductance with applied electric field due to ME coupling. In a magnetostrictive–piezoelectric composite, the converse ME effect is observed, which consists in inducing the magnetization in the applied electric field.

The use of ME composites in controlled inductive elements makes it possible to get a smooth variation of the inductance with input voltage. This device has advantages over previously used analogs, in particular, high Q, miniature sizes, and a relatively simple design. In addition, this type of inductance has a high temperature stability and radiation resistance.

As a means of simulating the inductive element, op-amp electronic circuits can be used in the low-frequency region for low powers. However, simulated inductors cannot completely supplant the real inductors as they do not reproduce all the distinctive parameters of physical objects.

Magnetoelectric Composites
Mirza I. Bichurin, Vladimir M. Petrov, Roman V. Petrov, and Alexander S. Tatarenko
Copyright © 2019 Pan Stanford Publishing Pte. Ltd.
ISBN 978-981-4800-04-4 (Hardcover), 978-0-429-48867-2 (eBook)
www.panstanford.com

2.1.1 Theoretical Model of the Device

As a model, an inductor is considered, which consists of a core based on ME composite and a coil wound on it. Theoretical estimates are presented for the laminates of the PZT and Metglas. Figure 2.1 shows a controlled inductive element based on longitudinal strains. For this case, the inductor core is a bilayer of PZT and Metglas. PZT is assumed to be polarized perpendicular to the sample plane. An inductive element with flexural deformations consists of magnetostrictive layer and piezoelectric bimorph. The inductor core is implemented as an asymmetric structure of Metglas layer and piezoelectric bimorph consisting of two PZT layers with opposite poling directions.

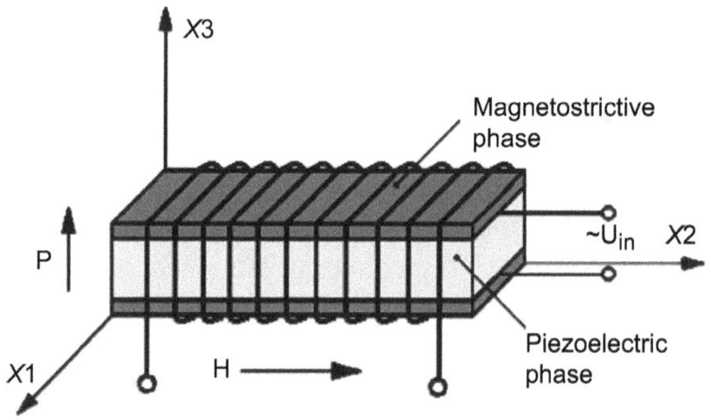

Figure 2.1 Electrically controlled inductance based on magnetostrictive–piezoelectric laminate with longitudinal deformations.

We assume the core to be a symmetrical structure with length much greater compared to the width and thickness. The composite layers are in the (X_1, X_2) plane, and X_3 direction is perpendicular to the sample plane. The poling direction of the piezoelectric layer is parallel to the X_3 axis. The applied electric field is also parallel to the X_3 axis. In this case, the ME effect is referred to as transverse one.

The application of electric voltage to the electrodes gives rise to a strain. The strain of piezoelectric layer ${}^{p}S$ is transferred to the magnetostrictive phase. The mechanical stress ${}^{m}T$ of magnetostrictive phase gives rise to variation in permeability μ and thus in coil inductance L. For the free-standing piezoelectric element, the tension–compression strains are not limited in any direction. When the piezoelectric and piezomagnetic layers are bonded together, the piezomagnetic layer restricts the strain of the piezoelectric one in an external electric field. The shearing forces in the layers produce turning moments in respect to the X_2 axis since the forces are not applied centrally. As a result, a flexural deformation appears which can be described by radius of curvature R. For taking into account the flexural deformations, the longitudinal axial strains of layers are considered as linear functions of X_3.

The magnetoelastic energy is determined by the saturation magnetization and stresses. Elastic stresses acting on a ferromagnet lead to a change in the orientation of the magnetic moments of the domains. This leads to a change in the magnetization of the ferromagnet. The magnetoelastic energy is directly related to magnetostriction. The equilibrium directions of magnetization correspond to minimum value of the magnetic energy which includes three components: (1) the energy of the magnetic anisotropy; (2) magnetoelastic energy; (3) the energy of the external magnetic field.

$$W_{\mathrm{MY}} = -\frac{3}{2}\lambda_{\mathrm{s}}{}^{\mathrm{m}}T_1 \cos^2\phi \tag{2.1}$$

where ϕ is the angle between the stress vector and the magnetization vector and λ_s is the saturation magnetostriction constant.

Effective permeability can be expressed in terms of effective magnetic field:

$$\mu_{\mathrm{eff}} = \frac{M_{\mathrm{s}}}{H_{\mathrm{eff}}} + 1 \tag{2.2}$$

The effective magnetic field equals the sum of the external magnetic field produced by the current I in the coil and magnetic anisotropy field H_{ME} due to the magnetoelastic effect:

$$H_{\mathrm{eff}} = H_{\mathrm{BH}} + H_{\mathrm{MY}} = \frac{NI}{l} + \frac{3\lambda_{\mathrm{s}}{}^{\mathrm{m}}T_1}{\mu_0 M_{\mathrm{s}}} \tag{2.3}$$

where N and L are the number of turns and the length of the coil.

At low frequencies, the effective magnetic permeability can be calculated by the following expression:

$$\mu_{\mathrm{eff}} = \frac{\mu_0 M_{\mathrm{s}}}{\dfrac{3\lambda_{\mathrm{s}}{}^{\mathrm{m}}T_1}{\mu_0 M_{\mathrm{s}}} + \dfrac{IN}{l}} + 1 \tag{2.4}$$

At low frequency the inductance is estimated as

$$L = \frac{(\mu_{\text{eff}} t_{\text{m}} + t_{\text{p}})\mu_0 d N^2}{l} \tag{2.5}$$

where t_{m} and t_{p} are the thicknesses of magnetostriction phase and piezoelectric phase, d is the core width. The fractional variation in inductance equals

$$\delta L = \frac{t_{\text{m}}(\mu_{U=0} - \mu_{\text{eff}})}{t_{\text{m}}\mu_{U=0} + t_{\text{p}}} \times 100\% \tag{2.6}$$

with $\mu_{U=0}$ denoting the magnetic permeability at $U = 0$.

The stress component can be found from elasticity equation. We assume that the surfaces of the plate are free. In this case, the normal components of the stress tensor on them are zero, that is, ${}^{\text{m}}T_1 = {}^{\text{m}}T_2 = {}^{\text{m}}T_3 = 0$ on the surfaces of the plate. Since the plate is thin and narrow, we can assume that ${}^{\text{m}}T_2 = {}^{\text{m}}T_3 = 0$ not only on the surface but also in the entire volume of the plate, and only the ${}^{\text{m}}T_1$ component is nonzero. The expression for the mechanical stresses in the magnetostrictive phase in the case of tension–compression strain has the form:

$${}^{\text{m}}T_1 = \frac{d_{31}U - q_{11}H_1\,{}^{\text{p}}t}{{}^{\text{p}}s_{11}\,{}^{\text{m}}t + {}^{\text{m}}s_{11}\,{}^{\text{p}}t}, \tag{2.7}$$

The expression for mechanical stress in the magnetostrictive phase in the case of bending deformations can be written as follows:

$${}^{\text{m}}T_1 = \frac{1}{{}^{\text{m}}s_{11}}\left({}^{\text{m}}S_{10} + \frac{{}^{\text{m}}t}{R} - {}^{\text{m}}q_{11}H_1\right) \tag{2.8}$$

where

$${}^{\text{m}}S_{10} = \frac{s_1}{t}\left[{}^{\text{m}}Y\int_{-{}^{\text{m}}t/2}^{{}^{\text{m}}t/2} {}^{\text{m}}q_{11}\,{}^{\text{m}}H_1 dz_{\text{m}} + {}^{\text{p}}Y\int_{-{}^{\text{p}}t/2}^{0}\left(-{}^{\text{p}}d_{31}\,{}^{\text{p}}E_3 + \frac{h_{\text{m}}\,{}^{\text{p}}t}{R}\right)dz_{\text{p}} + {}^{\text{p}}Y\int_{0}^{{}^{\text{p}}t/2}\left({}^{\text{p}}d_{31}\,{}^{\text{p}}E_3 + \frac{h_{\text{m}}\,{}^{\text{p}}t}{R}\right)dz_{\text{p}}\right]$$

is longitudinal deformations of the neutral axis for the magnetostrictive layer, R is the radius of curvature, $h_{\text{m}} = (t_{\text{p}} + t_{\text{m}})/2$, $t = {}^{\text{m}}t + {}^{\text{p}}t$, ${}^{\text{m}}Y$ and ${}^{\text{p}}Y$ are the elastic moduli of the piezomagnetic and piezoelectric components, respectively, z_i is the distance from the i-th layer of the structure to the neutral layer.

Next, we write down the expression for the ME susceptibility α_B, which is defined as the ratio of the induced magnetic induction dB to the applied voltage dU. In the case of the ME effect due to tensile–compression deformations, the ME susceptibility is expressed as

$$\alpha_B = \frac{d_{31}q_{11}}{{}^{\mathrm{p}}s_{11}{}^{\mathrm{m}}t + {}^{\mathrm{m}}s_{11}{}^{\mathrm{p}}t}, \tag{2.9}$$

In the case of bending deformations, the expression for the ME susceptibility takes a more complicated form:

$$\alpha_B = \frac{3d_{31}q_{11}t_{\mathrm{p}}\left(t_{\mathrm{p}}h_1{}^{\mathrm{m}}s_{11} + t_{\mathrm{m}}g/2\right)}{12{}^{\mathrm{p}}s_{11}{}^{\mathrm{m}}s_{11}t_{\mathrm{m}}t_{\mathrm{p}}h_1 + t_{\mathrm{m}}^3 g{}^{\mathrm{p}}s_{11} + t_{\mathrm{p}}^3 g{}^{\mathrm{m}}s_{11}}, \tag{2.10}$$

where

$$g = {}^{\mathrm{p}}s_{11}{}^{\mathrm{m}}t + {}^{\mathrm{m}}s_{11}{}^{\mathrm{p}}t$$

In Fig 2.2, the applied voltage dependence of relative change in inductance $\Delta L/L$ is shown. The results are presented for a PZT thickness of 0.2 mm and Metglas thickness of 40 μm.

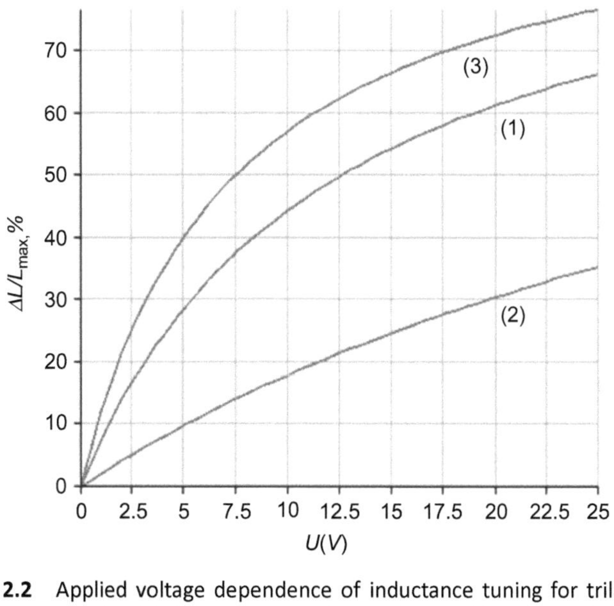

Figure 2.2 Applied voltage dependence of inductance tuning for trilayer of Metglas-PZT-Metglas (1), bilayer of Metglas-PZT (2), and laminate of Metglas-PZT-bimorph (3).

When applying voltage to the electrodes of the piezoelectric plate, a decrease in inductance is observed. Thus, the initial value of the inductance (without an external electric field) is 1.27 mH, and with voltage of 25 V, the inductance drops to approximately 0.83 mH, 0.55 mH and 0.29 mH for samples based on PZT-piezoelectric bimorph, Metglas-PZT-Metglas, and Metglas-PZT, respectively. The decrease in inductance with increasing voltage continues until the Metglas reaches saturation and the effective magnetic permeability becomes equal to 1. It is clear that for a sample based on a bimorph piezoelectric structure, the inductance tuning range will be higher than for the remaining structures with the same value of the control voltage.

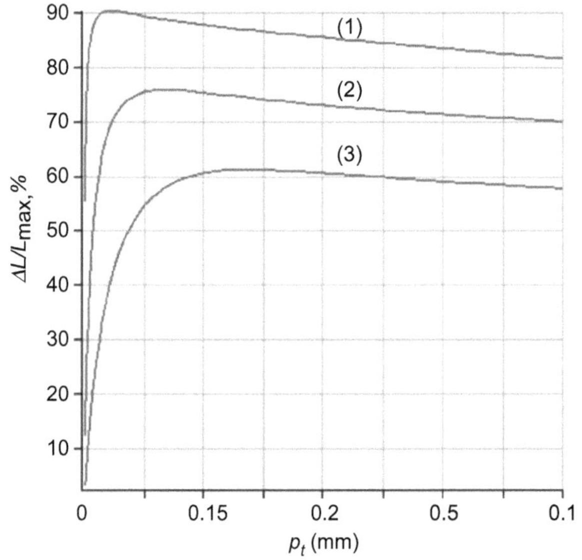

Figure 2.3 PZT thickness dependence of inductance tuning for inductive element based on the PZT bimorph and Metglas for Metglas layer thickness of 20 μm (1), 60 μm (2), and 120 μm (3).

Figure 2.3 shows the variation of inductance tuning with PZT thickness for several values of Metglas thickness. According to Fig. 2.3, the inductance tuning increases with the increase in the PZT thickness and reaches a maximum. Then it slowly decreases. The figure shows that the inductance tuning is maximal at PZT thickness of approximately 0.02 mm and Metglas thickness of 20 μm. For example, the inductance tuning equals 82%, 70%, and 58% for PZT thickness of 0.2 mm and for Metglas thickness of 20 μm, 60 μm, and 120 μm, respectively. It follows that the inductance tuning is stronger for thin Metglas layers.

The coil inductance strongly depends on the magnetostrictive phase thickness. It is clear that the inductance value increases with increasing Metglas thickness. Figure 2.4 shows the PZT thickness dependence of the ME susceptibility. The peak ME susceptibility is observed at specified piezoelectric and magnetic phases thicknesses.

The following material parameters were used for theoretical estimates: $q_{11} = 11000 \times 10^{-12}$ m/A; $d_{31} = -175 \times 10^{-12}$ m/V; $^{p}s_{11} = 15.3 \times 10^{-12}$ m^2/N; $^{m}s_{11} = 10 \times 10^{-12}$ m^2/N; $l = 0.01$ м; $\mu/\mu_0 = 2.9 \times 10^5$, $\lambda_s = 35 \times 10^{-6}$, $4\pi M_s = 18$ kG.

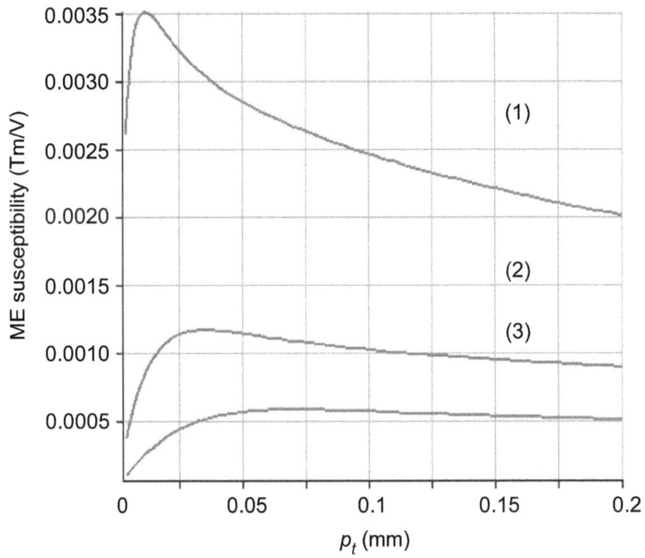

Figure 2.4 PZT thickness dependence of ME susceptibility for inductive element based on the PZT bimorph and Metglas for Metglas layer thickness of 20 μm (1), 60 μm (2), and 120 μm (3).

2.1.2 Comparison of Theoretical and Experimental Data

The manufacturing technique of a controlled inductive element based on the ME effect consists of the following steps:

(1) Fabrication of layered magnetostrictive–piezoelectric structure, which is the core of this device.

For the manufacture of the core, piezoelectric polarized plates and Metglas AMAG 170 plates are used. The most important stage in core fabrication is a proper mechanical coupling between the layers since it affects the ME interaction strength. We used epoxy compound which, on the one hand, provides a reliable mechanical adhesion, on the other hand, does not give the clamping effect.

(2) Fabrication of electrodes

The electrodes were connected to the piezoelectric layer by soldering. Since piezoelectric plates are metalized with silver, for a better connection, an active flux based on soldering acid was used.

(3) Manufacture of the coil

Since the initial value of the inductance does not influence its relative variation, a coil with a single-layer winding was used in the experiment. It was found that a decrease in inductance occurs at a higher frequency due to strong demagnetizing effect of eddy currents in the Metglas plate. That is, the inductance value noticeably decreases with increasing applied electric field at low frequencies. At higher frequencies, the inductance practically does not change. Therefore, experimental studies were conducted at frequency of 1 kHz.

At this stage, the measuring results for the inductance and its tuning are given for different values of the applied voltage for the piezoelectric–magnetostrictive bilayer with piezoelectric layer thickness of 1 mm and magnetostrictive layer thickness of 40 μm (2 Metglas layers), for Metglas-PZT-Metglas trilayer, and for Metglas-PZT bimorph structure.

According to the graphs in Fig. 2.5, the most effective sample is Metglas-PZT bimorph structure that confirms the theoretical findings.

Below are the results of measuring the inductance tuning when applying voltage to the 75 V electrodes for a sample based on a bimorph piezoelectric structure, depending on the number of layers of the Metglas.

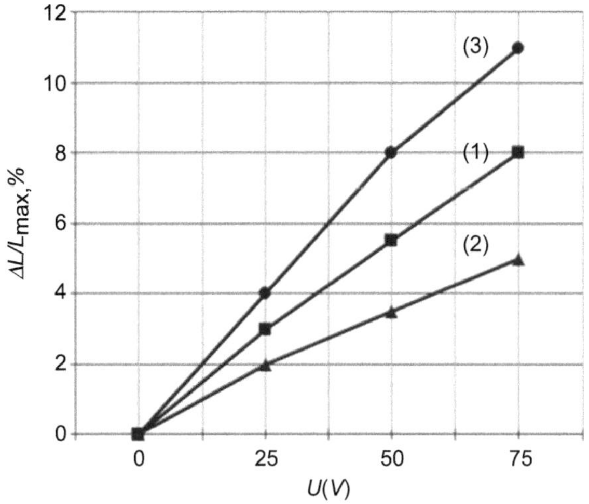

Figure 2.5 Experimental applied voltage dependence of inductance tuning for samples based on Metglas-PZT-Metglas (1), Metglas-PZT (2), and Metglas-PZT bimorph (3).

According to the graph above, the most effective sample is based on a bimorph piezoelectric structure with a magnetostrictive phase thickness of 20 μm (corresponding to 1 layer of Metglas). The data are in good agreement with theoretical findings.

In this chapter, we have considered possible applications of layered magnetostrictive–piezoelectric materials based on inhomogeneous components in the ME effect–based devices.

A theoretical model has been developed that makes it possible to determine the basic parameters of the inductive element, such as inductance and its relative tuning, ME susceptibility, and effective permeability. Expressions for the basic parameters of the inductive element are obtained for symmetric magnetostrictive–piezoelectric structure with longitudinal deformation, and for structure based on magnetostrictive layer and piezoelectric bimorph taking into account bending strain.

Numerical estimates of the inductance and its variation with applied voltage are presented for an inductive element based on the symmetric Metglas-PZT-Metglas trilayer, asymmetric Metglas-PZT bimorph, and Metglas-PZT structures. A voltage applied across the electrodes of piezoelectric plate induces a decrease in inductance. It was found that the inductance tuning range for the sample based on piezoelectric bimorph is higher than for the remaining structures for the same applied voltage.

Numerical estimates of inductance and variation of magnetic susceptibility due to ME coupling with PZT thickness are given for the inductive element based on piezoelectric bimorph for different thicknesses of the Metglas. The dependence in both cases has the form of a curve increasing to a certain peak value, and then slowly decreasing.

To confirm the theoretical studies, prototypes of inductive elements based on the ME effect were made and the inductance values were measured. The measuring results for the inductance and its tuning at different values of the applied voltage are given for Metglas-PZT-Metglas trilayer, Metglas-PZT bimorph, and Metglas-PZT structures with piezoelectric layer thickness of 1 mm and magnetostrictive layer thickness of 40 μm (2 Metglas layers).

Based on obtained results, one can conclude that the most effective sample is the magnetostrictive–piezoelectric bimorph structure.

2.2 ME Sensors

There are many different ways of measuring physical quantities. The most promising are methods using a magnetic field. To solve these problems magnetoelectric (ME) sensors are well suited.

Key features of ME sensors for measuring the magnetic field, electric current and crankshaft position sensor are discussed. ME sensors are shown to have advantages over semiconductor ones in the sensitivity, low price, and radiation resistance. To predict the feasibility of a composite for sensor application, we propose the nomograph method based on given parameters of the composite components. The sensor sensitivity depends on the construction and the materials parameters of the ME composite and bias magnetic field. ME laminates offer opportunities for low frequency (10^{-2}–10^{3} Hz) detection of low magnetic fields (10^{-12} Tesla or below) at room temperature in a passive mode of operation. Any other magnetic sensor does not reveal such combinations of characteristics. Current sensing based on ME effect is a good choice for many applications due to galvanic isolation between the current and measuring circuit. For increasing the sensor sensitivity one needs to use the ME composite based on materials with high magnetostriction and strong piezoelectric coupling. Proposed ME crankshaft position sensor would have several advantages against the traditional sensors such as smaller size and weight. Magnetostrictive–piezoelectric layered structure based on piezoceramic PZT plate with dimensions 0.5 mm of thickness, 20 mm of length, and 10 mm of wide was used as sensor.

In this chapter, under the ME sensors, we understand the devices recording the magnetic field, current in conductor, microwave power, and so on, at that the ME composites are the working material of these devices. In ME composites ferromagnetism and ferroelectricity occur simultaneously and coupling between the two is enabled and connected with the ME effect. The ME effect is defined as the dielectric polarization response of a material to an applied magnetic field, or an induced magnetization change upon application of an external electric field.

2.2.1 Magnetic Field Sensor

The magnetic field sensors are widely used in various fields of science and technology. For example, in physics they are used to measure the magnetic field, in medical technology as a measure of human magnetic fields, electrical heart signals, etc. More often to measure the magnetic field used sensors based on galvanomagnetic phenomena (the Hall effect), the Helmholtz coil, the magnetoresistive effect, and other.

A sensor is known to be a device that detects changes in quantities and provides a corresponding output. The ME sensor represents a structure with ME coupling with two electrodes for connecting to the voltmeter. The action of the sensor is based on the magnetoelectric effect. A composite of magnetostrictive and piezoelectric materials is expected to be magnetoelectric since a deformation of the magnetostrictive phase in an applied magnetic field induces an electric field via piezoelectric effect.

The ME effect in composites of magnetostrictive and piezoelectric phase is determined by the applied dc magnetic field, electrical resistivity, volume fraction of components and mechanical coupling between the two phases. The ME interaction is a result of magnetomechanical and electromechanical coupling in the magnetostrictive and piezoelectric phases, and stress transfer through the interface between these two phases. It should be noted that both the magnetomechanical response in magnetostrictive phase and electromechanical resonance in piezoelectric phases are possible origin of ME output peaks.

2.2.1.1 Principle of operation

The working principal of the ME sensor is that an input magnetic field changes the length of the Metglas via magnetostriction, and because the PZT fibers are elastically bound to the Metglas layers through an epoxy interfacial layer, the PZT fibers also change their length and generate an output voltage via piezoelectricity.

The static ME effect can be measured by using the electromagnet with Helmholtz coils between the electromagnet poles. The Helmholtz coils generate the dc magnetic field in the ME composite. Applying the dc magnetic field to the ME layered structure induces the output dc voltage across the piezoelectric layer. The relation between output voltage and magnetic fields can be described by sensitivity. The sensitivity of the ME sensor is determined by the expression: $S = \alpha_E \cdot {}^{p}t$ where $\alpha_E = \Delta E/\Delta H$ is ME voltage coefficient and ${}^{p}t$ is the piezoelectric layer thickness with ΔE and ΔH denoting the induced voltage and applied magnetic field.

The variation of static ME sensitivity, $(\Delta E/\Delta H)_H$, as a function of magnetic field H, at room temperature is shown in Fig. 2.6. It is observed that the sensitivity S initially increases up to a certain magnetic field and finally attains a maximum value and then decreases with an increase in applied dc magnetic field. This is because the magnetostrictive coefficient reaches saturation at a certain value of magnetic field. Beyond saturation the magnetostriction and the strain thus produced would also produce a constant electric field in the piezoelectric phase making the sensitivity decrease with increasing magnetic field.

The increases in $(\Delta E/\Delta H)_H$ with magnetic field is attributed to the fact that the magnetostriction reaches its saturation value at the time of magnetic poling and produces a constant electric field in the ferroelectric phase. Therefore, beyond a certain field, the magnetostriction and the strain thus produced would produce a constant electric field in the piezoelectric phase. Also, the possible reason for decrease in $(\Delta E/\Delta H)_H$ after certain magnetic field is attributed to the saturation state of the magnetic phase which does not show any response to the increased applied magnetic field, therefore, stress transfer through the interface between magnetic and electric phase decreases with magnetoelectric coupling $(\Delta E/\Delta H)$.

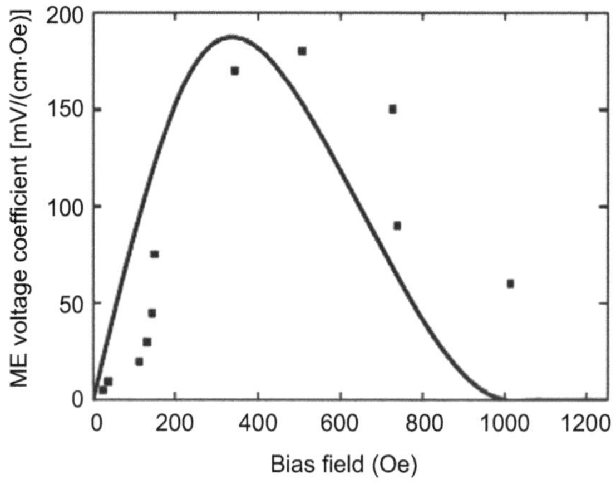

Figure 2.6 Theoretical ME voltage coefficient versus *H* and data for bilayer of NFO and PZT.

2.2.1.2 Equivalent circuit

To obtain the maximal ME output, the bias and ac magnetic fields should be simultaneously applied to the sample. For measuring either of these fields, the value of second field should be specified. In an ac magnetic field sensor, the reference bias magnetic field can be generated by both permanent magnet and electromagnet [2]. Making a dc (ac) magnetic field sensor implies using the additional magnetic system to produce the ac (dc) reference field as in Fig. 2.7.

Recent investigations of ME laminates sensors have shown that they have remarkable potential to detect changes in magnetic fields. It was shown that the feasibility of detecting magnetic field changes at the order of 10^{-12} T at near-quasi-static frequencies of

$f > 1$ Hz. This is an important achievement because the ME sensor does not itself require powering; rather it can harvest magnetic energy from inductances as a stored charge across a capacitor. Thus, ME laminates are small, passive magnetic field sensors with the potential of picotesla sensitivity at low frequencies while operated at room temperature. The potential for ME sensors resides in the fact that there are no other generations of magnetic sensors present that have the following key requirements [3, 4]: (i) extreme sensitivity (~pT $Hz^{-1/2}$), allowing for better magnetic anomaly detection; (ii) zero power consumption to foster long-term operation; (iii) operation at low frequencies, $f \sim 1$ Hz; (iv) miniature size, enabling deployment of arrays; (v) passivity; and (vi) low cost. It should be noted that ME laminate sensors are the only ones with the potential to achieve all key requirements. However, in spite of this potential, there are no available technologies that can fulfill requirements referred to above. The integration of ME laminates into an appropriate detection scheme is yet to be achieved. This detection scheme must be simple and capable of detecting anomalies in the time domain capture mode without either signal averaging or phase referencing.

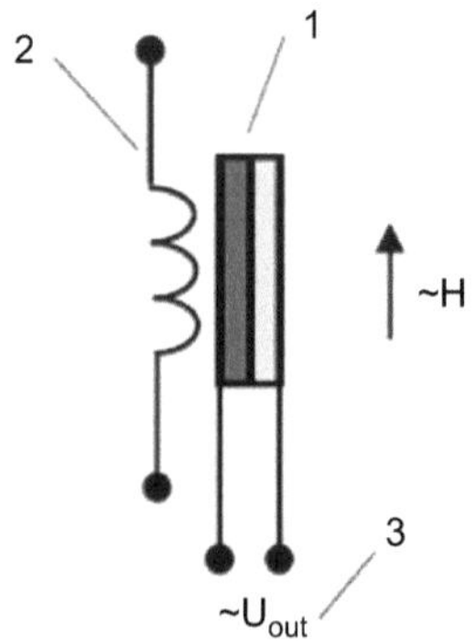

Figure 2.7 The equivalent circuit of ac (dc) magnetic field sensor. 1 and 2 are the ME composite sample and dc (ac) electromagnet.

Commonly, noise is defined as any undesirable disturbance that obstructs the relevant signal passage. It is of importance in the measurement of minute signals. Reducing the noise effect on the detection device is important since the sensitivity of a sensor is often limited by noise level. We will consider some simple ways to reduce noise.

The sensor itself and the measurement circuit contribute some inherent noise. This kind of noise cannot be removed since it comes from stochastic phenomena: thermal and radiation fluctuations between sensor and environment, generation and recombination of electron–hole pair, and current flows across a potential energy barrier in materials.

Development in the noise reduction of magnetostrictive/piezoelectric laminate sensors has been carried out in the past decade. Particularly, a 1 Hz equivalent magnetic noise of 5.1 pT $Hz^{-1/2}$ has been obtained, which is close to that of the optically pumped ultralow magnetic field sensors [5]. First of all, this was enabled by improved methods of interfacial bonding that can decrease the equivalent magnetic noise floor up to 2.7×10^{-11} T $Hz^{-1/2}$ [6]. Then, optimal poling conditions for the piezoelectric phase result in an increase in ME voltage coefficient by a factor of 1.4. The equivalent magnetic noise at $f = 1$ Hz was reported to be equal 13 to 8 pT $Hz^{-1/2}$ [7].

Magnetic flux concentration was found to enhance the ME coefficient of an ME sensor. A dumbbell-shaped sensor with an enhanced ME coefficient and reduced equivalent magnetic noise was reported [8], in which the dumbbell shape leads to concentration of magnetic flux. ME laminates with dumbbell-shaped Metglas layers exhibited 1.4 times lower required dc magnetic bias fields and 1.6 times higher magnetic field sensitivities than traditional rectangular-shaped ME laminates.

It was found that Mn-doped PMN-PT single crystals have the advantages of high piezoelectric coefficient and extremely low tan δ. Experimentally, an ultralow equivalent magnetic noise of 6.2 pT $Hz^{-1/2}$ was obtained at 1 Hz of the multi-push–pull mode for Metglas/PMN-PT single crystals [9].

2.2.1.3 Design

The combination of magnetostrictive amorphous ferromagnetic ribbons with piezoelectric materials allows obtaining magnetoelectric laminated composites that show an extremely high sensitivity for magnetic field detection. Magnetic alloys epoxyed to polyvinylidene fluoride (PVDF) piezoelectric polymer result in magnetoelectric coefficients above 80 V/(cm·Oe). Also, high-temperature new piezopolymers as polyimides can be used for magnetoelectric detection at temperatures as high as 100 °C.

This ME sensor is a Metglas/PZT-fiber laminate with a 100 circle coil wrapped tightly around it. The PZT fibers were 200 μm in thickness and were laminated between four layers of Metglas by use of a thin layer epoxy; the thickness of each Metglas layer was 25 μm and the total dimensions of the laminates were $10 \times 6 \times 0.48$ mm^3.

ME three-layer can be constructed sandwich-like with longitudinal magnetostrictive operation and transverse piezoelectric response-laminated composites by gluing two equal magnetostrictive ribbons to opposite sides of polymer piezoelectric films with an adhesive epoxy resin [3]. Magnetostrictive ribbons belonging to the family of Fe–Co–Ni–Si–B, Fe-rich metallic glasses have a measured magnetostriction that ranges between $\lambda s \approx 8$–30 ppm and maximum value for the piezomagnetic coefficient $d_{33} = d\lambda/dH$ of about 0.6–1.5 × 10^{-6}/Oe. This last parameter will modulate the magnetoelectric response of the composite as a function of the applied bias magnetic field. For the piezoelectric material, we firstly used the well-known polymer PVDF, which has glass transition and melting temperatures of about 35 °C and 171 °C, respectively, but a Curie temperature of about 100 °C. This makes its piezoelectric response to decay quickly above 70 °C. To develop a ME device being able to operate at higher temperatures, new amorphous piezoelectric polymers of the family of the polyimides were tested. It should be noted that its main parameters are a glass transition temperature (T_g) of about 200 °C and a degradation temperature (T_d) of about 510 °C; temperatures that make these polyimides suitable for our purposes. Taking advantage of the magnetoelastic resonance effect that enhances the magnetostrictive response, all measurements have been taken at resonance. For that, the static magnetic field H_{DC} necessary to induce the maximum amplitude of that resonance was first determined. The induced magnetoelectric voltage in the sandwich laminate (through two small silver ink contacts located at both opposite magnetostrictive ribbons) was measured by the following procedure: under an H_{AC} magnetic excitation applied along the length direction, the magnetostrictive ribbons will elongate and shrink along the same direction. This will make the piezoelectric polymer film to undergo an ac longitudinal strain, inducing a dielectric polarization change in its transverse direction. Thus, we can determine simultaneously the ME response dependence as the bias field H_{DC} changes and, at the H_{DC} value for the maximum magnetoelastic resonance amplitude, the ME voltage dependence *vs.* the applied ac magnetic excitation.

The highest ME response has been reported for laminated magnetostrictive–piezoelectric polymer composites. ME voltage coefficient of 21.5 V/(cm·Oe) for a Metglas 2605 SA1®/PVDF (Metglas, Conway, SC, USA) laminate was achieved at nonresonance frequencies and is, so far, the highest response obtained at sub-resonance frequencies [3]. At the longitudinal resonance mode, energy transference from magnetic to elastic, and vice versa, is maximum. This energy conversion at the resonance turns out to be very sharp for ME laminates while frequency bandwidth for applications based in this EMR enhancement effect remains limited. ME voltage coefficient of 383 V/(cm·Oe) on cross-linked *P* (VDF-TrFE)/Metglas 2605 SA1 is the highest reported to date. In order to avoid the observed sensitivity decrease when increasing the temperature, the same *L*–*T* structured magnetoelectric laminate was fabricated with the same magnetostrictive constituents but using a 40/60 copolyimide as high temperature piezoelectric constituent.

Efforts to get wider bandwidths for EMR and ME applications have been mainly based on magnetic field tuning procedures either in bimorph or tri-layered structures, but the maximum achieved frequency of operation has been some tenths of kHz. Another way to get high frequencies of operation can be based on the relationship between length and resonant frequency value of magnetostrictive ribbons at the magnetoelastic resonance. So, our efforts are now focused on fabricating short magnetoelectric *L*–*T* type laminates showing good magnetoelectric response at high frequencies. Nevertheless, the higher the resonant frequency, the lower the amplitude of the resonance and as a first consequence, the magnetoelectric response will be also decreased. It is clear that a compromise between the length of the device and therefore the working frequency and the induced magnetoelectric signal, must be achieved. Thus, a device 1 cm long for which the resonant (working) frequency rises to 230 kHz was developed. The measured magnetoelectric voltage coefficient is about 15 V/(cm·Oe) when PVDF is used as piezoelectric constituent. Thus a 0.5 cm long device that will work at a resonant frequency about 500 kHz is expected to be constructed. This fact, combined with the use of a high temperature piezopolymer as the polyimides previously described, can lead to a very useful class of magnetoelectric laminates working simultaneously at high temperature and within the radiofrequency range, both characteristics of great interest for low distance near field communications in aggressive environments (e.g., in a desert, in a tunnel, or while fighting a fire).

Combining the excellent magnetoelastic response of magnetostrictive amorphous ferromagnetic ribbons with piezoelectric polymers, the short length magnetoelectric laminated composites that show an extremely high sensitivity for magnetic field detection were fabricated.

2.2.1.4 Discussions

The magnetic sensors based on magnetoelectric composites for the practical purposes including the use in biomagnetic imaging have been of considerable interest in recent years [10–21]. Migratory animals are capable of sensing variations in geomagnetic fields as a source of guidance information during long-distance migration. It is well known that geomagnetic fields are on the order of 0.4–0.6 Oe and have different inclinations at different locations. The earth's mean field and its inclinations at many points over much of the earth's surface are known to be tabulated. Accordingly, geomagnetic field sensors could be used for guidance and positional location. There are many types of magnetic sensors: for example, superconducting quantum interference devices or giant magnetoresistance spin valves. However, these sensors require very low operational temperatures of liquid nitrogen in order to achieve high sensitivity. Fluxgate sensors based on exciting coil have been investigated for many years to detect dc magnetic and geomagnetic fields. This widely used sensor is relatively cheap and temperature independent; however, its magnetic hysteresis, offset value under zero magnetic field, and large demagnetization factor restrict design considerations. Recently, new types of passive ac and active dc magnetic field sensors have been developed based on a giant magnetoelectric ME effect. They are simple devices that work at room temperature. Using a Metglas/PZT-fiber ME sensor enables one to detect precisely both geomagnetic fields and their inclinations along various axes of a globe. Detection of the earth's magnetic field was performed by applying of 1 kHz ac magnetic field H_{AC} via about 10 mA ac input to the coil and by measuring the dc voltage and its phase induced in the PZT fibers by a lock-in amplifier SR-850. Over the range of $-1.5 < H_{DC} < 1.5$ Oe, V_{ME} was linearly proportional to H_{DC} and equal to 300 mV under a $H_{DC} = 1$ Oe. When the sign of H_{DC} was changed, a dramatic 180° phase shift was found. This shift could be used to distinguish the direction along which changes in H_{DC} occur with respect to the length-long axis of the sensor. This is an important advantage compared to fluxgate. Previously, it was reported that VME from a Metglas/PZT fiber laminates was strongly anisotropic, offering good sensitivity to magnetic field variations only along its length direction. In the other two perpendicular directions, only very weak signals were found with changes in H_{DC}. These unique properties of Metglas/PZT-fiber ME sensors are due to the ultrahigh relative permeability r of Metglas, which is 103 times larger than that of Terfenol-D or nickel ferrite. Correspondingly, the high r of Metglas results in an ultrasmall demagnetization field, enabling a high effective-piezomagnetic coefficient at low biases.

Thus sensitivities of a few picotesla to hundreds of femtotesla for 1–30 MHz magnetic fields are required for use in biomagnetic imaging. A possible approach for achieving such sensitivities is a bilayer ME sensor operating under frequency modulation at bending resonance [3]. It is of interest to compare the low-frequency and resonance ME voltage coefficients in representative bilayer composite systems. One of the best values for low-frequency ME voltage coefficient, ~52 V/(cm·Oe), was measured in samples of Metglas and a piezofiber and was attributed to high q value for Metglas and excellent magnetic field confinement field due to high permeability. A recent study compared the low-frequency and resonance ME effects in bilayers of composites with permendur and ferroelectric PZT and PMN-PT and piezoelectric langatate and quartz [20]. The highest ME voltage coefficient of 1000 V/(cm·Oe) at bending resonance among these systems was measured for a permendur-langatate bilayer. But the highest resonance ME voltage coefficient obserbed to date is 20 kV/(cm·Oe), which was reported for AlN-FeCoSiB for measurements under vacuum that reduces damping of bending resonance in air [10, 11]. A very high ME sensitivity was also reported under bending resonance in a cantilever of FeCoSiB and PZT with inter-digital electrodes.

ME sensitivity optimization should take into account the environmental or external noise sources, such as thermal fluctuation and mechanical vibration. These external noises are dominating factors that affect the sensor's sensitivity in practical applications. For ME sensors, the dominant ones are the thermal fluctuation and mechanical vibration sources. Thermal fluctuation noise is pyroelectric in origin, where the spontaneous polarization of the piezoelectric phase is temperature dependent, resulting in a dielectric displacement current in response to temperature changes; whereas the vibrational noise is piezoelectric in origin, where the spontaneous polarization is coupled to pressure and stress changes via piezoelectricity. As for all magnetic field sensors, it is important that ME sensors be designed by such a means that optimizes its abilities to cancel these external noise.

In the case of the push-pull laminate, the extreme enhancement in the sensitivity limits ($\sim 10^{-15}$ T $Hz^{-1/2}$) at EMR is nearly equivalent to that of a SQUID sensor operated a 4 K and 15 mA. ME laminates offer much potential for low frequency (10^{-2}–10^{3}) Hz detection of minute magnetic fields (10^{-12} T or below) at room temperature in a passive mode of operation; such combinations of characteristics are not available in any other magnetic sensor.

2.2.2 Current Sensor

The current sensors are very important type of devices. There are a large number of current sensors operating on different physical principles. The most common sensors are current transformer and magnetoresistance and Hall sensors [22, 23]. Despite the fact that ME current sensors have small size and weight and high sensitivity, it received little attention in the publications, compared to the ME magnetic field sensors [24]. The first attempt to measure the direct current (DC) based on the ME effect was made in [25]. There are also a number of articles about ME current sensors [34, 35]. Evaluation of the value of current flowing in the conductor is possible by measuring the magnetic field. The ME current sensor uses the ME effect as a basis of its measurements. As is well-known, the ME effect is a polarization response to an applied magnetic field, or conversely a magnetization response to an applied electric field. ME behavior exists as a composite effect in multiphase systems of piezoelectric and magnetostrictive materials. In a magnetostrictive–piezoelectric layered structure the interaction between magnetic and electric subsystems occurs through mechanical deformation. It means that the ME effect is much stronger at frequencies corresponding to electromechanical resonance range. In current sensor applications, the induced ME voltage coefficient is more important than the induced ME electric field coefficient, as voltage is the physical quantity measured. ME current sensors can work in different regimes. In the first case, ME element works in the regime of non-resonant mode, and in the second case, in the resonant one. As a sensitive element of the sensor in both cases, the same ME element can be used. The properties of the nonresonant ME current sensor were considered in Refs. [26, 27] and the resonant one in Ref. [28].

2.2.2.1 Nonresonant current sensor

2.2.2.1.1 *Principle of operation*

Magnetoelectric current sensor consists of a ME element made in the form of a piezoelectric plate and the side surfaces of the magnetostrictive plates are glued. The principle of operation of ME sensor is shown in Fig. 2.8. Here a current conductor or a current coil replaces a bias magnetic field for inducing the ME effect. The modulating coil is needed to define the frequency range. Then everyone can choose to operate the linear portion of the ME voltage coefficient dependence and get a result of the sensitivity dependence for the ME sensor, that is, the dependence of the output voltage on the input measuring current.

2.2.2.1.2 *Equivalent circuit*

For a better understanding of the current processes, it is suggested to consider the magneto-electro-mechanical circuit, equivalent to nonresonant current sensor (Fig. 2.9) [28, 29]. The magnetic section (I) comprises an AC voltage source and an inductance coil. Flowing in the magnetic section (I) an electric current by means of an ideal transformer causes in mechanically section (II) the occurrence of mechanical current $\bar{U}$. Mechanical section (II) is connected with the second ideal transformer with an electric section (III), thus the mechanical current results in flow of electric current i in an electric section. The output voltage V is taken from the capacitor C_0 of electrical section.

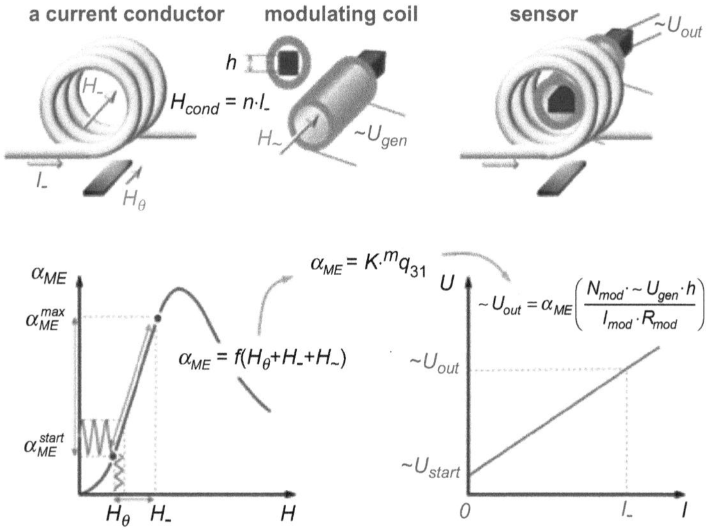

Figure 2.8 Magnetoelectric current sensor: principle of operation.

For the calculation we used the value of coil inductance (L_s = 5.83 × 10^{-3} H) and active resistance (R_s = 510 Ohm) obtained from the experimental data. It is necessary to make the resistance Z as much high as possible, because then the mechanical circuit will not distort the input current of the solenoid circuit. And so as not to distort the mechanical connection and the output circuit capacity, C_0 must be reduced as much resistance Z has been increased.

$$Z = -\frac{1}{2} jM\rho v A_1 \cot\frac{kl}{2} \tag{2.11}$$

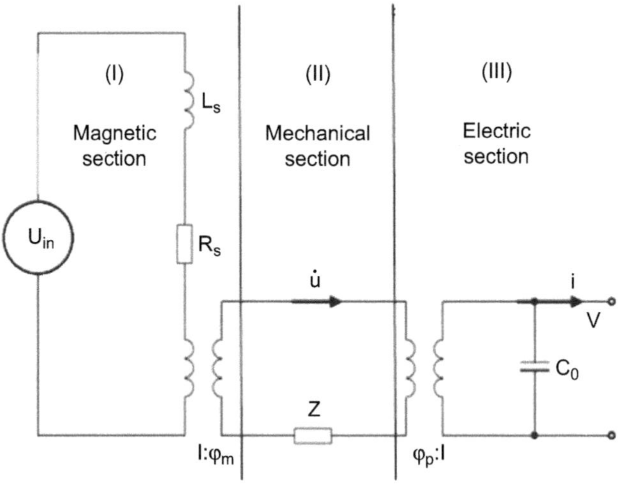

Figure 2.9 The equivalent circuit of ME nonresonant sensor DC.

$$C_0 = \frac{lb\left(\varepsilon_{33} - \frac{d_{31}^2}{{}^{\mathrm{p}}s_{11}}\right)}{M\,{}^{\mathrm{p}}t} \qquad M >> 1 \tag{2.12}$$

In [30], authors used the input section with the magnetic field source. In our case the input section contains an electric current. Therefore, the ideal ratio of transformation has the following form:

$$\varphi_{\mathrm{m}} = \frac{{}^{\mathrm{m}}\nu q_{11}}{{}^{\mathrm{m}}s_{11}} A_1 n \tag{2.13}$$

For this equivalent circuit the equations are as follows:

$$\begin{aligned} &R_s J + j\omega L_s J = U_{\mathrm{in}} + \varphi_{\mathrm{m}}\dot{u} \\ &Z\dot{u} = \varphi_{\mathrm{m}} J + \varphi_{\mathrm{p}} V \\ &J_{\mathrm{disp}} = j\omega C_0 V - \varphi_{\mathrm{p}}\dot{u} \end{aligned} \tag{2.14}$$

On using open-circuit conditions $J_{\mathrm{disp}} = 0$, where $J_{\mathrm{disp}} = (\delta D_3/\delta t) = 0$, the following equation for the output voltage was obtained:

$$U_{\mathrm{out}} = \frac{\varphi_{\mathrm{m}}\varphi_{\mathrm{p}}}{j\omega C_0\left[Z\left(R_s + j\omega L_s\right) - \varphi_{\mathrm{m}}^2\right] - \varphi_{\mathrm{p}}^2\left(R_s + j\omega L_s\right)} U_{\mathrm{in}} \tag{2.15}$$

In order for match the theoretical and experimental data, we introduced the coefficient due to manufacturing technology

K = 0.087. In the graph below, the theoretical curve and the experimental points of output voltage for a low-frequency sensor are presented (Fig. 2.10).

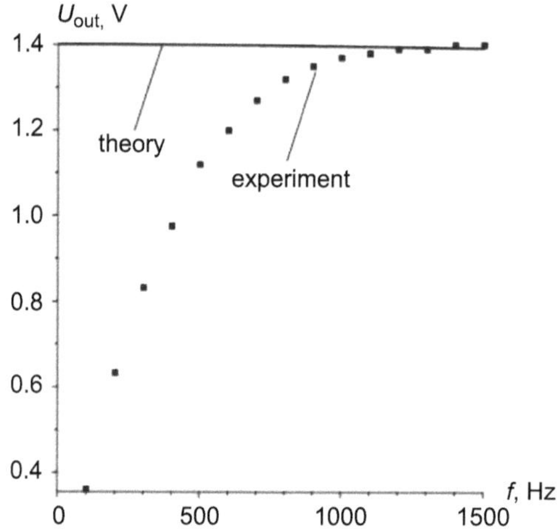

Figure 2.10 Output voltage depending on the frequency for nonresonant current sensor.

Study of the equivalent circuit for the low-frequency mode of operation allowed to find the coefficient due to manufacturing technology K, necessary for the further consideration of the resonant mode of operation.

2.2.2.1.3 *Design*

(a) Sensor design

The ME nonresonant current sensor consists of three major parts: the sensing head (generator, inductance coil, ME element, current coil), power supply, and the signal processing (amplifier, peak detector). Each part affects to performance of the current sensor. The block diagram of the ME nonresonant current sensor is shown in Fig. 2.11.

The principle of operation of the sensor is based on measuring the electrical voltage appearing at the output of sensitive element due to ME effect as a result of the influence of alternating and bias magnetic fields.

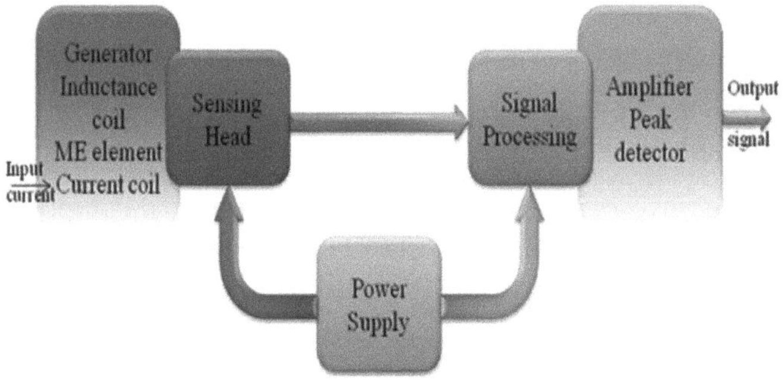

Figure 2.11 Block diagram of ME nonresonant current sensor.

(b) Sensing head

The ME element, shown in Fig. 2.12, is the primary part of the whole current sensor, and its function is to convert the measuring current into the voltage signal that can be measured easily.

ME element is the sensitive part of ME current sensor and in our case consists of piezoelectric and magnetostrictive layers. Layered structure based on piezoceramic PZT plate in our case had thickness of 0.38 mm, length of 10 mm, and width of 1 mm [26, 27]. Piezoelectric was polarized in the thickness direction. The electrodes are applied on two sides of the magnetostrictive plates. Thickness of one layer of Metglas was 0.02 mm. Metglas (1) and PZT (3) plates are connected by gluing (2). Different types of glue can be used. Thickness of the adhesive layer does not exceed 10 microns. To increase the sensitivity, different number of Metglas layers may be used. The electrical signal is taken from the electrodes (4) and then passes the signal processing.

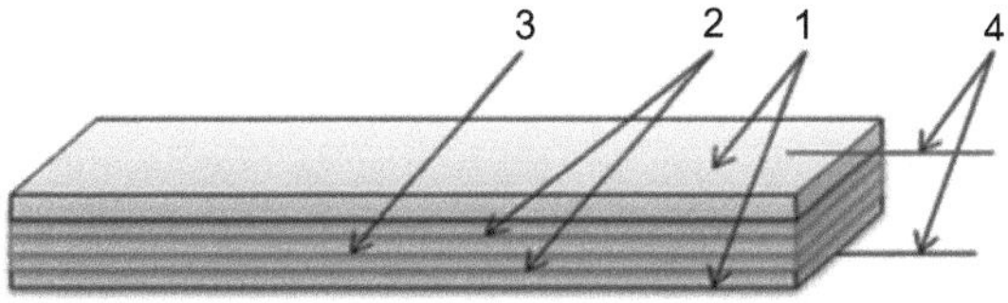

Figure 2.12 Magnetoelectric element: (1) magnetostrictive plates (Metglas), (2) adhesive layer, (3) piezoelectric plate (PZT), (4) electrodes.

The sensitive element of the sensor is a magnetostrictive–piezoelectric composite and its magnetostriction constant is a function of magnetic field and in general can be written as $\lambda = f(H_0)$ [31, 34]. In turn the piezomagnetic module depends on DC magnetic field as follows:

$$q = d\lambda/dH \qquad (2.16)$$

For the transverse ME voltage coefficient of layered composite, it is necessary to use the equation (1.30) from Chapter 1.

ME element is placed into the current and inductance coils, where a bias magnetic field and a variable modulation magnetic field are created, respectively (Fig. 2.13).

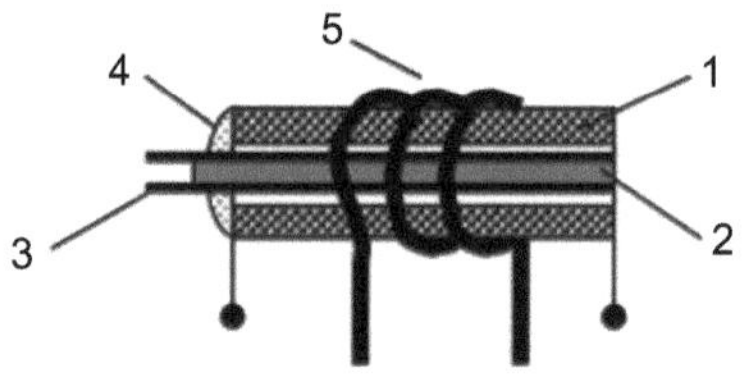

Figure 2.13 Design of hard core of magnetoelectirc sensor: (1) inductance coil, (2) ME element, (3) leading-out wire of ME element, (4) glue, (5) current coil.

An important part is the fixation of ME element in the inductance coil. ME element was fixed by the glue on one end of sample in order to avoid jamming of ME element surface.

(c) Signal processing

The signal processing is used to process the voltage signal from the sensing head, the block scheme is shown in Fig. 2.14.

Figure 2.14 Schematic diagram of signal processing.

In nonresonant case a signal processing scheme of current sensor consists of a generator that is tuned to the frequency of 500 Hz. A generator is connected to the inductance coil, then the output signal from the ME element is amplified and provided to a peak detector. The current coil creates a bias magnetic field proportional to the current in the coil. A microprocessor can be used for signal conversion.

In resonant case the signal processing scheme of current sensor similar with one, but sensitivities of the scheme increases from ten to hundred times. If the level output signal is enough for estimation and further processing, then it is possible to work without the amplifier.

(d) Construction

The ME current sensor is a system consisting of the ME composite (3), a generator (4), an amplifier, a peak detector, two coils (1, 2) placed into one another and a body. The amplifier and the peak detector are attached to the back of the generator. The construction of the DC ME sensor is shown in Fig. 2.15.

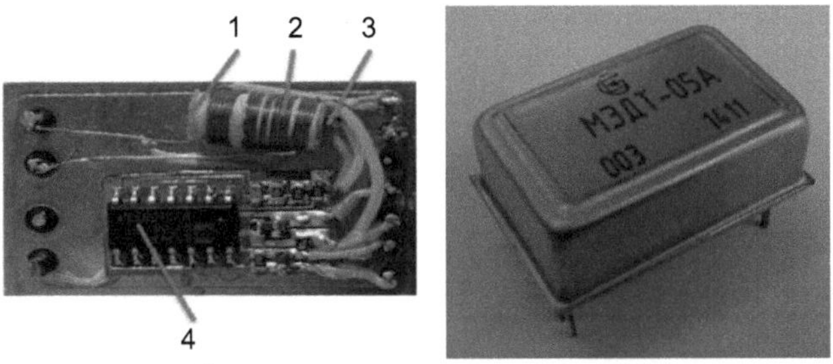

(a) (b)

Figure 2.15 Design of magnetoelectirc nonresonant current sensor's prototype: (a) nonresonant current sensor, (b) current sensor in the body.

It can be noted that the design of AC sensors is similar to the design of magnetic field sensors and differs in the presence of the conductor near the ME element. AC sensors can also be used like DC sensors only without the generator and modulating coil. The sensitivity of ME sensors will mainly depend on ME properties and maximal sensitivity should be at frequency of electromechanical resonance.

(e) Characteristics

The theoretical curve and the experimental points of the output characteristic of ME nonresonant current sensor are shown in Fig. 2.16. The output voltage of the sensor was calculated by the following equation:

$$U_{\text{out}} = \frac{K_{\text{amp}} K^{\text{m}} v^{\text{p}} L q_{11} \left(H_0 + \dfrac{I_-}{d} \right) d_{31} n U_{\text{in}}}{{}^{\text{m}}s_{11}\, {}^{\text{p}}s_{11} \varepsilon \varepsilon_0 c_{11} R_{\text{s}}} \infty$$

where K_{amp} is coefficient of amplifier,

K is coefficient due to manufacturing technology,

U_{in} is input amplitude voltage,

q_{11} is piezomagnetic module,

c_{11} is effective longitudinal stiffness coefficient of the composite, $c_{11} = {}^{p}v/{}^{p}s_{11} + {}^{m}v/{}^{m}s_{11}$,

${}^{p}s_{11}$, ${}^{m}s_{11}$ are coefficients of compliance for piezoelectric and magnetostrictive phase at constant electric field,

${}^{m}v$, ${}^{p}v$ are volume fractions of magnetostrictive and piezoelectric phases, respectively,

${}^{m}v = {}^{m}L/({}^{p}L + {}^{m}L)$ and ${}^{p}v = {}^{p}L/({}^{p}L + {}^{m}L)$, ${}^{m}L$, ${}^{p}L$ are thicknesses of magnetostrictive and piezoelectric layers, respectively,

d_{31} is piezoelectric module,

n is number of windings per unit length of the AC solenoid,

R_s is the active resistance of the AC solenoid,

ε is dielectric permittivity of piezoelectric phase,

ε_0 is electric constant.

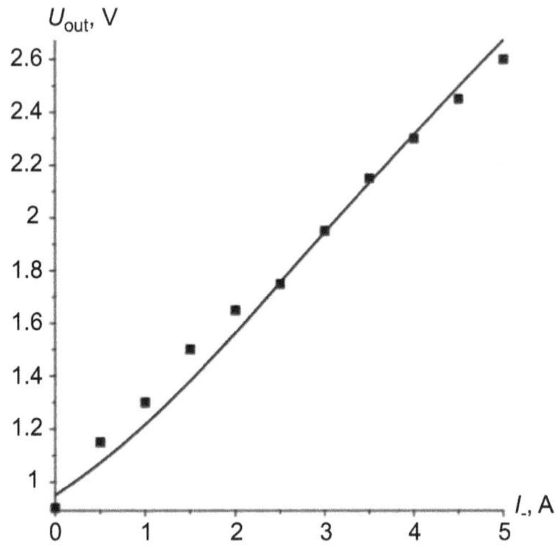

Figure 2.16 Output characteristic of ME nonresonant current sensor.

The calculations showed allowable matching between theory and experiment. Consequently, all the parameters of the sensor are chosen quite optimally.

In our case, the magnetostrictive material Metglas based on iron alloy was used, and the dependence of ME coefficient on bias magnetic field is shown in Fig. 2.8. The plot of the curve from α_{ME}^{start} to α_{ME}^{max} can be used for designing the sensor with high linearity. Measurements of Metglas parameters and calculations necessary to determine the values of bias magnetic field and the maximal measured current were made. The sensitivity of the sensor reached 0.34 V/A. The working range of currents was set up to 5 A. The nonlinearity of output characteristic was within 1%.

2.2.2.2 Resonant current sensor

2.2.2.2.1 *Principle of operation*

Magnetoelectric current sensor consists of a ME element made in the form of a piezoelectric plate and the side surfaces of the magnetostrictive plates are glued. The principle of operation of ME sensor is shown in Here a current conductor or a current coil replaces a bias magnetic field for inducing the ME effect. The modulating coil needs for definition of frequency range. Then everyone can choose to operate the linear portion of the ME voltage coefficient dependence and get a result of the sensitivity dependence for the ME sensor, that is the dependence of the output voltage on the input measuring current.

2.2.2.2.2 *Equivalent circuit*

When considering the ME equivalent circuits, one generally follows the method developed in [29]. In this paper, the authors use magnetic circuit of the magnetic field source directly, without specifying how the magnetic field is created. In contrast, we consider the appearance of the magnetic field during the flow of electric current through the inductor. It allows to consider the resulting frequency response of the sensor of our current dependence on the complex resistance of the coil of the alternating voltage source.

In the study resonant mode of operation of the sensor it can be considered completely simplified electrical equivalent circuit (Fig. 2.17). The input voltage is U_{in} whereby the current through the coil flows. This gives rise to a current in the resonant circuit containing the dynamic inductance of L, resistance R, and capacitance C. The output voltage is removed from the capacitor C_1 constituting together with the C_2 capacity, full operating capacity.

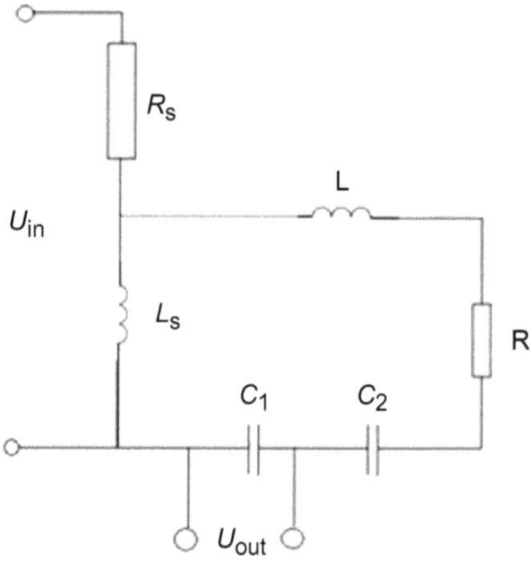

Figure 2.17 The equivalent circuit of ME resonant DC sensor

ME voltage coefficient is given by the following equation:

$$\alpha_E = \frac{\overline{E}}{h_1} = -\frac{{}^{m}v\,{}^{p}v q_{11} d_{31}\,{}^{p}s_{11} \tan(\eta)}{{}^{m}s_{11}\left(\varepsilon\varepsilon_0\,{}^{p}s_{11}^{2} c_{11}\eta + d_{31}^{2}\left[\tan(\eta) - c_{11}\,{}^{p}s_{11}\eta\right]\right)} \quad (2.17)$$

where

$$\eta = \frac{kl}{2} \quad (2.18)$$

The magnetic field inside a long solenoid:

$$h_1 = nJ \quad (2.19)$$

where n is the number of coils per unit length, J is a current.

Current in the solenoid:

$$J = \frac{U_{\text{in}}}{R_s + j\omega L_s} \quad (2.20)$$

Output voltage:

$$U_{\text{out}} = \bar{E}(^{\text{p}}L + {}^{\text{m}}L) = \alpha_{\text{E}} h_1 (^{\text{p}}L + {}^{\text{m}}L)$$

$$= -\frac{^{\text{m}}\nu\, {}^{\text{p}}L q_{11} d_{31}\, {}^{\text{p}}s_{11} \tan(\eta)}{^{\text{m}}s_{11}\left(\varepsilon\varepsilon_0\, {}^{\text{p}}s_{11}^2 c_{11}\eta + d_{31}^2[\tan(\eta) - c_{11}\, {}^{\text{p}}s_{11}\eta]\right)} \tag{2.21}$$

$$nJ = -\frac{^{\text{m}}\nu\, {}^{\text{p}}L q_{11} d_{31}\, {}^{\text{p}}s_{11} n \tan(\eta) U_{\text{in}}}{^{\text{m}}s_{11}\left(\varepsilon\varepsilon_0\, {}^{\text{p}}s_{11}^2 c_{11}\eta + d_{31}^2[\tan(\eta) - c_{11}\, {}^{\text{p}}s_{11}\eta]\right)(R_{\text{s}} + i\omega L_{\text{s}})}$$

Taking into account the calibration coefficient K, the following ratio of voltages was found:

$$\frac{U_{\text{out}}}{U_{\text{in}}} = -\frac{K\, {}^{\text{m}}\nu\, {}^{\text{p}}L q_{11} d_{31}\, {}^{\text{p}}s_{11} n \tan(\eta)}{^{\text{m}}s_{11}\left(\varepsilon\varepsilon_0\, {}^{\text{p}}s_{11}^2 c_{11}\eta + d_{31}^2[\tan(\eta) - c_{11}\, {}^{\text{p}}s_{11}\eta]\right)(R_{\text{s}} + i\omega L_{\text{s}})} \tag{2.22}$$

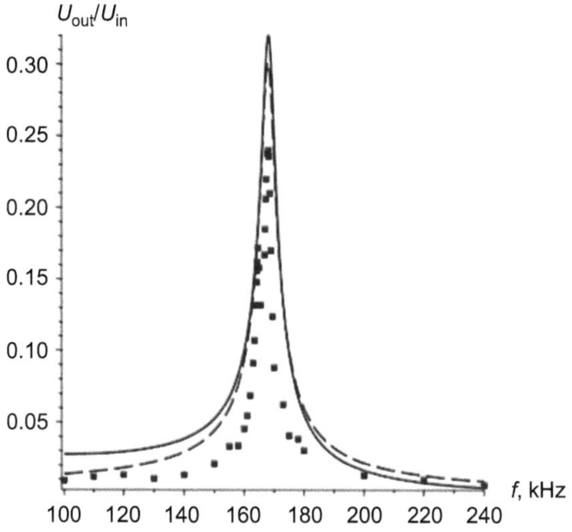

Figure 2.18 Amplitude frequency characteristic of resonant current sensor.

The results of calculation and experimental data are presented in Fig. 2.18. The value of the resonant frequency calculated by Eq. 2.23 is f_{res} = 168 kHz, and it agrees with the experimental data. The solid line and dotted line show the results of calculation by equation and the equivalent circuit, respectively. Considered equivalent circuit adequately describes the current operation of the sensor in the resonant mode because the frequency response of the equivalent circuit is in good agreement with theoretical frequency response and accurate experimental data.

2.2.2.2.3 *Design*

(a) Sensor design

The resonant current sensor has a similar design with nonresonant one. It also consists of a generator, inductance coil, ME composite, permanent magnet, current coil (sensing head), power supply, and rectifier (signal processing). The block diagram of the resonant current sensor is shown in Fig. 2.19.

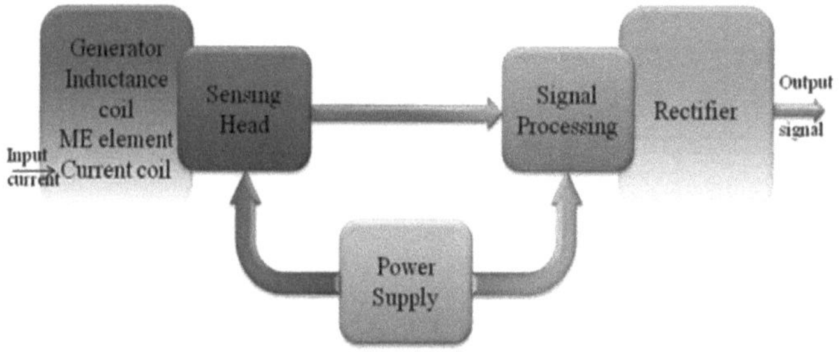

Figure 2.19 Block diagram of resonant current sensor.

The principle of operation of the sensor is based on measuring the electromotive force appearing at the output of sensitive element due to ME effect. As a result of influence the alternating magnetic field and a bias magnetic field. The amplifier here is not applicable because the output signal level is enough for estimation. In this case the rectifier (diode bridge) is sufficient.

(b) Construction

The sensing head is similar as in the nonresonant current sensor. We used the same composite, but with a slight difference. It consists of piezoelectric and magnetostrictive layers as shown in Fig. 2.12. The ME current sensor shown in Fig. 2.20 consists of the ME composite (3), a generator (4), a rectifier (5), a permanent magnet, inductance (2) and current coils (1) (coils placed into one another).

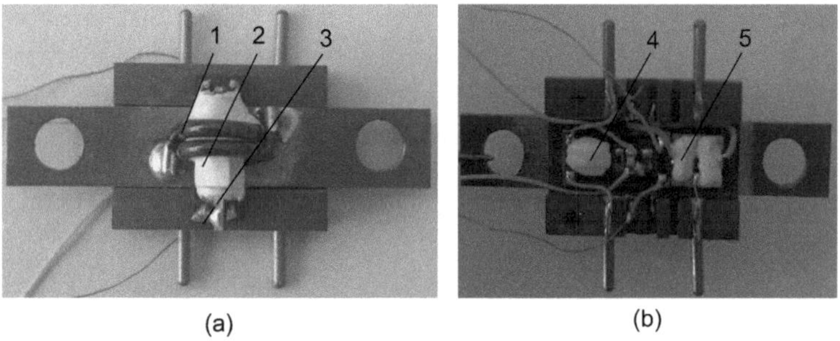

Figure 2.20 Design of magnetoelectric resonant current sensor prototype: (a) view from above; (b) bottom view.

The ME composite is a layered structure which includes a thin piezoceramic plate of PZT placed between two magnetostrictive layers of Metglas performing a function of electrodes. The final dimensions of the composite were 6 × 1 × 0.62 mm with the PZT concentration of 0.8. The PZT layer was poled in an electric field in the thickness direction. The ME laminate operates in a regime of the transverse mode in which applied magnetic fields ($H_{\sim}$, H_0) are parallel to each other and are located in the plane of laminate.

2.2.2.2.4 *Construction*

The resonant ME current sensors were designed. The sensor works in the electromechanical range with the resonance frequency f_{res} = 168 kHz. The sensors sensitivity and ME voltage coefficient dependencies were investigated using the designed measurement setup. The measurement stand includes power supply APS-7315 (Aktakom) and the power supply HMP4040 (HAMEG instr.), the multimeter HM 8112-3 (HAMEG instr.), the oscilloscope HMO722 (HAMEG instr.), and the electromagnet GMW 5403 (Magnet Systems). The graph below shows the output characteristic of the resonant current sensor.

Tables are numbered serially within each chapter following Fig. 2.21 shows the theoretical curve and the experimental points of the output characteristic of the resonant current sensor. As can be seen from the graph, the calculated and experimental results are in good agreement. This means that the selected parameters of the sensor and materials are most optimal. The maximum linearity is reached at the additional bias field of 10 Oe. Also it is known that sensitivity is described by the slope of the output characteristic. The sensitivity of the sensor reached 0.53 V/A. The working range of currents was set up to 5 A. The nonlinearity of output characteristic was within 0.5%. The resonant frequency of composite for the first axial mode was calculated by the following equation:

$$f_{\text{res}} = \frac{1}{2L}\sqrt{\frac{s_{11}^{\text{p}} + rs_{11}^{\text{m}}}{s_{11}^{\text{p}} s_{11}^{\text{m}} (r\,{}^{\text{p}}\rho + {}^{\text{m}}\rho)}} \tag{2.23}$$

where L is plate length, ${}^{\text{p}}\rho$ and ${}^{\text{m}}\rho$ the density of the piezoelectric and magnetic phases, respectively, and r is ratio of thicknesses, $r = {}^{\text{m}}L/{}^{\text{p}}L$.

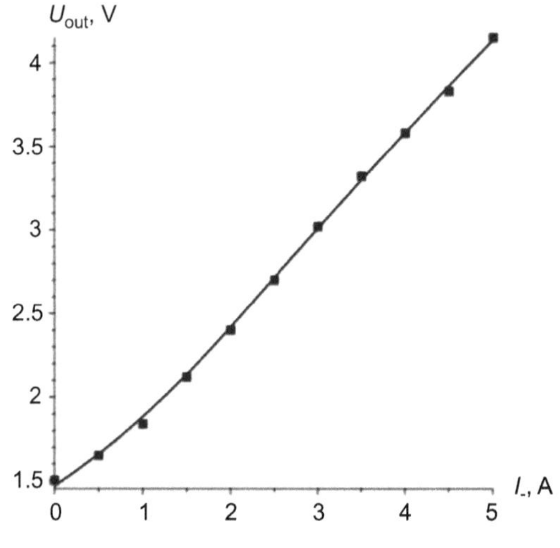

Figure 2.21 Output characteristic of ME resonant current sensor.

For calculation of the output characteristics the following equation was obtained:

$$U_{\text{out}} = \frac{8KQ\,{}^{\text{m}}v\,{}^{\text{p}}Lq_{11}\left(H_0 + \dfrac{I_{-}}{d}\right)d_{31}nU_{\text{in}}}{{}^{\text{m}}s_{11}\,{}^{\text{p}}s_{11}\varepsilon\varepsilon_0 c_{11}\pi^2(R_{\text{s}} + i2\pi f_{\text{res}}L_{\text{s}})} \tag{2.24}$$

where Q is quality factor of resonance, f_{res} the resonance frequency, and L_{s} the inductance of the AC solenoid.

2.2.2.2.5 *Discussions*

Table 2.1 shows the comparative characteristics of DC current sensors. The information is about the sensors' production of LEM Holding SA, Honeywell Inc., and Allegro MicroSystems. As can be seen from the table below the ME current sensors have higher sensitivity and lower current consumption.

Table 2.1 Performance comparison of current sensors

Sensors	**HO8-NP**	**CSLW6B5**	**ACS712ELCTR-05B-T**	**Non-resonant magneto-electric sensor**
Measuring principle	Hall-effect measuring principle	Miniature ratiometric linear Hall-effect sensor	Hall-effect sensor	Magneto-electric effect
Primary current, measuring range, I_{pm} (A)	0–20	±5	±5	0–5
Sensitivity (V/A)	0.1	0.2	0.185	0.53
Supply voltage (V)	5±10 %	4.5–10.5	5 ± 10%	5 ± 10 %
Current consumptions (mA)	19	9	10	2.5
Accuracy (%)	1	0.5	1.5	0.5
Output voltage range U_{out} (V)	2.5–0.5	2.7-3.7	2.5–4.5	0.3 - 2.6
Size (mm)	24 × 12 × 12	16.2 × 14 × 10	6 × 5 × 1.75	30 × 20 × 10

Known current transformers (CT), with high breakdown voltage, are the most widely applied for AC sensing in traditional power systems. CT can't be used as a DC sensor because it is based on Faraday's law of induction. Disadvantages of CT are: large size, high price, limited bandwidth, and large consumption of metal resources, it is only used in power stations and substations.

Shunts are more often applied in DC converter stations and power electronics. Disadvantages of shunt are: the measured current has to be interrupted in the sensor, an over current may permanently damage it, and the intrinsic inductance limits the accuracy and bandwidth.

Hall-effect current sensors are mainly applied in non-contact current measurements. The problems of the Hall-effect sensor are: low sensitivity, low breakdown voltage, and susceptibility to temperature, which limit it to applications in high-voltage power systems [23].

Compared with the current sensors presented above, the ME current sensors has advantages of higher sensitivity, higher linearity, lower cost, simple structure of sensor, which make it the most promising for current measurements.

Eventually the characterization of nonresonant current sensor showed that in the operation range upto 5 A, its sensitivity was 0.34 V/A and nonlinearity less than 1%, and for resonant one in the same operation range, the sensitivity was 0.53 V/A and non-linearity less than 0.5%. The resonant current sensor has higher sensitivity compared with nonresonant current sensor, therefore it is recommended for the detection of very low currents, for example leakage currents.

Finally, ME current sensors can be used in measuring equipment, power nets and control systems, security and safety systems, and metal detectors; in the automotive industry; in rail transport; in wireless electricity metering systems; and in space technology and robotics.

2.2.3 Crankshaft Position Sensor

Industrial progress in the modern era demands the invention of new technical devices and their components. The automotive industry is one of the important sectors of the economy that undergoes unceasing updating, in looking for competitive advantages even in small things, and more so using breakthrough technology. Electronics, computers, and sensors have become an integral part of any modern car. For example, almost every auto is equipped with a crankshaft position sensor (CKP sensor) or sometimes also a camshaft position sensor (CMP sensor) [33]. CKP sensor is used to determine the angular position of the crankshaft of the engine, synchronization of the control unit with the workflow engine, and the rotational speed of the shaft. This gives significant advantages and such autos generally have a high consumer demand. The principle of operation of the CKP and CMP sensors is based on the phenomenon of electromagnetic induction or the Hall effect. Magnetoelectric sensors developed recently are a good alternative to inductive sensors and Hall sensors.

2.2.3.1 Principle of operation

The crankshaft position sensor based on magnetoelectric effect works as follows. Figure 2.22 shows the scheme of ME CKP sensor. ME element (2) should be located in the immediate vicinity of the steel toothed ring (1). The constant magnetic field H_0 (bias field) is created using the permanent magnet (3) located near ME element. $Nd_2Fe_{14}B$ magnet with dimensions of 20 mm × 10 mm × 2 mm with a residual flux density of 1 T was used for the prototype. Steel-toothed ring has one type of label determined by successive identical combinations of "tooth/interval between the teeth" and the distinguishable labels with increased intervals between the teeth. The magnetoelectric element is experiencing the action of alternating magnetic field $H_{\sim}$, caused by the rotation of the steel toothed ring (1) due to the alternation during the rotation of plots with high magnetic permeability—"tooth" and low magnetic permeability—the "interval between teeth".

ME sensor for each passing combinations "the tooth/interval between the teeth" generates a pulse with the different levels. The signal has a high level in the locations of teeth, and low level at the locations of all intervals between teeth. The alternating magnetic field $H_{\sim}$ and the bias field H_0 are collinear to each other and perpendicular to the polarization vector P of the piezoelectric layer of ME elements. The conversion of magnetic energy into electric potential in ME element is due to the magnetoelectric effect, which is a result of the interaction of magnetostrictive and piezoelectric components of ME material. Alternating electrical signal proportional to the rotational speed of the crankshaft and depending on the angle of rotation occurs at the electrodes (4) of ME element. ME element and the magnet are placed in the case (5).

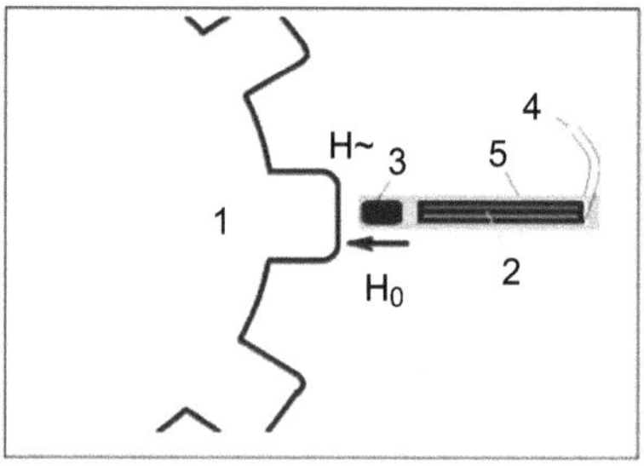

Figure 2.22 Scheme of ME CKP sensor.

2.2.3.2 Design

We applied for the manufacture of CKP sensor magnetostrictive–piezoelectric layered structure. Layered structure based on piezoceramic PZT plate in our case had 0.5 mm of thickness, 30 mm of length, and 10 mm of width. Piezoelectric was polarized in the thickness direction. The electrodes are applied on two sides of the piezoelectric plate. The electrodes are made from three layers of Metglas and correspond in size to the PZT plate. Thickness of one layer of Metglas was about 0.02 mm. Joining of layered design was done by gluing. Thus, the total number of thin layers in Metglas was 6 with a total thickness of magnetostrictive phase about 0.12 mm. The layered structure is symmetrical. The electrical signal is taken from the surface of Metglas plates. Low-frequency magnetoelectric coefficient in this ME magnetostrictive–piezoelectric structure, measured at the frequency about 20 Hz, was 0.75 V/(cm·Oe). The size of ME element, technology, materials, and manner of fastening have a significant influence on the value of ME coefficient.

The simulation stand consists of HMF2550 generator; oscilloscope HMO722; magnetometer DX-180; solenoid coil with a width of 3 cm, internal diameter of 2 cm, outer diameter of 4 cm, number of turns about 2000; the impedance of 120 Ohms; the core of a soft magnetic steel alloy; and ME element.

Measurement of characteristics is performed as follows. The signal from the generator is fed to the solenoid coil. The alternating magnetic field is formed in the coil at a preset frequency. This alternating magnetic field through the steel core is supplied to ME sensor. The distance between ME sensor and the steel core can be adjusted. The simulation of the steel-toothed ring is performed in this way. This method makes it possible to develop ME sensor design without using the actual hardware and to estimate limiting characteristics of the sensor. The signal generator with the frequency of 20 Hz and amplitude 10 V was supplied on the coil for study of the characteristics of ME prototype sensor. Solenoid coil causes a slight distortion in the signal, simulating the work of real devices. ME sensor converts the signal and then this signal goes to the oscilloscope. The magnetometer measures the amplitude of constant and alternating magnetic fields. For the sensor described in Section II, we obtained the following characteristics. Figure 2.23(a) shows the characteristic of ME sensor for the case when the gap between the steel core and the sensor was 1 mm, Fig. 2.23(b) shows the same characteristic with a gap of 3.5 mm, Fig. 2.23(c) shows the characteristic when steel core was absent in the coil. As can be seen from the figures, ME sensor has a high sensitivity to AC magnetic field. Maximum output peak-to-peak value voltage was about 270 mV for the above parameters. Data obtained by measurements on this stand allowed us to refine the sensor design.

The measuring stand also includes an electric motor with mounting system for a pulley; oscilloscope HMO722; CKP sensor, company Start-Volts, type VS-CS0112; and ME CKP sensor. Standard sensor VS-CS0112 and ME sensor were tested and the comparative data was obtained.

Figure 2.24(a) shows the oscillogram of the standard CKP sensor of type VS-CS0112 and Fig. 2.24(b) shows the curve shape of a crankshaft position marker.

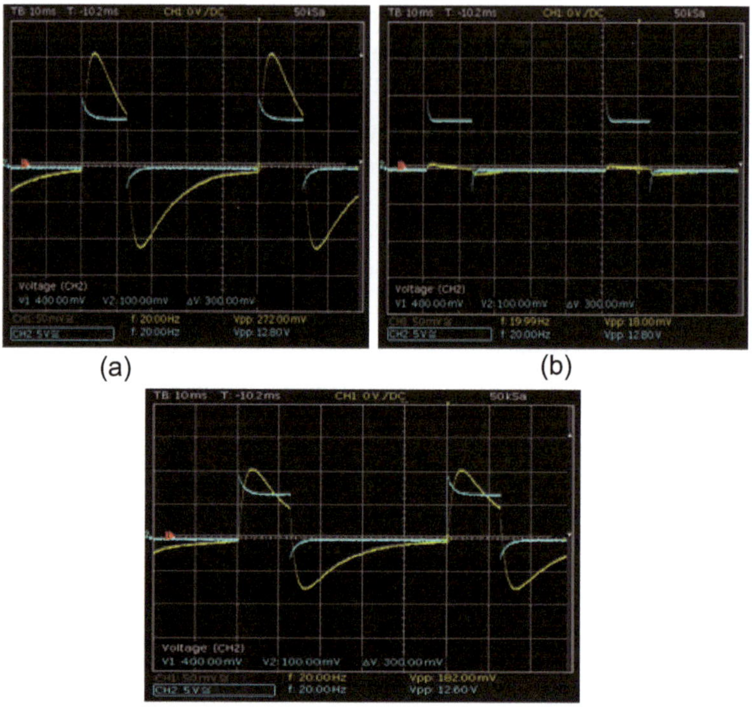

(a) (b)

(c)

Figure 2.23 Dependence of the sensor output voltage on the pulse signal in coil when the gap between sensor and steel core of: (a) 1 mm; (b) 3,5 mm; (c) absence of steel core.

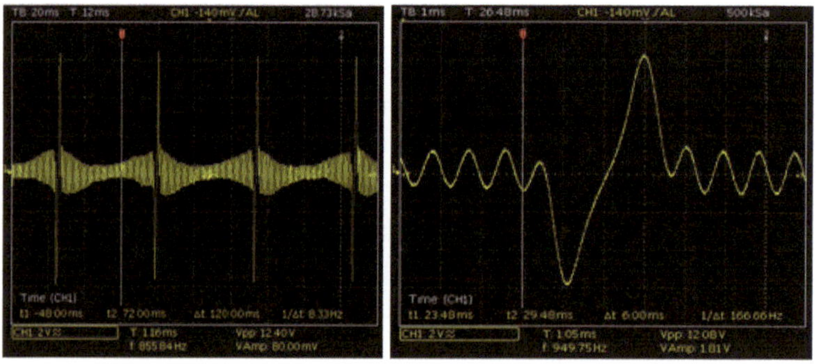

(a) (b)

Figure 2.24 The output characteristics of sensor VS-CS0112: (a) oscillogram; (b) shape of the curve marker.

The developed ME CKP sensor was installed at an angle of 45° to the standard sensor. Its oscillogram is shown in Fig. 2.25(a). The shape of the curve pulse from the tag of the pulley is shown in Fig. 2.25(b). From these figures it is seen that the signal of ME sensor is significantly smaller in amplitude. If the signal of the standard sensor gave the value of the peak-to-peak voltage as 12 V, then ME sensor gave the signal value only as 130 mV. Then, the curve shape of ME sensor is significantly different from the curve shape of the standard sensor. Although it should be noted that the experiments show the ability to change the shape of the curve in certain limits.

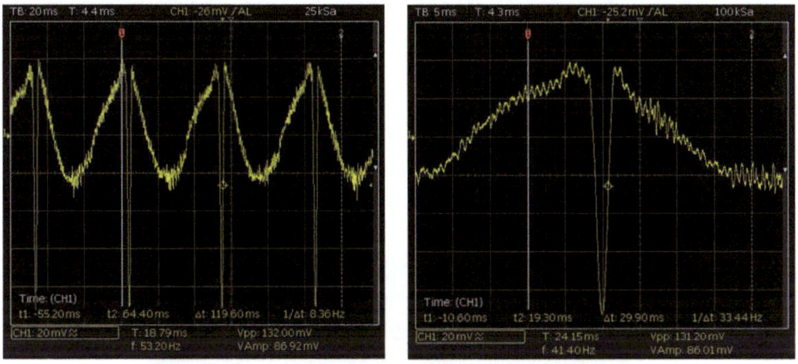

(a) (b)

Figure 2.25 The output characteristics of ME sensor: (a) oscillogram, (b) shape of the curve marker.

To obtain a more complete picture of the capabilities of ME sensor the following experiment was carried out. The permanent magnet which is part of the ME sensor was moved from the design and fixed at the marker surface of the pulley. Previous works on the development of ME alternator demonstrated the effectiveness of this design. Accordingly, Fig. 2.26(a) shows the oscillogram of work of ME sensor with magnet attached to the pulley, and Fig. 2.26(b) shows the curve shape of the marker from the rotating magnet. ME sensor reacts very clearly to the impact of the magnet. With a maximum pulse amplitude of 1.4 V, the resulting pulse shape is a single tooth (peak). So, this is the second possible option of ME CKP sensor design. This option has the explicit advantage over previous option where amplitude of the signal at the sensor output was essentially lower. The peak value of the voltage on the sensor was about 1.4 V, which is sufficient for reception and further processing of the signal in the electronic control unit (ECU) of the auto.

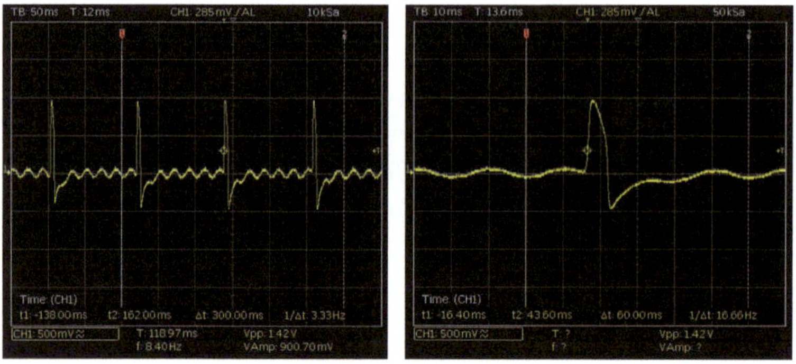

(a) (b)

Figure 2.26 (a) Oscillogram of work of ME sensor with magnet attached to the pulley, (b) the curve shape of the marker from the rotating magnet.

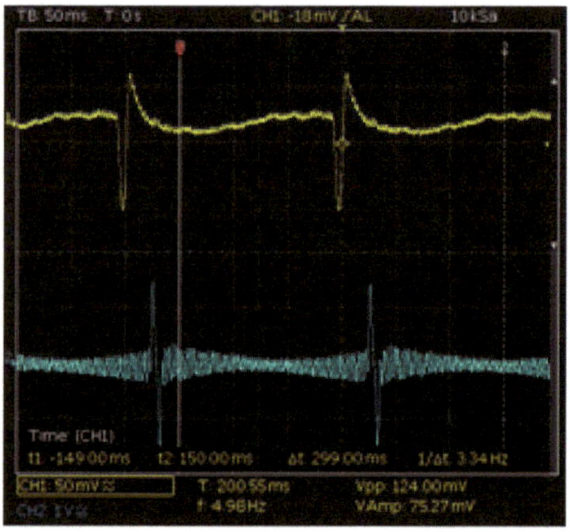

Figure 2.27 Comparison of oscillogram of CKP sensors.

Signals received from a standard sensor VS-CS0112 and ME sensor were simultaneously measured to compare the obtained characteristics. Figure 2.27 shows the oscillogram corresponding to the carried-out measurements. As the transducers are mounted at an angle of 45° relative to the axis of the shaft and to each other, the signal from one is ahead of another by one-eighth period, as can be seen on the oscillogram. In general it is important to note the prospects of the developed sensors. As a meaningful result it can be noted that success was achieved in providing lower manufacturing cost of ME sensors along with smaller size and weight compared with standard sensors. It is obvious that the first results can be significantly improved and sensors with a larger output voltage and a more acceptable form of the output impulse can be obtained.

2.2.3.3 Discussions

The well-known expression for the calculation of the magnetomotive force in the electrical devices can be used for approximate calculation of parameters of ME CKP sensor:

$$F = H_1 l_1 + H_2 l_2 + H_3 l_3 + H_4 l_4 + H_5 l_5 \tag{2.25}$$

where F is the magnetomotive force, H_i is the magnetic field strength on the corresponding area of the magnetic circuit, l_i is the length of the magnetic circuit.

In this case, the magnetic flux is determined by the residual magnetic induction of the permanent magnet, which is used in the sensor. Magnetomotive force is defined as the product of the magnetic flux on the magnetic resistance of the subcircuit. The magnetic flux passes through the metal magnetic paths and the air and is redistributed between the layer of Metglas of the sensor and the steel of pulley. The magnetic flux is close to the steel pulley when the pulley tooth is next to the sensor. The location of the tooth greatly increases the magnetic field strength in all parts of the magnetic circuit. The gap between the pulley and the sensor is increased by approximately 3.5 mm at the moment of passing markers. This causes significant changes of the magnetic flux, which passes through the sensor. The flow is decreased and this causes a strong change of the characteristics of the sensor. Thus, the label of the pulley produces a strong voltage pulse on ME sensor. The value of the output voltage on ME sensor can be calculated by using an expression for ME coefficient [30]:

$$U = E \cdot d = \alpha_{\mathrm{ME}} \cdot H_4 \cdot d \tag{2.26}$$

where E is the electric field in the piezoelectric material, d is the thickness of the piezoelectric material, H_4 is the magnetic field strength at the location of the sensor, α_{ME} is the magnetoelectric coefficient.

A more accurate expression can be obtained taking into account the location of the sensor relative to the magnet and relative to the pulley, the properties of the materials used, and design features of the sensor.

The proposed sensor would have the following advantages against the traditional sensors such as smaller size and weight. The designed sensor showed the peak voltage pulse of 1.4 V. The design of ME sensor is made of mechanical solid components and provides greater reliability of the sensor. ME sensor is sensitive to strong magnetic fields, therefore, it is recommended to provide the magnetic shielding.

2.3 ME Harvesters

Energy harvesting devices occupy a significant niche in the energy mix of many countries. Their further improvement should lead to the development of the local energy systems independent from the global networks. Along with well-known devices such as solar panels, thermal sources, and wind turbines, these are proposed and investigated for installations on other physical principles. The magnetoelectric (ME) effect [34] is largely prospective for the development of energy harvesting systems. ME energy harvester structure that can simultaneously scavenge magnetic and vibration energy in the absence of DC magnetic field, as shown in Ref. [35–38]. This structure consisted of a piezoelectric macro-fiber composite bonded to a Ni cantilever. Large ME coefficient ~50V/(cm·Oe) and power density ~4.5 mW/cm^3 (1 g acceleration) were observed at the resonance frequency in this work. Iteratively coupled finite element scheme for ME devices was presented in the paper [39]. The studies that propose the use of piezoelectric and ME thick films for fabricating power sources in wireless sensor nodes was published [40]. Recent advances in phase-transition transduction [41] enabled the design of a nonresonant broadband mechanical energy harvester that is capable of delivering an energy density per cycle up to two orders of magnitude larger than resonant cantilever piezoelectric type generators. One of the energy harvesting devices are ME energy generators [42] based on the ME effect. ME generator, unlike the piezoelectric generator, uses a double-power conversion: first the external mechanical energy is converted into the change of magnetic energy and as a result in the deformation of the magnetostrictive layer, and secondly, the deformation of the magnetostrictive layer causes deformation of the piezoelectric layer which causes the appearance of electric charge on the electrodes of ME structure. General dimensions of the generator were as follows: diameter of 12 cm and thickness of 2.4 cm. The model of generator comprising eight ME elements with dimensions of one element of 40 × 10 × 0.5 mm at the frequency of the alternating magnetic field of 38 Hz provides the output constant voltage of 1.12 V and current of 3.82 microamps. Variable voltage before the rectifier was of 1.7 V. Total generated power was of 4.28 μW. Also, studies of the energy characteristics of ME elements on the axial mode was performed [43]. This paper investigated an element with dimensions 40 × 10 × 0.5 mm and composition of PZT-Metglas. Low-frequency magnetoelectric coefficient was of 1.24 V/(cm·Oe) at an output current of 2.6 microamps and at the resonance frequency of 41 kHz the magnetoelectric coefficient was of 1.32 V/(cm·Oe) at an output current of 205 microamps. The application of vibrations in the tens of kHz in practice for real devices is a difficult task and it requires a design of ME elements that is capable of operating in the range of hundreds and tens of Hertz. This is the subject of the study presented in this paper. In the study, the low-frequency range for bending mode, which is the lowest resonance mode, have been realized to further improve the effectiveness of ME devices. The goal was to study the characteristics of ME element in the bending mode oscillations and to give practical recommendations for further research in this area.

2.3.1 ME Elements Design

ME element for energy harvesting devices can be made, for example, of magnetostrictive and piezoelectric layers [34]. A few samples of ME elements were made with different sizes to obtain comparative characteristics. The ME sample is a two-phase structure [Fig. 2.28(a)], the first phase of piezoelectric PZT type with 0.5 mm thickness is polarized along the thickness with electrodes on the surface, and the second phase is two layers of foil Metglas, the thickness of each layer is 20 μm. Layers of Metglas and PZT was fixed with glue.

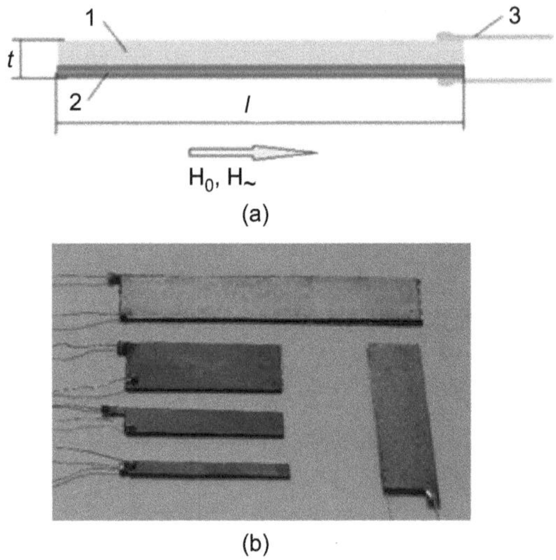

Figure 2.28 ME elements. (a) Design: 1 is PZT, 2 is Metglas layer, 3 is electrodes; (b) Photo of the samples.

Four elements of different length and width [Fig. 2.28(b)] were made. The parameters of PZT are the elastic modulus ^{p}Y = 70 GPa, density ρ = 7.4 kg/m^3, dielectric permittivity $^{p}\varepsilon_{33}$ = 1750, piezocoefficient $^{p}d_{31}$ = -175×10^{-12} m/V. The parameters of Metglas is the elastic modulus ^{m}Y = 110 GPa, density ρ = 7.18 g/cm^3, magnetic permeability $\mu = 10^5$, piezomagnetic coefficients $^{m}q_{11} = 11 \times 10^{-9}$ m/A. The elements' sizes, length (l) × width (h) × thickness (t), are as follows: 60 × 10 × 0.54 mm, 30 × 10 × 0.54 mm, 30 × 5 × 0.54 mm, 30 × 2.5 × 0.54 mm. These dimensions allowed us to precisely identify the resonant frequencies and to trace the dynamics of changes of the output power. The wire electrodes were soldered to the outer surfaces of the element.

2.3.2 Measurement Stand

The measuring stand for measuring the ME coefficient and AC current are shown in Fig. 2.29. The measuring stand included generator HMF2550, oscilloscope HMO722, gauss meter DX-180, multimeter HM8112-3, solenoid, permanent magnet, and ME element. Measurement of the amplitude–frequency characteristics were carried out by the oscilloscope and the generated voltage and current by the multimeter.

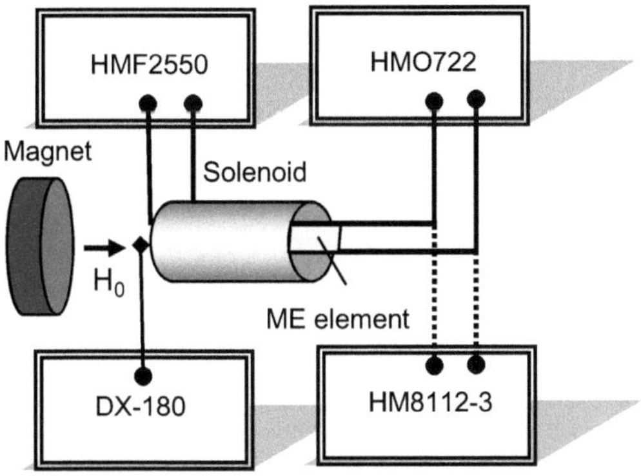

Figure 2.29 Measuring stand.

The gauss meter allows to measure the values of AC and DC magnetic fields. AC is supplied from the generator to the solenoid and creates AC magnetic field, which acts on ME element. The value of AC magnetic field inside the solenoid was 1.8 Oe and the value of DC magnetic field inside the solenoid was about 50 Oe. Via ME effect, the alternating electric potential and current in ME element were initiated and then it was observed by the oscilloscope and multimeter. Bias magnetic field was applied to ME element for correct work. ME element was located in the solenoid freely, without fixation.

2.3.3 Measurement Data

Studies of the frequency characteristics of ME elements were carried out. The characteristic of the output voltage and output current has a resonant character. Resonance occurs at the frequency of the electro-mechanical resonance of the piezoelectric phase in the composite. At the resonant frequencies for bending and axial modes, a large increase in the voltage and current is observed.

The characteristics of output voltage and current for ME element depending on the frequency at bias magnetic field of 50 Oe is shown in Fig. 2.30. The frequency of bending mode for ME element with length 3 cm located around 2 kHz.

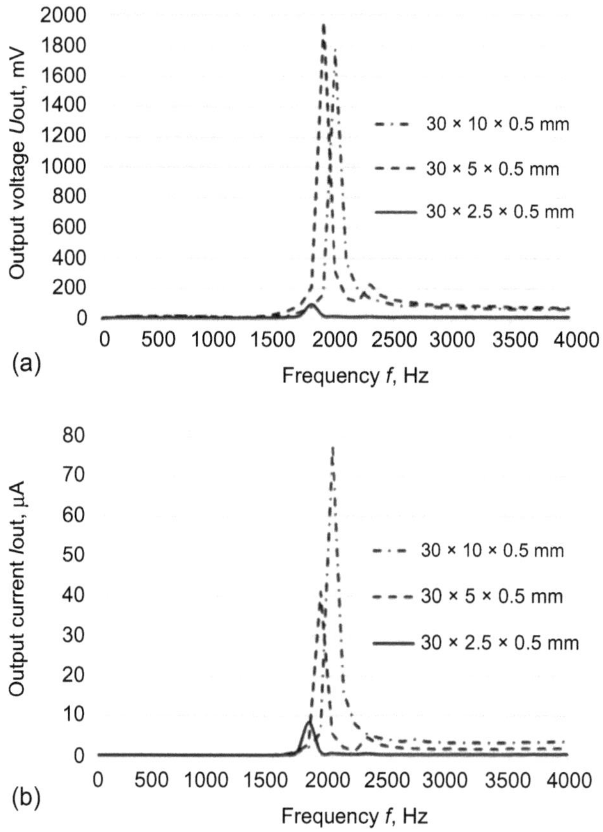

Figure 2.30 The frequency characteristics of ME element depending on width for the bending mode: (a) frequency *vs.* output voltage, (b) frequency *vs.* output current.

A comparison of the output characteristics for ME element in bending vibration for different lengths 6 cm and 3 cm are shown in Fig. 2.31. For ME element with a length of 6 cm, the frequency of the bending resonance was reduced to 500 Hz.

Figures 2.32 and 2.33 depict the frequency characteristics for the axial mode of ME element. The values of the output voltage and current for the axial modes are much higher than for the bending modes.

Figure 2.34 shows comparative characteristics of the output power *vs.* the width for the bending mode [Fig. 2.34(a)] and the axial mode [Fig. 2.34(b)]. The output power of ME element increases with an increase in element width in bending mode as well as in axial mode.

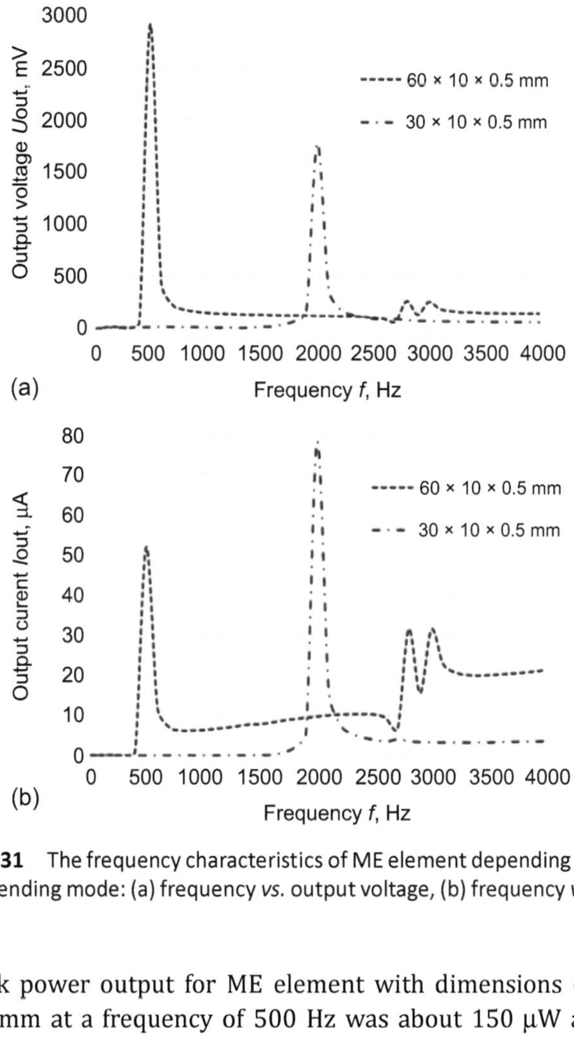

Figure 2.31 The frequency characteristics of ME element depending on length for the bending mode: (a) frequency *vs.* output voltage, (b) frequency *vs.* output current.

Peak power output for ME element with dimensions 60 × 10 × 0.54 mm at a frequency of 500 Hz was about 150 µW and at a frequency of 27 kHz was about 60 mW; for 30 × 10 × 0.54 mm at a frequency of 2 kHz was about 140 µW and at a frequency of 52 kHz was about 15 mW; for 30 × 5 × 0.54 mm at a frequency of 2 kHz was about 80 µW and at a frequency of 53 kHz was about 6 mW; and for 30 × 2.5 × 0.54 mm at a frequency of 2 kHz was about 0.8 µW and at a frequency of 52 kHz amounted to about 0.3 mW.

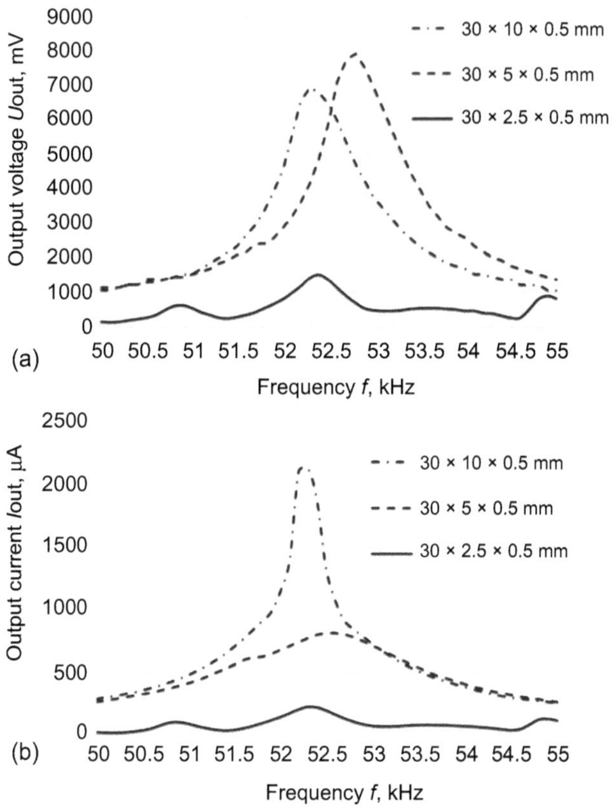

Figure 2.32 The frequency characteristics of ME element depending on width for the axial mode: (a) frequency *vs.* output voltage, (b) frequency *vs.* output current.

Frequency dependence of ME coefficient has a resonance character. ME coefficient for ME element with dimensions 60 × 10 × 0.54 mm for bending mode was about 32 V/(cm·Oe); for 30 × 10 × 0.54 mm was about 20 V/(cm·Oe); 30 × 5 × 0.54 mm was about 21 V/(cm·Oe); and 30 × 2.5 × 0.54 mm was about 1 V/(cm·Oe). The use of tens of ME elements with a length of 60 mm, it is possible to obtain a peak power of 1.5 mW, which is technically relevant and valid.

ME coefficient for ME element with dimensions 60 × 10 × 0.54 mm for axial mode was about 160 V/(cm·Oe); for 30 × 10 × 0.54 mm was about 77 V/(cm·Oe); 30 × 5 × 0.54 mm was about 88 V/(cm·Oe); and 30 × 2.5 × 0.54 mm was about 16 V/(cm·Oe).

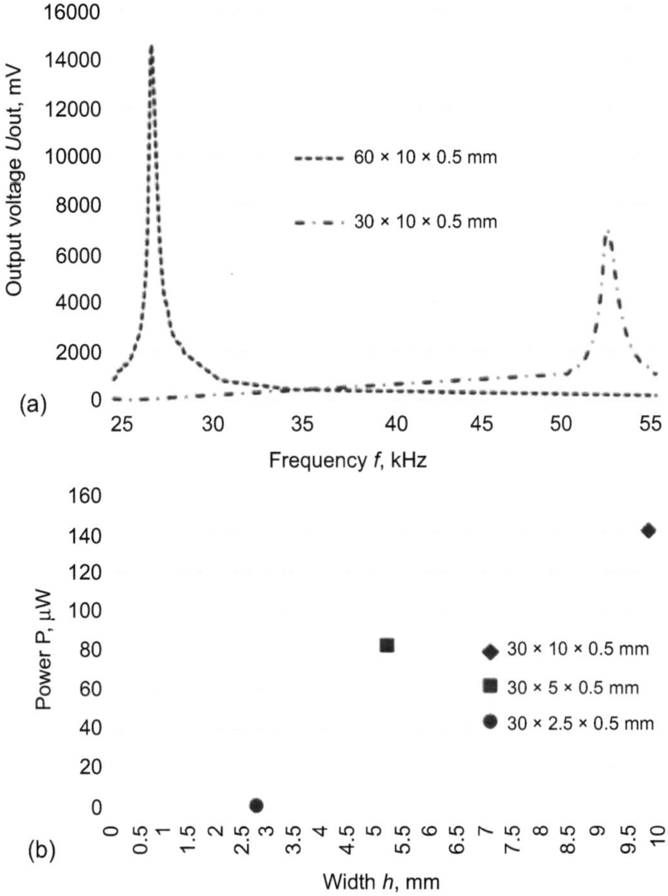

Figure 2.33 The frequency characteristics of ME element depending on length for the axial mode: (a) frequency *vs.* output voltage, (b) frequency *vs.* output current.

2.3.4 Theoretical Approach

The formula that allows to calculate the frequency of bending vibrations are well known and widely practised [44]:

$$f = \frac{\lambda t}{2\pi l^2 2\sqrt{3}}\sqrt{\frac{E}{\rho}} \tag{2.27}$$

where λ is a coefficient depending on the shape of the oscillations, t is thickness of the element, l is the length of element, E is the elastic modulus, and ρ is the density. The coefficient λ depends on the method of fixation of the element. It is equal to 22.4 for freely placed element and 3.52 for the element clamped on one end. Thus significant reduction in the frequency of bending resonance is possible using the fastening ME element. Frequencies of resonances of ME elements, which are calculated by the formula (1) showed good agreement with the obtained experimental data.

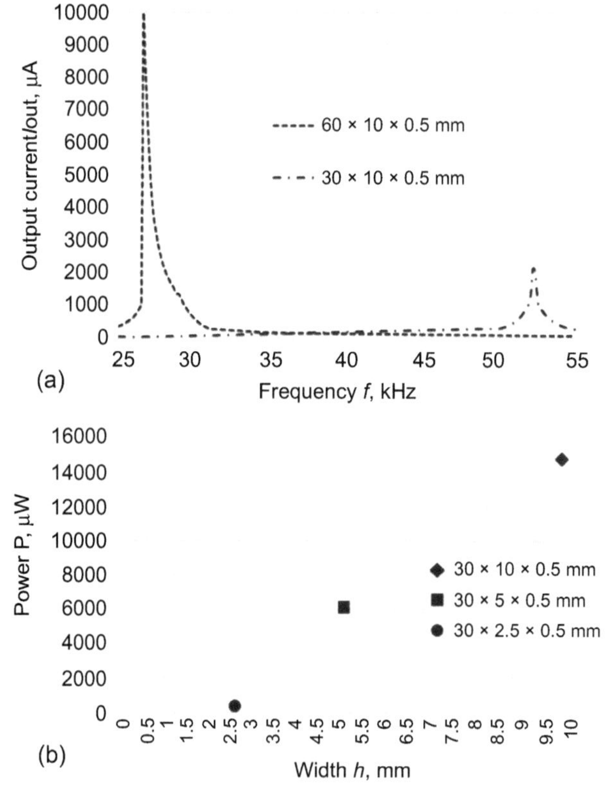

Figure 2.34 Peak output power *vs.* width of ME elements: (a) bending mode, (b) axial mode.

The calculation of the ME coefficient of the proposed layered structure can be carried out according to the following equation [45]:

$$\alpha_{E,31} = \frac{E_3}{H_1} = \frac{A \cdot {}^{p}d_{31} \cdot {}^{p}Y \cdot {}^{m}q_{11}}{\Delta \cdot {}^{p}\varepsilon_{33}} F \quad (2.28)$$

where

E_3 and H_1 are the intensities of the electric and magnetic fields,

A, F, and Δ are the coefficients that depend on the geometric dimensions of the element and the original material parameters,

${}^{p}d_{31}$ is the piezoelectric coefficient of the piezoelectric phase,

${}^{p}Y$ is the elastic modulus of the piezoelectric phase,

${}^{m}q_{11}$ is the piezomagnetic coefficient of the magnetic phase, and

${}^{p}\varepsilon_{33}$ is permittivity of the piezoelectric material.

The output voltage for ME element can be calculated by the known ME voltage coefficient of the material, which is determined experimentally or theoretically, if the magnitude of the alternating magnetic field is known ($H_1 = H_{\sim}$):

$$U_{out} = \alpha_{E,31} \cdot H_1 \cdot d \quad (2.29)$$

where d is the thickness of ME element.

The power generated by ME element can be calculated by the well-known formula:

$$P = U_{out} \cdot I_{out} \quad (2.30)$$

where I_{out} is the output current of ME element.

According to the calculations, the power output of the studied samples can reach values of tens of milliwatts. Output power of about 1 watt can be achieved if you combine in one design the hundreds of ME elements such as shown in Ref. [38].

Figure 2.35 shows the calculated characteristic resonance frequencies for the bending modes of vibration depending on the length of ME element for the three cases. The oscillation frequency can be significantly reduced by increasing the length of the element, reducing its thickness or by using the low-frequency bending vibration, which appears on the element with the clamped end. Oscillations of the clamped element in comparison to the freely placed one will be 6.36 times below. Thus, oscillations with a frequency of about 10 Hz can be generated by increasing the length of the element up to 12 cm, reducing the element thickness to 0.25 mm, and fixing it with one end.

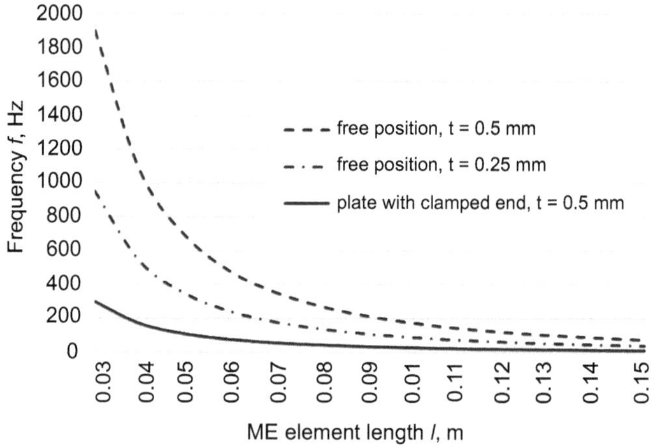

Figure 2.35 Frequency of bending resonance mode depending on ME element length.

2.3.5 Generator

2.3.5.1 Design

The design of the generator is based on the use of ME elements. ME elements may be made of different materials. In our case we use a magnetostrictive-piezoelectric layered material, which consists of a piezoelectric material PZT and magnetostrictive material Metglas [37]. Metglas layers were glued to a piezoelectric material with two sides and used as electrodes for induced electrical voltage. Schematic drawing and photo of ME element is shown in Fig. 2.36. The size of the ME element was of $40 \times 10 \times 0.5$ mm.

The ME element should be correctly positioned in the DC and AC magnetic fields for efficient operation of the device. Transverse direction of ME element magnetization is one of the structurally optimal variants of implementation of the design, when DC and AC magnetic fields are directed at an angle of 90 degrees to the direction of polarization, as shown in Fig. 2.36. Deviations in the direction of the magnetic fields can significantly reduce ME effect and the value of the output voltage should be considered while designing the generator.

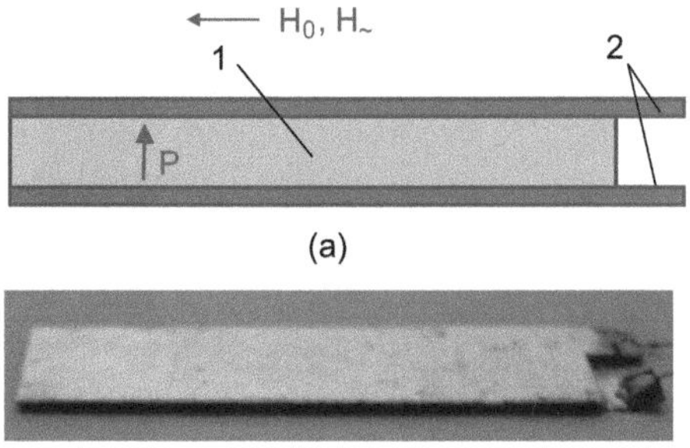

Figure 2.36 ME element. (1) Piezoelectric PZT, (2) Magnetostrictive material Metglas. (**a**) Structure of ME sensor, (**b**) Photo of ME sensor.

The location of ME elements can be arbitrary, but for the best result, one needs to place the elements according to axes of symmetry with a maximum packing density. Consider two variants of this disposition with the forms of chamomile and the roll. Figure 2.37(a, b) shows two variants of the design. Figure 2.37(a) shows a longitudinal and cross-sectional view of the three-phase synchronous ME generator in the form of the roll. Figure 2.37(b) shows the structure in the form of the chamomile.

The device consists of a stator and a rotor. The stator consists of frame 1, in the plane in which the ME elements 2 and the bias magnets 3 were located. The rotor consists of axis 4 mounted in the bearings of the stator 5, fixed on the axis by means of constructional holders 6, and magnets 7. The electrical potential is induced at the electrodes 8.

The device operates as follows. On the axis 4 of rotor the torque is transmitted from an external source movement. The axis rotates the magnets 7 fixed by the holders 6. The magnets are designed for the creation of alternating magnetic fields on the ME elements of the stator 2 on rotation. Permanent bias magnets 3 are installed in the stator to create the necessary offset in ME element. The alternating magnetic field induces an alternating electric potential on ME element via the ME effect: the alternating magnetic field acts on Metglas that, as a result of the magnetostriction, changes its dimensions, which leads to pressure on the piezoelectric phase of the ME element, and this in turn induces the appearance of the electric potential on the electrodes of the ME element due to the action of the piezoelectric effect. The ME elements may be interconnected, for example, as shown in Fig. 2.38, or separate. If the polarized piezoelectric material is used for the manufacture of ME elements, it is necessary to consider the polarity of these elements. Finally, the electric potential is induced on electrodes 8. In addition, the speed of the rotor at a constant frequency of magnetic field the stator ME elements is kept constant and does not depend on the output load on the shaft, i.e., the mode of operation of the generator is synchronous.

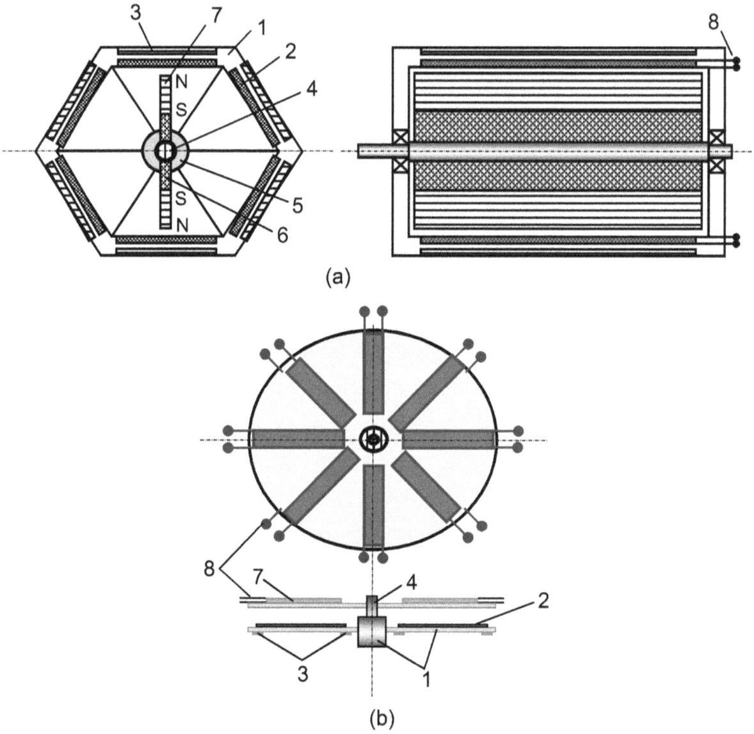

Figure 2.37 Examples of structures of ME generators: (a) Three-phase synchronous ME generator, (b) Single-phase synchronous ME generator

2.3.5.2 Prototype of generator

Designs shown in Fig. 2.37 were manufactured in prototypes. The generator shown in Fig. 2.37(b) is considered in more detail. In the design of the device shown in Fig. 2.37(b) the ME elements are placed and fixed symmetrically on the stator in the form of a disk. General view of the prototype is shown in Fig. 2.39. Permanent bias magnetic field acting on ME elements is created by the magnets fixed on the stator. The rotor in the form of the disc is rotated by an external source of motion from a small electric motor. Two niobium magnets of size 24 × 12 × 4 mm were installed on the rotor. Permanent magnets create a variable magnetic field in the area of ME elements due to the rotation of the rotor, which leads to the emergence of the electric potential in them due to ME effect. The energy received from ME elements is rectified and then stored. General overall dimensions of the generator's prototype amounted to a diameter of 12 cm and thickness of 2.4 cm. Eight ME elements with size of 40 × 10 × 0.5 mm of the composition of Metglas and PZT were presented in the structure of the prototype. All eight elements were combined into a single circuit through the rectifier bridges.

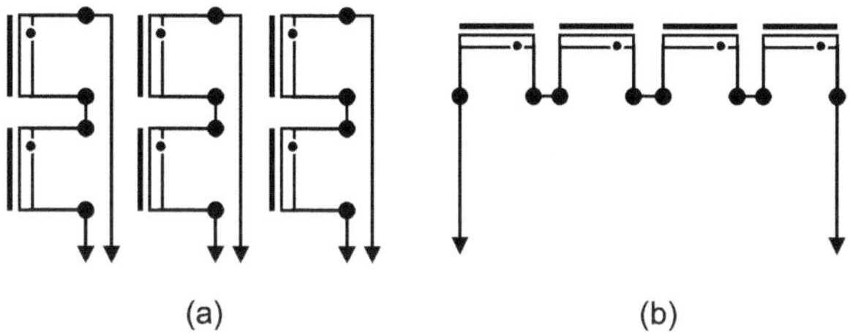

Figure 2.38 (a) Diagram of connections for the three-phase synchronous ME generator is shown in Fig. 2.37(a, b). Diagram of connections for the single-phase synchronous ME generator is shown in Fig. 2.37(b).

Figure 2.39 Prototype of generator.

2.3.5.3 Measuring stand

Figure 2.40 shows a schematic diagram of the measuring stand; photo of the stand is shown in Fig. 2.41. The measuring stand consists of the ME generator, rectifier bridge, oscilloscope HMO722, multimeter HM8112-3, power supply HY3002, gauss meter DX-180, and the collector electric motor, the speed of which is controlled by the voltage from the power supply. Gauss meter DX-180 was used to measure the variable and constant magnetic fields. The frequency of rotation of the motor is regulated by means of a power supply. The electric motor rotated the rotor of ME generator. Measurement of the potential at ME elements was carried out using the oscilloscope. The multimeter provided accurate measurement of the voltage and the current in both AC- and DC-rectified signal.

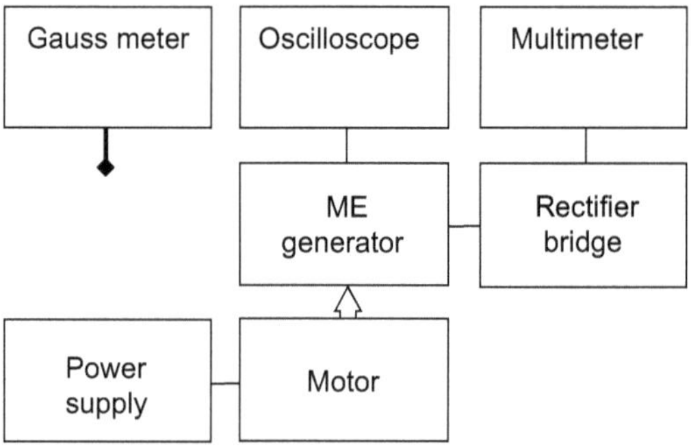

Figure 2.40 Diagram of ME generator measuring stand.

Figure 2.41 Prototype of ME generator.

2.3.5.4 Characteristics of ME element

Figure 2.42 shows the characteristics of ME element depending on the frequency when bias field is near 40 Oe. The optimum mode of operation in which the ME coefficient reaches a maximum of 2.64 V/(cm·Oe) is about 50 Hz at low frequency and 198 V/(cm·Oe) and about 41 kHz at the resonance mode. The real designs of synchronous generators can't work at kilohertz frequency range. Therefore, the oscillator frequency tends to decrease to the optimum for the ME element. The decrease of the resonance frequency is possible by increasing the length of ME element and the excitation of oscillations at the bending mode of resonance frequency, which can be below 1 kHz.

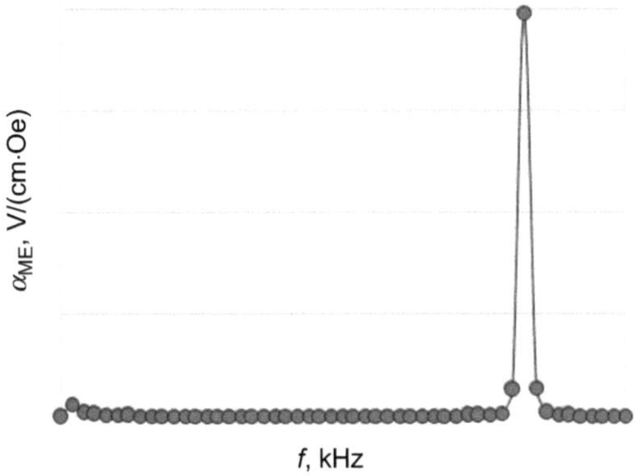

Figure 2.42 Frequency characteristics of ME element.

Figure 2.43 shows the characteristics of ME element depending on the applied DC magnetic field at a frequency of about 50 Hz and the amplitude of the AC magnetic field of 1 Oe. Figure 2.43 shows that ME coefficient is strongly dependent on the presence of a constant magnetic field and the optimum field is about 65 Oe when ME coefficient reaches a value of 2.64 V/(cm·Oe).

Figure 2.44 shows the bias magnetic field dependence of the output current at frequency of 50 kHz. The current reaches a value of 0.01 μA at a bias magnetic field of 65 Oe and an alternating magnetic field of 1 Oe at a frequency of 38 Hz. The current reached its maximum of 10 μA at a resonance frequency of 41 kHz.

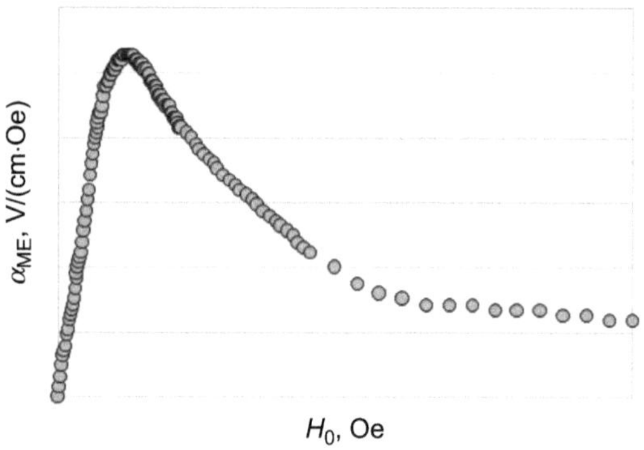

Figure 2.43 DC magnetic field characteristics of ME element.

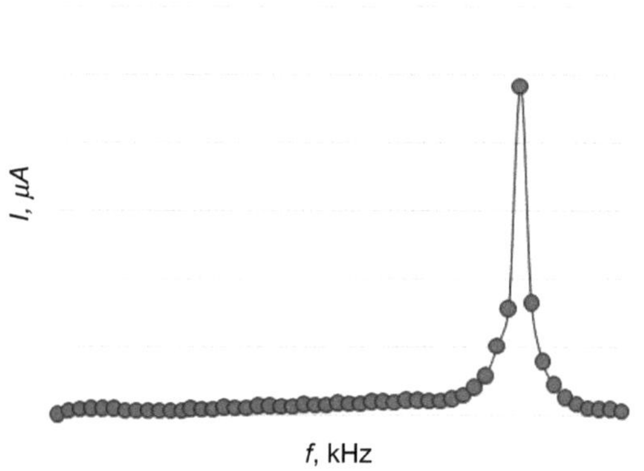

Figure 2.44 Frequency characteristics of the output current of ME element.

The value of the output power is defined as a product of voltage and current. Figure 2.45 shows the dependence of output power *vs.* frequency in the resonance region. The maximum of output power was of 0.55 μW with a bias magnetic field of 65 Oe and an alternating magnetic field of 1 Oe at resonance frequency. At the same time, the value of the output power in the low frequency region amounted to 0.55 nW. Figure 2.46 shows the characteristic of the output current of ME element at a frequency of 38 Hz and the bias magnetic field of 65 Oe in the range of alternating magnetic fields from 0 to 20 Oe.

The nonresonant low-frequency ME effect was used in the generator's design. Resonance effect has the undoubted advantage of power output, but so far, the implementation of its design is planned in the future. For use of the resonance effect, it is necessary to reduce the resonant frequency of the element increasing the length of ME element to the required value or it is also possible to use bending vibrations [46]. However, in our opinion, the study of the design of the generator with the use of low-frequency ME effect is of considerable interest for the design of energy harvesting devices.

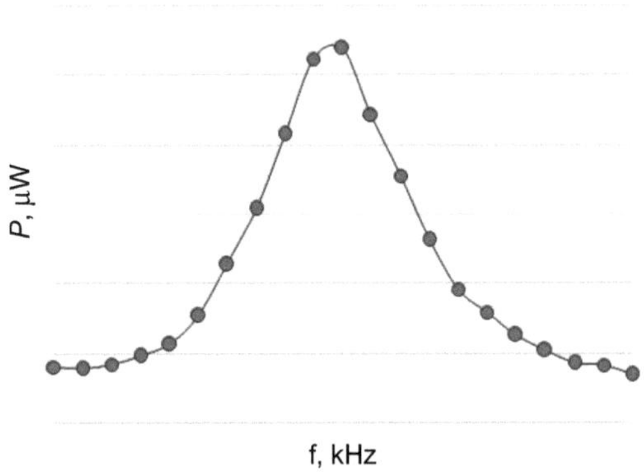

Figure 2.45 Frequency characteristics of the output power of ME element.

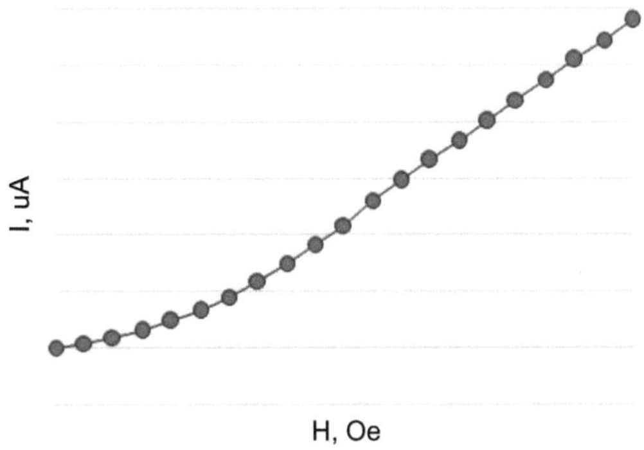

Figure 2.46 Characteristics of the output current of ME element depending on the amplitude of the alternating magnetic field.

2.3.5.5 Characteristics of generator

The prototype of the generator contained eight of the ME elements connected according to the scheme shown in Fig. 2.47. The elements were grouped by two for increase of the output voltage. For connecting the ME elements in pairs, one needs to take into account their polarization. The obtained four groups of elements then connected to diode bridges. Further, the rectified voltage is summed and the output characteristics of the generator is measured.

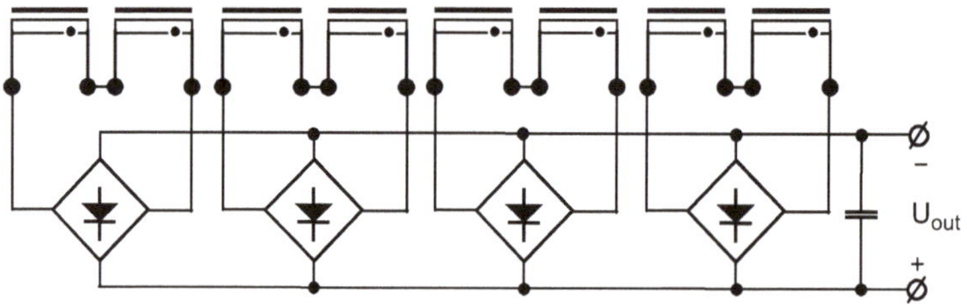

Figure 2.47 Schematic diagram of ME generator prototype.

A constant voltage of 1.12 V and a current of 3.8 μA was observed as the generator output when the rotation frequency of the rotor and the alternating magnetic field was 38 Hz. Alternating voltage until the rectifier was of 1.7 V. Total generator power amounted to 4.28 μW. The mode of operation of the generator was nonresonant. Figure 2.48 shows the waveform at the output of ME generator.

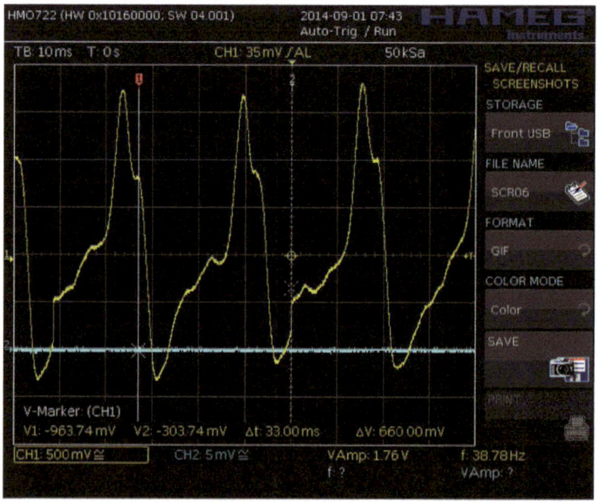

Figure 2.48 Oscillogram of the output voltage of ME generator.

2.3.5.6 Configuration of the magnetic field generator

The results of measuring the magnetic fields of the stator are shown in Fig. 2.49. The magnetic field measurements were carried out on the model of the generator by using a magnetometer.

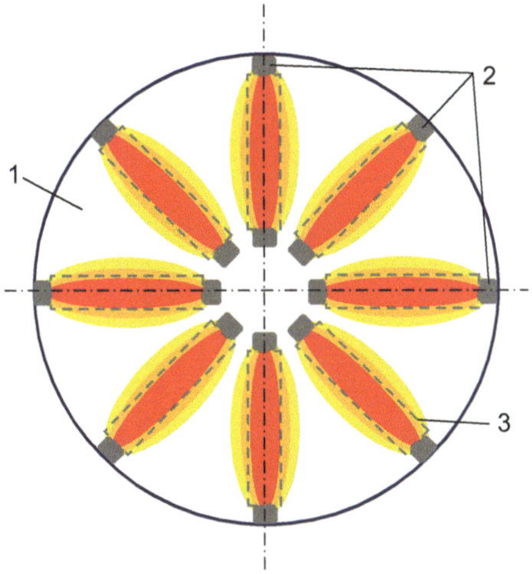

Figure 2.49 Configuration of the magnetic fields of the stator. Red zone between the magnets area is zone with a field of 20 Oe. (1) Stator, (2) Permanent magnet, (3) Contour of ME element.

The results of measuring the magnetic fields of the rotor are shown in Fig. 2.50. The influence of the magnetic field of the rotor on ME elements fixed on the stator was measured. The magnitude of the field decreases on increasing the distance from the magnet. Therefore, the magnetic field of the rotor was measured at a distance between the rotor and the stator equal to 2 cm.

The configuration of the magnetic field of the rotor was studied in order to determine the optimal location of ME elements and to find the best distance between the disks of the rotor and the stator. The rotor was rotated with a frequency of 20 Hz, and because two magnets were installed on the rotor, the observed frequency of the output signal was of 40 Hz. The frequency of the generated signal will increase if more magnets are installed on the rotor. The efficiency of the device depends on the correct ratio of magnetic fields that are applied to the ME element. The field of the magnets on the stator form the necessary bias for ME elements in linear mode, and the variable magnetic field of the rotor causes the alternating potential at the ME element. Correctly calculating and configuring the magnetic fields make it possible to achieve better results, therefore, this calculation as well as the choice and location of the magnets in the generator design should be given great attention.

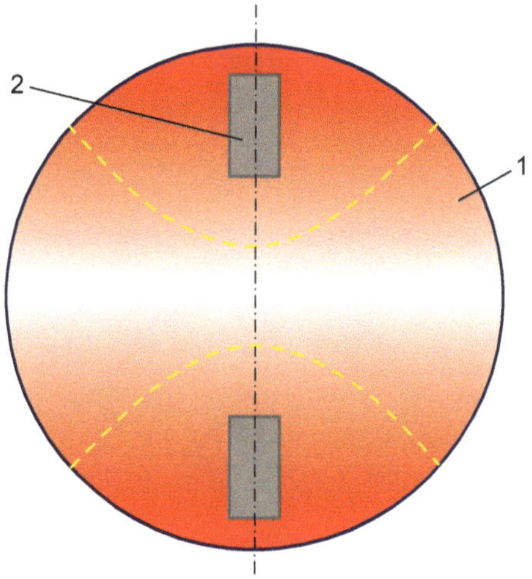

Figure 2.50 The configuration of the magnetic field at a distance of 2 cm from the rotor. Red zone is the zone of maximum magnetic field with a magnitude of 100 Oe, white is the zone of the field with magnitude about 0 Oe. Yellow dotted lines are isomagnetic lines with the magnitude of the field of 30 Oe. (1) Rotor disk, (2) Permanent magnets.

2.3.5.7 Calculation of ME coefficient

For a successful analysis of the characteristics of the generator it is sufficient to know the level of the output voltage obtained at each of the ME elements. The method of calculation of the output voltage is reduced to obtaining expressions for the ME coefficient. The expression for the transverse ME coefficient at low frequencies is presented in Eq. (1.30).

The output voltage of the ME element can be calculated by the known ME voltage coefficient of the material, which is determined either experimentally or theoretically, if the magnitude of the alternating magnetic field is known:

$$U_{\text{out}} = \alpha_{\text{E},31} \cdot H_1 d\,, \tag{2.31}$$

where d is the thickness of the ME element.

2.3.5.8 Outlook for increasing output power of the ME generator

One ME element generates very little power in the nonresonant mode. The power of one ME element may be significantly higher by several orders in magnitude in the resonant range. In addition, the output power can be increased by increasing the density of the space-filling generator by the ME elements. According to our estimates, the power generated can be up to several watts with the same geometric dimensions of ME generator if the density of the space-filling generator by the ME elements is increased and a resonant mode is used. In this case, the ME generator can be used as an effective energy harvesting device.

2.4 ME Microwave Resonators

With a growing interest in area of microwave and millimeter wave communication and radar systems, there is a need for high-frequency tunable resonant devices. Layered ferrite–ferroelectric structures are ideal for studies directed at fundamental understanding of high frequency magnetoelectric effects and for a variety of device applications. Typical magnetic materials used for magnetoelectric (ME) bilayers are ferrites, such as yttrium iron garnet (YIG) and barium ferrite (BaM) or strontium ferrite (SrM), due to their small magnetic losses, as well as ferromagnetic alloys, such as Fe_3O_4 and FeGaB, due to their large magnetostriction. Typical piezoelectric materials are lead zirconium titanate (PZT), lead magnesium niobate–lead titanate (PMN–PT), and lead zinc niobate–lead titanate (PZN–PT) due to their high electromechanical properties.

The ME effect in the single-phase materials [47] arises from the long-range interaction between the atomic moments and electric dipoles in ordered magnetic and ferroelectric sublattices. The realizable ME coefficient in single-phase materials is, however, very small [1–20 mV/(cm·Oe)] and not sufficient for practical applications.

Moreover, ME effect in most of these single-phase materials is observed only at low temperatures as either ferromagnetic (or antiferromagnetic) or ferroelectric transition temperature is very low. For the two-phase systems, the physical properties are determined by the interaction between the constituents as well as by their individual properties. Some effects, which are already present in the constituents may be averaged or enhanced for the overall system. However, the ME effect is among the novel effects that arises from the product properties originating through the interaction between the two phases.

Composite materials, on the other hand, provide an alternative strategy which makes use of indirect coupling via mechanical strain between the materials of two different phases: ferroelectric and ferromagnetic (or antiferromagnetic or ferrimagnetic such as ferrites). ME coupling (ME effect) in bonded bilayers composite have been studied so far. An electric field E applied to the composite produces mechanical deformation in the piezoelectric phase that in turn is coupled to the magnetic phase, resulting in a shift in ferromagnetic resonance (FMR) or magnetic (or magnetostatic) modes. The interaction is based on the fact that ferromagnetic resonance frequency is strain dependent and that the magnitude of this strain dependence is determined by the piezoelectric and magnetoelastic constants [47]. The strength of the interactions is measured from the frequency or field shift in FMR.

A composite of ferrite and piezoelectric is attractive for use in dual magnetic and electric field–tunable components for microwave, millimeter wave, and terahertz devices. For example, microwave resonators based on ferrite–piezoelectric-layered structures have rapid, narrowband electric tuning and slow broadband magnetic tuning. The magnetic tuning is realized through the variation in the magnetic permeability of the ferrite layer by changing the bias magnetic field. The electric tuning is realized through the ME effect. Ferrite materials having either very weak magnetocrystalline anisotropy, namely, YIG, or strong uniaxial anisotropy, namely, BaM or SrM, are usually utilized in the devices.

This part describes the fabrication of the ferrite–piezoelectric bilayer structures and the measurement procedure. The bilayers were fabricated in the form of ferrite–piezoelectric resonators.

The operation principle of ME resonators is based on microwave ME effect. The frequency of the ferromagnetic resonance with a high degree of linearity is proportional to the external magnetic or electric field, which allows designing of the resonator, and does not depend on the sample size, which makes the resonator very small. A uniform precession of the magnetization of the ferrite component of the resonator is usually used as a working mode of oscillations in ME resonators, the excitation conditions of which are ensured in ferrite samples that are small in comparison with the microwave wavelength and have the shape of an ellipsoid of revolution. The most widespread in microwave devices are ferrite resonators of spherical shape from high-quality single crystals of YIG. Along with the YIG spheres, epitaxial YIG films have been used for many years to design various microwave devices. One of the advantages of the devices on YIG films is the planarity of the design, which makes it possible to create a filter topology in a single technological cycle and is more convenient for technical applications. Due to the strong dependence of the resonance on the shape, the resonance element is usually formed in the form of a sphere or disk with a diameter of 0.5–5 mm, with a well-treated surface.

As single crystal YIG has been shown to exhibit the narrowest ferromagnetic resonance linewidth, most of the efforts on ME coupling–based devices have focused on incorporating single-crystal YIG with different types of ferroelectric layers, either in single-crystal or ceramic form [50].

A tunable YIG/PZT microwave resonator based on ferromagnetic resonance for YIG has been demonstrated [48, 49]. The schematics of the YIG–PZT bilayer resonator is shown in Fig. 2.51.

For wide-band ME measurements, a microstripline structure as shown in Fig. 2.51 was used. The ferrite–piezoelectric resonator was placed on a microstrip transducer, 50 μm in width and 3 mm in length on an alumina substrate. The microstrip transducer was connected to a standard 50 Ω microstrip transmission line, open at one end and shorted at another end. The device structure was placed in between the poles of an electromagnet so that a bias magnetic field H could be applied parallel or perpendicular to the resonator plane. A dc electrical field E = 0–10 kV/cm was generated across piezoelectric by applying a voltage. Microwave measurements were carried out using an Agilent vector network analyzer. A cw input signal with f = 2–10 GHz and power P_{in} = 0.1 mW was applied to the microstrip transducer. The experimental investigation of reflected power $P_{ref}(f)$ *vs.* f was based on the measurements of the resonator structures' frequency responses for different values of applied electric field E and bias magnetic field H. Namely, reflection spectra $S_{11}(f) = 10\log[P_{ref}(f)/P_{in}(f)]$, where $P_{in}(f)$ is an incident power, $P_{ref}(f)$ is a reflected power, and f is the excitation frequency, were measured in the microwave and millimeter frequency range.

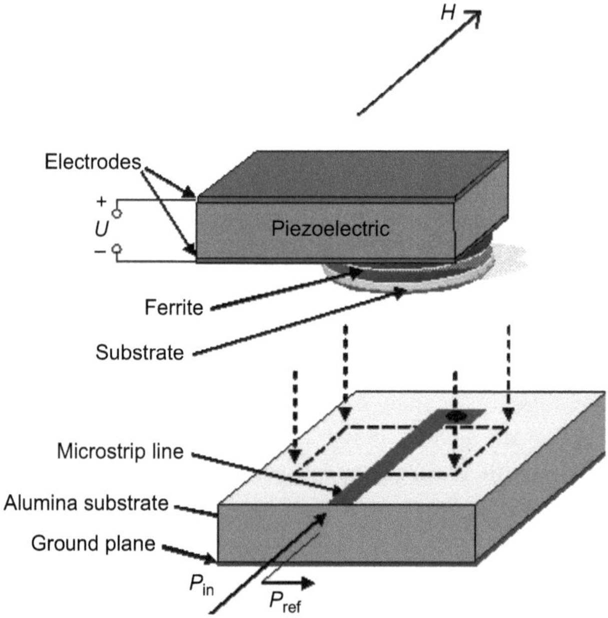

Figure 2.51 A tunable YIG/PZT microwave resonator.

The FMR frequencies for in-plane (ω_1) and out-of-plane (ω_2) applied magnetic fields are given by

$$\omega_1 = \gamma\left[H_0(H_0 + 4\pi M_s)\right]^{1/2} \text{ and } \omega_2 = \gamma(H_0 - 4\pi M_s) \qquad (2.32)$$

The finite lateral dimensions of the ferrite film, metallization on the ferroelectric surfaces, and the anisotropy fields in the ferrite may result in slight deviations from the theoretical estimates; however, FMR frequencies obtained from Eq. 2.33 were shown to be in good agreement with measurements.

The range of the frequency tuning by electric field could potentially be increased by either decreasing the thickness of the GGG substrate or by using a ferroelectric with higher piezoelectric coefficients.

Hexaferrites have high uniaxial anisotropy fields H_a and high saturation magnetization $4\pi Ms$. This type of ferrites are promising for operation at frequencies as high as 40–70 GHz for external fields H_0 = 5–8 kOe, and higher for higher magnetic fields. M-type barium hexaferrite $BaFe_{12}O_{19}$ with H_a = 17 kOe and saturation magnetization $4\pi Ms$ = 4.8 kG is appropriate for studies of millimeter-wave ME interactions.

2.4.1 ME Microwave Devices

Ferrites are an important class of materials for use in microwave passive devices such as phase shifters, circulators, filters, isolators, and resonators. Depending on the frequency of operation, different types of ferrite materials are needed and used. Because ferrites are magnetic dielectric materials (unlike magnetic metals which are conductive), they permit electromagnetic penetration and thus interaction between the electromagnetic wave and magnetization within the ferrite. Ferrite devices permit the control of microwave propagation by a static or switchable DC magnetic field. The devices can be reciprocal or nonreciprocal, and linear or nonlinear. What is certain is that their development requires a good deal of knowledge of magnetic materials, electromagnetic theory, and microwave circuit theory.

Miniaturization of microwave devices is one of the fundamental requirements in communication systems. Therefore, small-size and high-performance devices are always necessary to reduce cost and enhance system performance. Today there are diversified production engineering methods for forming electric circuits that have small losses and manifest frequency-dependent properties. It is possible to follow constructive alternatives.

Ferrites are used in tunable microwave and millimeter-wave devices and the tunability is traditionally realized through the variation of a bias magnetic field. This magnetic tuning could be achieved over a very wide frequency range, but is relatively slow, noisy, and requires high power for operation. Similar devices but with some unique advantages could be realized by replacing the ferrite with a ferrite–ferroelectric composite [51, 52]. Such heterostructures are magnetoelectric due to their response to elastic and electromagnetic force fields.

Layered structures based on the magnetostrictive and piezoelectric materials are prospective structures for new microwave devices. The given materials contain magnetostrictive and piezoelectric components and possess magnetoelectric interaction that will allow creating essentially new devices on their basis.

2.4.2 Magnetoelectric Band-Pass Filter

2.4.2.1 Characteristics

The filter is based on ferromagnetic resonance (FMR) of ferrite component. The device operating at 5–10 GHz can be tuned over a wide frequency band with a bias magnetic field and over a narrow band with a voltage applied across piezoelectric component. The voltage tuning of the device is possible through ME interactions that are mediated by mechanical deformation and manifests as a shift in FMR. Data on tuning range, insertion loss, and device characteristics are presented for filters with single- and double-ME resonators.

The objects used for magnetostrictive component was yttrium iron garnet (YIG) and for the piezoelectric component were lead zirconate titanate [PZT: $Pb(Zr,Ti)O_3$] or lead magnesium niobate–lead titanate [PMN-PT: $Pb(Mg_{1/3}Nb_{2/3})O_3$]. Designing band-pass filters utilizing ferrite–piezoelectric structure that is ME involves calculations of coupling of energy to the first ME resonator and of coupling of energy between resonators. Tunable resonators are widely used in microwave circuits as the frequency-controlling element in reference oscillators, tunable resonator tank for voltage-control oscillators (VCOs), and building elements for tunable filters. Traditional MMIC varactor-tuned resonators need to use GaAs MOSFETs or a *pin* diode as tuning elements, which will lead to high insertion loss and also small tuning range. Microelectromechanical systems (MEMS) technology has been successfully applied in developing a tunable resonator to achieve large tuning range from 3.5 to 7 GHz. However, large difference in S_{11} response is found in the whole tuning range. Tunable ME resonators are used in microwave range. The resonant frequency of the ME resonator can be shifted at a tunable range of 5 to 10 GHz. There are also various other considerations of importance such as temperature properties, unwanted mode suppression, signal handling capability, etc. What is not available is adequate information on coupling between ME resonators and means of designing filters for the minimum excitation and coupling of unwanted modes. One of the goals of this work was the determination of those physical parameters which affect the coupling of the fundamental mode.

We recently studied ME interactions at FMR in layered ferrite-ferroelectric composites [53, 54]. The studies were done on bilayers of single-crystal YIG films and PMN-PT. Microwave ME measurements at 9.3 GHz were performed using a traditional FMR spectrometer. An electric field E applied to the sample produced a mechanical deformation in the piezoelectric phase that in turn was coupled to the ferrite and manifested as a shift in FMR profiles. Profiles of FMR absorption *vs.* bias magnetic field H were obtained for a series of electric voltage across PMN-PT. With the application of E, we measured a shift in the profile by ΔH and the ME constant $A = \Delta H/E$ varied over the range 1–5 cm·Oe/kV, depending on the YIG film thickness. Since 1 Oe of field shift corresponds to a frequency shift of 2.8 MHz, frequencies of FMR-based microwave devices can be tuned with E. For a nominal field of 30 kV/cm, a YIG/PMN-PT resonator, for example, can be tuned by 80–400 MHz. Thus the ME effect would facilitate rapid voltage tuning for any FMR-based microwave device. Other advantages of microwave ME devices include miniaturization, near-zero power consumption, noise reduction, and compatibility with integrated circuit technology.

This work is on the fabrication and characterization of an electric field tunable microwave band-pass filter with single- and dual-ME resonators. Since the frequency of operation is 5–10 GHz, we choose low-loss single-crystal YIG films for the magnetic phase of the resonator and polycrystalline lead zirconate titanate (PZT) for the ferroelectric phase. Our studies indicate very good tunability due to strong ME interaction and an acceptable insertion loss. Theoretical models account very well for the observations.

2.4.2.2 Filter design

Both single- and dual-ME cavity filters were studied. Microstrip transducers fabricated by substrates of foil-clad microwave material FLAN-10 [permittivity at 10 GHz, not more than 10.0 ± 0.8; tangent of dielectric loss angle at 10 GHz, not more than 0.0045; sheet thickness (including copper foil) from 1.0 ± 0.1 mm]. It is a sheet microwave material manufactured from the composition on the basis of filled polyphe nyleneoxide with electrolytic galvanoresistant 35 micron copper foil bonded to both sides. It possesses a high-stability value of dielectric permittivity and low dielectric losses in the microwave range. The base of the band-pass filter is a microstrip transmission line on dielectric substrate and ME resonators. The single-cavity ME filter consisted of a 1 mm thick ground plane, input and output microstrips of nonresonance lengths, and an ME-element. The microstrip transducers, 1 mm in width and 18 mm in length, were separated by 1.5 mm. The input–output decoupling is determined by this gap between the microstrips. Power is coupled from input to output under FMR conditions in the ME element. A dual-cavity filter shown in Fig. 2.52 consisted of a 1 mm thick ground plane, input and output microstrips of nonresonance lengths equal to 15 mm, section of a nonresonance length microstrip (3 mm) that carries out the role of a coupling element between two resonators, and a two ME-element. Width of all line are 1 mm, gaps between micostrips are 1 mm also. The ME element consisted of epitaxial YIG film bonded to PZT. A YIG film grown by liquid-phase epitaxy on a (111) gadolinium gallium garnet (GGG) was used. The ME-element was made from YIG film 110 μm thick on GGG substrate 210 μm thick with diameter 1.5 mm and PZT plate (dimensions $4 \times 0.9 \times 0.5$ mm^3) with electrodes.

The film saturation induction $4\pi M$ was of 1750 G and the FMR linewidth was of 1 Oe. A PZT plate was initially poled in oil by heating up to 110 °C within 5 hours and cooling back to room temperature in an electric field of 4 kV/mm perpendicular to the sample plane. The layered structure was made by bonding the YIG film surface to PZT with a thin layer of ethyl cyanoacrylate or a fast-dry epoxy. The layered structure was placed between the transducers and was subjected to a field H parallel and perpendicular to the sample plane.

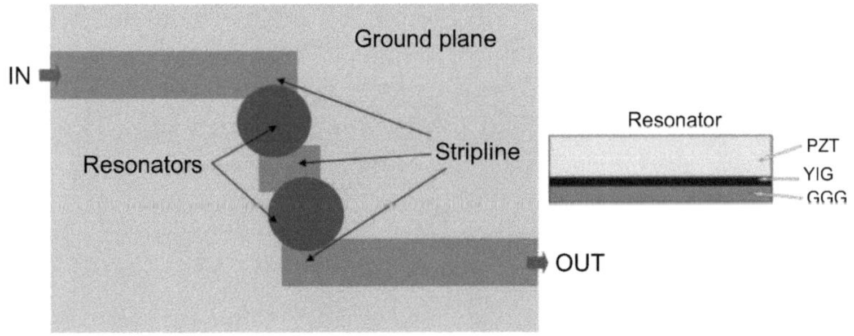

Figure 2.52 Schematic diagram showing a dual-cavity magnetoelectric filter and the ME resonators.

2.4.2.3 Results

The device characterization was carried out with a vector network analyzer (PNA E-8361). An input continuous-wave signal $P_{in}(f)$ = 1 mW was applied to the filter. The frequency f dependence of the insertion loss L, i.e., the transmitted power through the ME element, was measured at 4–10 GHz as a function of H and E applied across the PZT layer.

We are first interested in measuring the frequency dependences of the insertion losses as a function of magnetic field H (applied in the plane ferrite layer) and the electrical field E (applied across the piezoelectric layer). Next we will measure other characteristics of the filter such as bandwidth of pass-band (level of –3 dB), stop-band, shape factor – $K_p = BW_{stop-band}/BW_{pass-band}$, and Q-factor.

The insertion-loss dependences of the frequency (Fig. 2.53) were measured for one-resonator filter as compared to two-resonator filter. The data are for a central frequency f = 7.36 GHz (corresponding to H_0 = 1915 Oe, H is parallel to plane of resonator).

Characteristics of one-resonator and two-resonator band-pass filters are presented in the Table 2.2.

Figure 2.54 shows the dependence of the frequency shifts δf on YIG/PZT structure *vs.* intensity of E for cases in which the magnetic field is parallel and perpendicular to plane of resonator.

One observes a near-linear variation in δf with small hysteresis. The observable linear dependence of the shift may be explained by internal distortions of YIG films and it indicates the necessity of application in practice of thin YIG films.

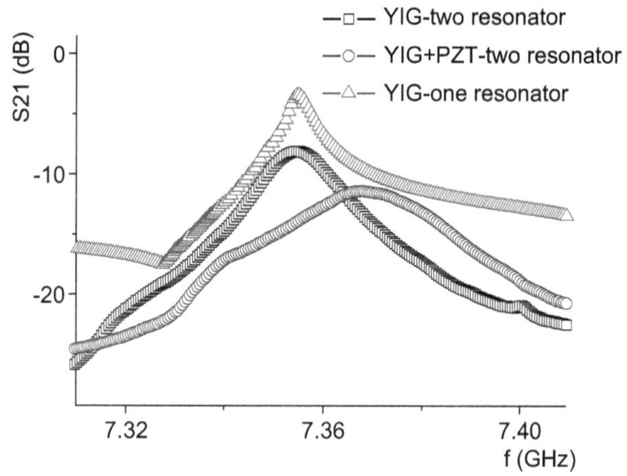

Figure 2.53 One-resonator filter as compared to two-resonator band-pass filter.

Table 2.2 Characteristics of one-resonator and two-resonators band-pass filter

Type of filter	$f_{central}$ (MHz)	$BW_{-3\ dB}$ (MHz)	IL_{pass} (dB)	IL_{stop} (dB)	K_p	Q
One-resonator YIG-filter	7356	8	3.0	12	3.0	660
Two-resonator YIG-filter	7355	16	8	24	3.0	450
Two-resonator YIG/PZT filter	7370	30	11.4	25	4.0	240

Thus, the electrically tunable band-pass filter designed in microstrip transmission line and with layer resonator based on YIG-film and PZT is described. First, we measured the insertion loss as a function of the magnetic field H (applied in the plane and across ferrite layer) and the electrical field E (applied across the piezoelectric layer). The minimum insertion loss is 3 dB for one-resonator YIG-filter and 8–11 dB for two-resonator band-pass filter in the frequency region 5–11 GHz. The insertion loss is up to 25 dB for out-of-band frequencies.

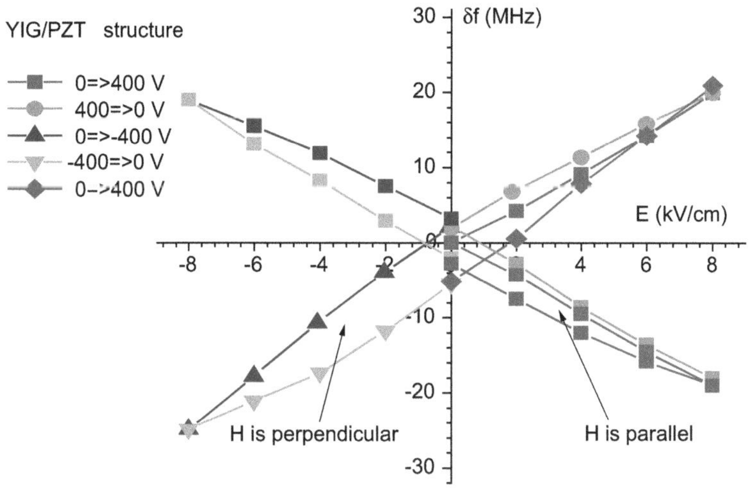

Figure 2.54 The dependence of the frequency shifts δf *vs*. intensity of E for in which magnetic field is parallel and perpendicular to plane of resonator.

Next, we measured the frequency shift δf on YIG-PZT layered structure for different electric field E (magnetic field H is parallel and perpendicular to plane of resonator). In a magnetoelectric composite, the frequency tuning is realized through the variation of the applied electric field that creates effect of magnetostriction. Tunability of the device's characteristics in the chosen working range can be executed by application of an electric field. This electric tuning is possible in a narrow-frequency range, but it is relatively fast and is not power-consuming. It will allow carrying out tuning in a range of 25 MHz. The electric tuning would facilitate high-speed operation, small size, and compatibility with integrated circuit technology.

2.4.3 Magnetoelectric Phase Shifter

Traditional ferrite phase shifter use magnetic tuning systems that are slow, demand high power, and are not miniature in size. Here we present the design and characteristics of a new electric field-tunable phase shifter based on ferrite–piezoelectric layer composite. The electrical control of the phase shifter is realized through microwave magnetoelectric effect. The phase shifter is capable of rapid tuning and is miniature in size. This microwave phase shifter is of interest for phased array antenna system. The base of the phase shifter is a microstrip transmission line on alumina substrate (ε = 9.8, thickness = 1 mm) and ME resonator.

2.4.3.1 Experiment

During the research, ME composite consisting of epitaxial single-crystal (111) YIG films on gadolinium gallium garnet (GGG) substrates and 0.44 mm thick (001) lead magnesium niobate-lead titanate $0.68PbMg_{1/3}Nb_{2/3}O_3–0.32PbTiO_3$ were used. Metal electrodes (200 nm in thickness of gold and 30 nm in thickness of chromium) were deposited on PMN-PT for electrical contacts and the crystal was initially poled by heating to 373 K and cooling it back to room temperature in E = 2 kV/cm. A thin layer (<0.08 mm) of an epoxy ethylcyanoacrylate was used to bond YIG to PMN-PT. Circular polarization of the magnetic field in volume of the magnetoelectric resonator results from the use of microstrip stubs of one-eighth and three-eighth wavelength. A bias field corresponding to FMR is applied to the ME resonator. Electrical tuning is realized with the application of a control voltage to electrodes leading to a shift of FMR line.

The ferrite film used is a 26.8 μm thick disk with a diameter of 4 mm. A PMN-PT plate with diameter 6 mm was used as a piezoelectric element. Device characterization was carried out with a vector network analyzer PNA series E8361.

Schematics of a ME microwave phase shifter and ME resonator are represented in Fig. 2.55.

Basis of a design of the phase shifter is the microstrip line on a substrate from an alumina and a magnetoelectric composite served by the resonator. In the volume of the resonator by means of microstrip loops of $\lambda/8$ and $3\lambda/8$ wavelengths a circular polarization of a magnetic field is created.

The resonator is installed in an area of circular polarization of a magnetic field, H, which is constant and is parallel to the sample plane. For reduction of losses the working point gets out on a dispersive curve outside the resonance area.

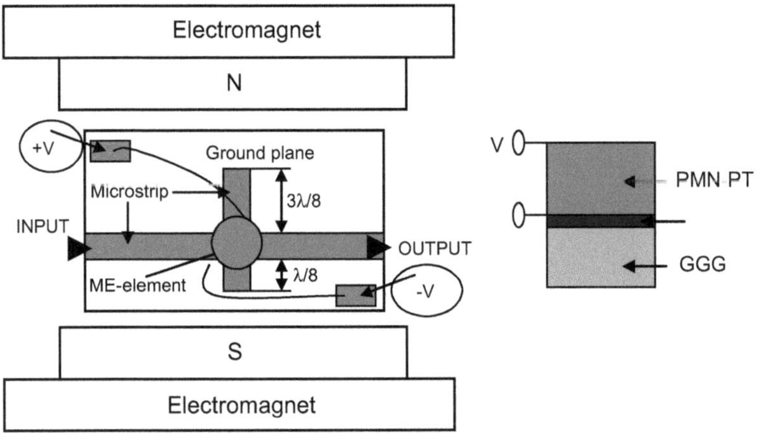

Figure 2.55 Schematics of a ME microwave phase shifter and ME resonator of YIG film on GGG bonded to PMN-PT.

Work of the phase shifter is based on the following: for a choice of the working point laying on a dispersive curve, to the resonator on frequency close the external magnetic field is put to a resonance. Under the influence of the operating voltage of the electrodes that are located at the end faces of the resonator and owing to microwave effect, there is a shift of line FMR due to which electric management in parameters of the phase shifter is realized.

Calculation is carried out for the phase shifter in the resonant frequency range of 6–9 GHz. The ME composite is included as heterogeneity in area of circular polarization of a magnetic making microwave of a field. Change of a phase corner is achieved by choosing a strong but not mutual connection of the resonator with a line of transfer. Thus regulation of a phase corner is achieved by a change of the operating voltage of the electrodes of the resonator.

The working condition of the phase shifter is chosen based on the dispersion curve of permeability in the external magnetic field. Figure 2.56 shows typical permeability μ *vs.* H profiles for YIG; FMR is expected for $H = H_r$. We choose $H_0 > H_r$, to obtain low-insertion loss and a linear or nonlinear variation of μ' with H.

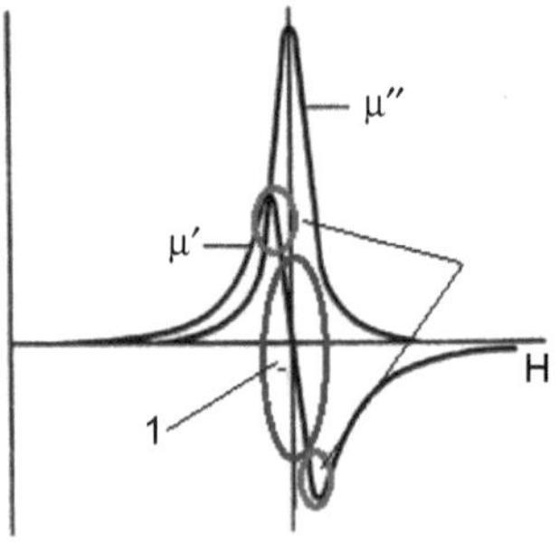

Figure 2.56 Typical ferromagnetic resonance regime for YIG showing the real and imaginary parts of the permeability as a function of bias field H_0. The change in the permeability and phase with H_0 or E is linear in region 1 and it is nonlinear in region 2.

2.4.3.2 Results

Representative data on *phase vs. frequency* for different frequency and magnetic field H are shown in Figs. 2.57 and 2.58. We measured the phase characteristics as a function of the electrical field E (applied across the piezoelectric layer) at frequency region 6–9 GHz. A magnetic field H was used for choice of working point. We used parallel magnetic field. The loss was 3–4 dB at the frequency region.

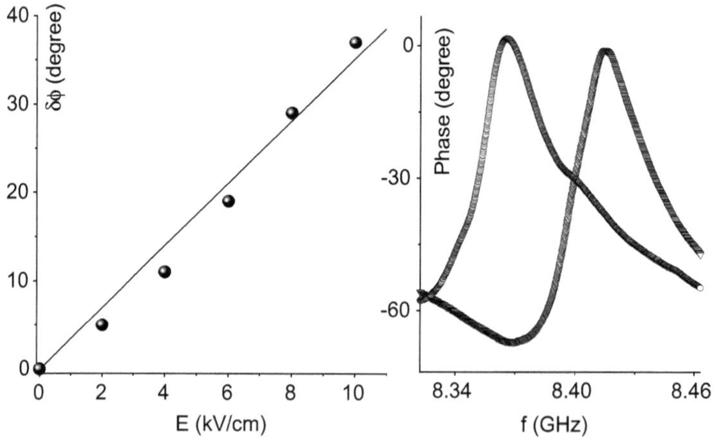

Figure 2.57 Differential phase shift *vs.* electric field E. H is parallel to sample $H_0 = 2277$ Oe; $L = -4$ dB; $f_0 = 8420$ MHz; $t_{YIG} = 26.8$ μm; $\phi = 4$ mm; $t_{GGG} = 440$ μm; $t_{PMN\text{-}PT} = 0.5$ mm; $\phi = 6$ mm.

The data in Figs. 2.58 and 2.59 constitute demonstration of the electrical tunability of the phase shift in a microwave ME microstrip device. The tunability is due to the influence of mechanical deformation on the piezoelectric phase. A near-linear variation in phase with E is evident from the data. A maximum phase shift is 30–40 degree at the frequency region 6–9 GHz. Obtained data allow to speak about an opportunity of practical application of our device. Optimization of the material, both ferrite and piezoelectric phase, and device parameters will allow a further reduction in the insertion loss and provide a high electric tunability.

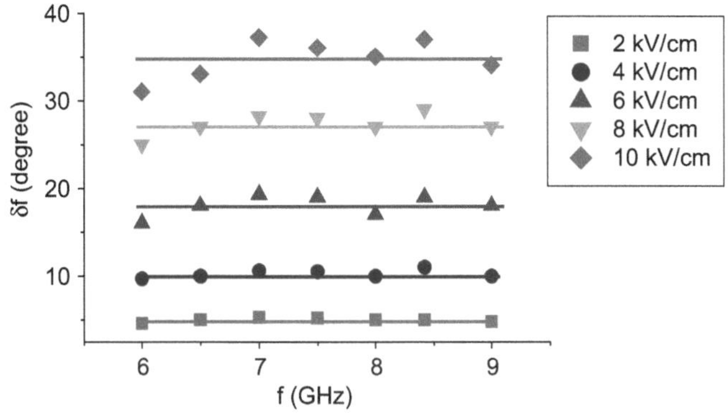

Figure 2.58 Differential phase shift *vs.* frequency at different electric field E. H is parallel to sample L = –3–4 dB; YIG = 26.8 μm; ϕ = 4 mm; t_{GGG} = 440 μm; PMN-PT = 0.5 mm; ϕ = 6 mm.

Thus, we designed, fabricated, and tested microwave ME phase shifter in a standard thin-film passive technology. Experimental results showed that 30–40 degree phase shift was achieved at frequency range of 6–9 GHz and at electric field change from 0 to 10 kV/cm. Our technology can significantly reduce the cost of phased array systems, where the phase shifter circuit is a large fraction of the overall antenna cost.

2.4.4 Magnetoelectric Microwave Isolator-Attenuator

Microwave devices, such as isolators, attenuators, and circulators are the devices which enable to protect the input of the transmitter from stray electromagnetic radiation acting in the operating frequency band. They are indispensable devices for television, radar system, broadcasting radio, and communications. The task of matching the microwave power source to the load is achieved by using nonreciprocal properties of ferrites. However, the nonreciprocal ferrite devices control the parameters realized by the magnetic system, therefore these devices have low speed of operation, high power consumption, and cannot be manufactured by integrated technology. ME nonreciprocal devices can reduce the disadvantages of existing ferrite devices: reduce the cost of devices; improve the speed of switching; reduce the power consumption in the control circuit; use integral technology manufacturing [57]. The difference between the ferrite devices and the proposed ME nonreciprocal devices is that the ferrite resonator with magnetic control system is replaced by the ME resonator and system of electrodes connected to the source of the control voltage.

The base of the ME isolator-attenuator is a microstrip or coplanar waveguide transmission line on dielectric substrate and ME resonator (Fig. 2.59). In case of a microstrip line, the design is same as the phase shifter but the working area, which depends on the magnetizing field, is different.

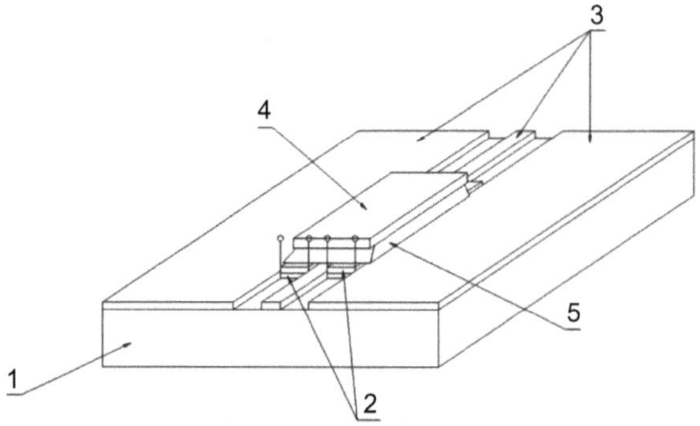

Figure 2.59 Microwave ME coplanar isolator-attenuator: 1 is the dielectric substrate; 2 is an ME resonator; 3 are metallic conductors forming the coplanar waveguide; 4 is the magnet; and 5 is the ceramic plate with high permittivity with beveled edges.

The principle of operation of the isolator-attenuator based on microwave ME effect is to shift the ferromagnetic resonance line under the influence of an electric field. ME layered structure in this case plays the role of a resonator. Shift of magnetic resonance line at an applied electric field is defined as $\delta H_E = AE$, where A is magnetoelectric coefficient and E is applied DC or AC electric field.

2.4.4.1 Results and discussion

First we describe the ME isolator–attenuator based on coplanar waveguide. The ME resonator based on layered structure of YIG has dimensions 13 × 4 × 0.9 mm and PZT has dimensions 15 × 4 × 0.5 mm. The next parameters of coplanar waveguide are the width of the gap S = 3 mm, distance between slits W = 2 mm, substrate thickness h = 2 mm, and dielectric permittivity of substrate ε = 10. The measurement of the transmission characteristics of the prototype were carried out on the measuring stand, consisting of an Agilent vector network analyzer PNA-L N5230C, electromagnet GMW5403, and DC current source Sorensen DLM 20-30. Figure 2.60 shows the experimental dependence of attenuation in the forward and reverse directions for the investigated structure. So, it is clear to see that the isolation ratio was equal around 5 for this case.

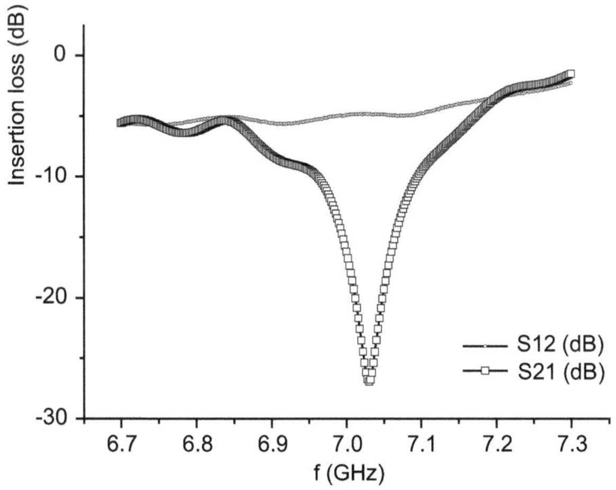

Figure 2.60 The experimental frequency dependence of attenuation. Constant magnetic field is H_0 = 1780 Oe.

Isolation ratio was about 20 dB. The shift of the resonance line corresponded ±30 MHz at an applied electric field E = ±10 kV/cm. Thus, the electric control allows to realize the tunable isolator or controllable isolator-attenuator. The magnitude of frequency tuning for the isolator was 60 MHz. The amount of attenuation change was about 20 dB.

The schematics of the YIG/PMN-PT bilayer and the microstrip line structure are shown in Fig. 2.55. A YIG film of thickness of 110 μm and lateral dimensions of 1×2.5 mm^2 grown by the liquid-phase epitaxy on one side of a 0.2 mm thick (111) gallium gadolinium garnet (GGG) substrate was used. The film had a saturation magnetization of 1750 G and FMR linewidth of ~0.6 Oe, measured at 7.25 GHz. A (001) PMN-PT plate of dimensions of $4 \times 4 \times 0.5$ mm^3 and with 30 nm thick gold electrodes on both sides was bonded to the YIG film with a fast-dry epoxy. The stripline structure consisted of microstrip transmission lines of nonresonant lengths on a 0.3 mm thick alumina ground plane. The device structure was placed in between the poles of an electromagnet so that a bias magnetic field H could be applied parallel or perpendicular to the bilayer plane. A dc electrical field E = 0–10 kV/cm was generated across PMN-PT by applying a voltage.

Microwave measurements were carried out using a vector network analyzer. A standard calibration procedure was performed before measurements. A cw input signal at 7.25 GHz and power P_{in} = 0.1 mW was applied to the microstrip transducer. Low input power was chosen to prevent heating of the sample due to power absorption at FMR. Profiles of $S_{21}(f) = 10\log[(P_{out}(f)/P_{in}(f)]$ *vs*. f, for an in-plane magnetic field H_0 = 1.91 kOe are shown in Fig. 2.61.

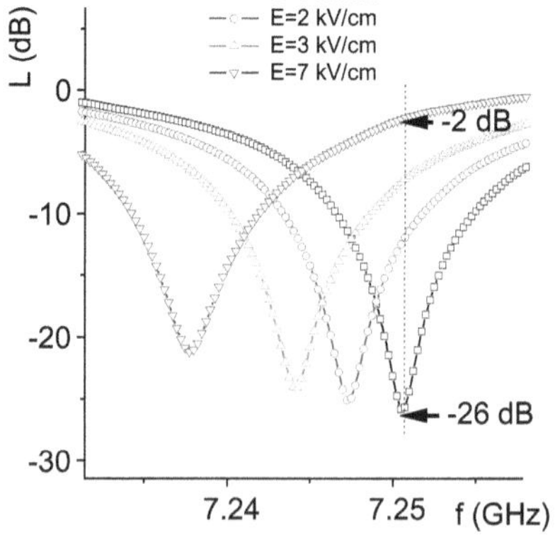

Figure 2.61 Profiles of S_{21} *vs*. f, for an in-plane magnetic field H_0.

For $E = 0$, the spectra contained a well-defined FMR absorption peak with a maximum insertion loss of –26 dB and a 3 dB linewidth $\Delta f = 2$ MHz. The off-resonance loss of 2 dB is due to ferroelectric and other losses as discussed later. With the application of $E = 2$ kV/cm (100 V across PMN-PT), the FMR peak shifted to a lower frequency by $\delta f = 4$ MHz and the ME coupling constant $A = \delta f/E = 2$ MHz cm/kV. With increasing E, Fig. 2.58 shows a linear increase in δf and the FMR profile becomes broader. Similar studies showed a variation in A over the range 1–8 MHz cm/kV, depending on the YIG film thickness, its crystallographic orientation, GGG thickness, and the thickness of epoxy bonding layer. Figure 2.62 shows the attenuation as a function of electric field E. These data were obtained by setting the bias field $H = H_0$ and by increasing E from 0 to 7 kV/cm. With increasing E, one observes a sharp decrease in attenuation up to 5 kV/cm and then it levels off at 2 dB.

The power absorbed by the ME element is given by

$$P(H) = P_0 \exp\left[-\gamma \frac{(H - H_0)^2}{2\sigma^2}\right], \tag{2.33}$$

where σ is the gyromagnetic ratio, H is the effective bias magnetic field, P_0 is absorption at the resonance field H_0, and σ is a parameter determined from the width of resonance line. The electric field leads to a frequency shift δf of the resonance profiles by $\delta f = \gamma(H - H_0) = AE$. Substituting for A in Eq. (2.34), it can be shown that the insertion loss of ME element can be expressed as:

$$L = 20\log[1 - (1 - |s_{21}(H_0)|^2)]\exp(-A^2E^2/(2\sigma^2)). \tag{2.34}$$

Here $S_{21}(H_0)$ is the S parameter at resonance and for $E = 0$.

For the estimation of dependence of insertion loss L *vs.* electric field E, we used measured values of $A = 2$ MHz·cm/kV and $S_{21}(H_0) = 0.22$ and choose $\sigma = 6$ MHz. The estimated loss is compared with data in Fig. 2.62 and there is excellent agreement between the two. The off-resonance insertion loss of the attenuator is the sum of losses in YIG and PMN-PT of about 1 dB, metal striplines (0.3 dB), dielectric ground plane (0.2 dB), and due to nonideal coupling between the resonator and transmission lines (0.5 dB) for a total of 2 dB. Estimated attenuation *vs.* electric field characteristics for a

YIG/PMN-PT bilayer in a stripline structures are in agreement with data.

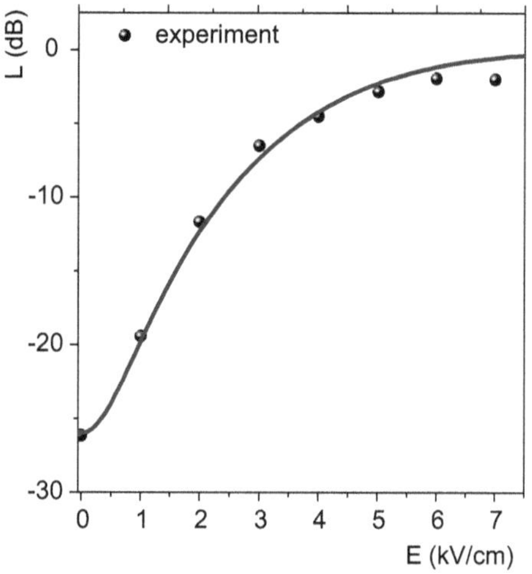

Figure 2.62 The attenuation as a function of electric field *E*.

The proposed models performs functions of both the isolator and the attenuator; are controlled by an electric field; can be made by integral technology; and have higher performance and lower power consumption on comparison with ferrite attenuators. To decrease the control voltage and increase the isolation ratio it is necessary to reduce the thicknesses of the piezoelectric and the ferrite. The use of ME structures in the resonant microwave devices allows electrical control of their parameters and opens promising opportunities of the designing the new devices for functional electronics.

2.4.5 Modeling of ME Microwave Devices

Magnetoelectric interactions in ferrite–ferroelectric composites have facilitated a new class of microwave signal processing devices. Such devices are based on either hybrid spin electromagnetic waves or mechanical force–mediated magnetoelectric interactions. When a ferrite–piezoelectric bilayer is driven to ferromagnetic resonance (FMR) and an electric field $\boldsymbol{E}$ is applied across piezoelectric (ferroelectric), the ME effect results in a frequency or field shift of FMR. Thus, devices based on FMR can be tuned with both electric field E and magnetic field H. Several dual tunable ME devices, including resonators, filters, attenuators, circulators, isolators, and phase shifters have been demonstrated so far.

Simulation of ME microwave devices by the modern computer program that calculate multimode S-parameters and the electromagnetic field in the three-dimensional passive structures greatly simplifies the selection of optimal parameters of such devices: the parameters of the transmission line (dimensions and relative substrate permittivity, the size of the conductors) and the resonator parameters (size, shape, material) [56].

As the industry turns to monolithic integrated/hybrid nonreciprocal microwave devices, planar geometries have to be used. This requires the development of planar elements, compatible with strip line and microstrip systems. As high-frequency systems are manufactured using monolithic microwave integrated circuit (MMIC) designs, the size of the ME resonator must be compatible with the MMIC chip technology.

The design of ME microwave device assumes the presence of ME resonator, which is placed on the microstrip line or circuit-resonator, slot line, or into waveguide using the circular polarization area of microwave field. The circular polarization of microwave field allows more effective use of composite component and allows an increase in the magnetic susceptibility. The working point is selected depending on the purpose of the device. For example, in case of attenuator or isolator, the device is tuned to the resonance absorption because the phase shifter selects the area near a resonance with the lowest absorption but maximal depth control.

Computation, design, and manufacturing technology of nonreciprocal microwave devices intended for application in receiving–transmitting modules of antenna array are of great interest in current time. Recently, a large program high-frequency system simulator (HFSS) has been developed by the company AnSoft, which is intended for the analysis of three-dimensional microwave structures, including antennas and nonreciprocal devices containing ferrites and ferroelectrics. Electromagnetic simulation in HFSS is based on the use of the finite element method (Finite Element Method, FEM).

Microstrip line [57], coplanar line, and slot line are used in the microwave range. The microstrip lines are the most widely used [58,

63]. However, for designing the nonreciprocal devices using ferrites microwave field of circular polarization is required. In microstrip line, this region is absent and additional elements are needed, for example, in the form of stubs to create an area of circular polarization. From this point of view, the slot and coplanar line are of interest. The structure of the microwave field in the slot line and coplanar waveguide is significantly different from the structure of the wave field in microstrip line. Coplanar waveguide (CPW) is a transmission line which consists of a center strip, two slots, and a semi-infinite ground plane on either side of it [59]. This type of waveguide offers several advantages over conventional microstrip line, namely, it facilitates easy shunt as well as series mounting of active and passive devices; it eliminates the need for wraparound and the holes and has low radiation loss. Another important advantage of CPW which has recently emerged is that CPW circuits render themselves to fast and inexpensive on-wafer characterization at frequencies as high as 50 GHz. Lastly, since the microwave magnetic fields in the CPW are elliptically polarized, nonreciprocal components such as ferrite circulators and isolators can be efficiently integrated with the feed network.

Figures 2.63(a–c) show computer models of ME devices on different types of transmission lines.

The transmission line structure in Fig. 2.63(a) consisted of microstrip lines of nonresonant lengths with two stubs of lengths 1/8 and 3/8 wavelengths on a dielectric substrate with ground plane on bottom side. The stubs are required for creating elliptically polarized microwave magnetic field.

The slot line transmission systems [Fig. 2.63(b)] [60, 61] have been shown to contain elliptically polarized ***H*** field regions which are required for producing nonreciprocal microwave devices. The development of such a device was dependent on being able to determine an ME composite slot line configuration that would yield good interaction between the ME resonator and the propagating mode of the slot line with a minimum of concurrent insertion loss. The microstrip to slot line transition is used to convert input microwave signals from a TEM mode to the required slot line mode. The slot width on the transition is designed so as to match into the slot line etched on one of the ME resonator inserts in the slot of the

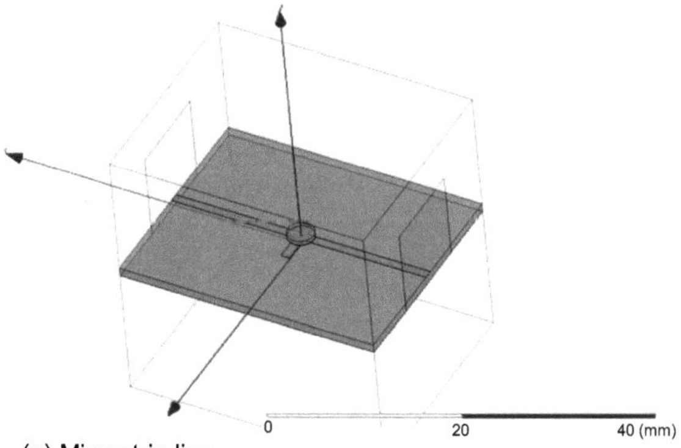

(a) Microstrip line

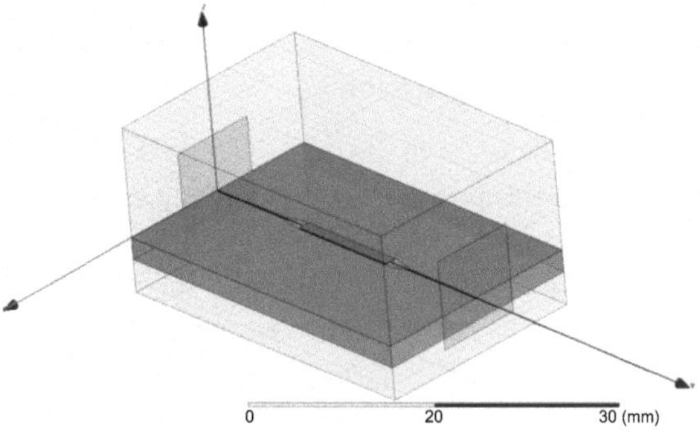

(b) Slot line

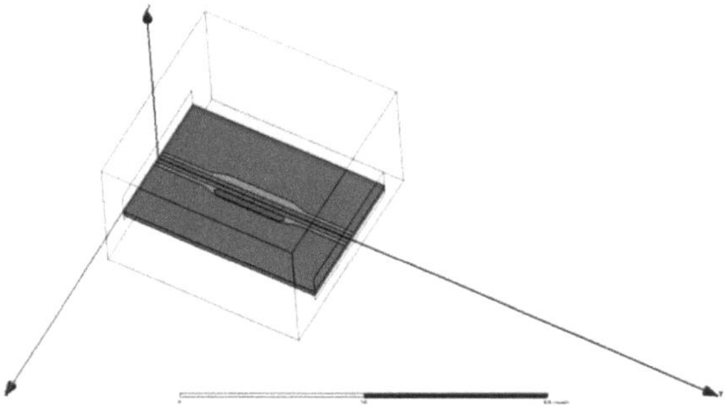

(c) Coplanar waveguide

Figure 2.63 Computer models of ME devices on different types of transmission lines: (a) microstrip line, (b) slot line, and (c) coplanar waveguide.

device. The pertinent characteristics of this type of transmission system such as field configurations, propagation constants, and impedance as functions of dielectric material characteristics, dielectric thickness, and slot width were derived. The slot line contained a microwave magnetic field configuration which was suitable for generating nonreciprocal ME devices. There existed regions within the slot line that contained circularly or elliptically polarized microwave magnetic field.

The use of modern simulation software allows the fast design of various types of nonreciprocal microwave devices. We conducted a simulation of various types of nonreciprocal magnetoelectric devices based on slot and coplanar lines by using the HFSS [Fig. 2.63(c)]. A comparison with similar devices based on the microstrip line was made.

2.4.5.1 Results and discussion

Simulation of the devices is made in the software environment of the HFSS program. *S* parameters in the frequency range are optimized for the investigated device. The characteristicsof the amplitude were investigated. Computer simulation results for different designs of ME microwave devices realized on the strip transmission lines are shown in the figures.

Figure 2.64 shows the frequency dependence of the microstrip line attenuation in the forward and reverse directions.

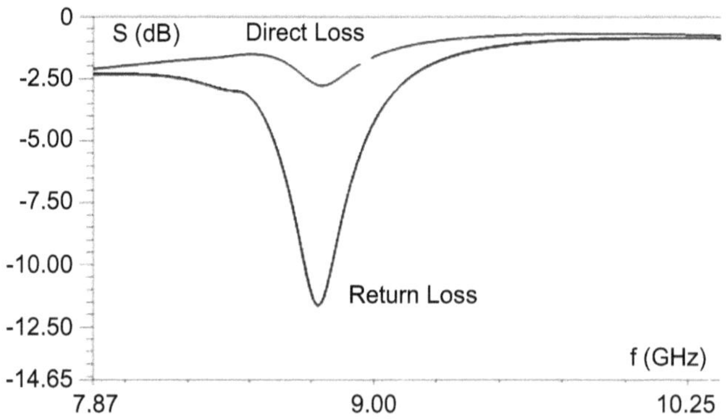

Figure 2.64 The microstrip transmission line. Dependence of attenuation *vs.* frequency. The resonators parameters with YIG disk: thickness is 0.1 mm on GGG substrate of thickness 0.44 mm and diameter is 3 mm; magnetizing field is 2700 Oe.

Figure 2.65 shows the frequency dependence of the slot transmission line attenuation in the forward and reverse directions.

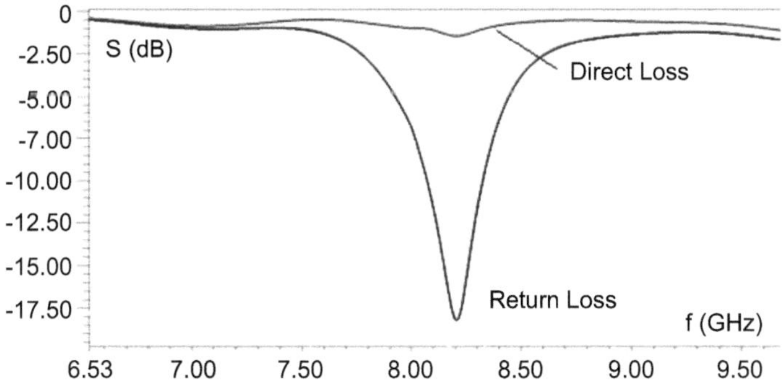

Figure 2.65 Slot transmission line.

The dependence of attenuation *vs.* frequency when resonator dimensions is 10 mm × 1 mm × 0.2 mm; slot linewidth is 0.62 mm, widening the gap to 1.2 mm; the relative permittivity of the substrate is 30, the substrate thickness is 2 mm; magnetizing field is 2514 Oe.

Figure 2.66 shows the frequency dependence of the coplanar transmission line attenuation in the forward and reverse directions.

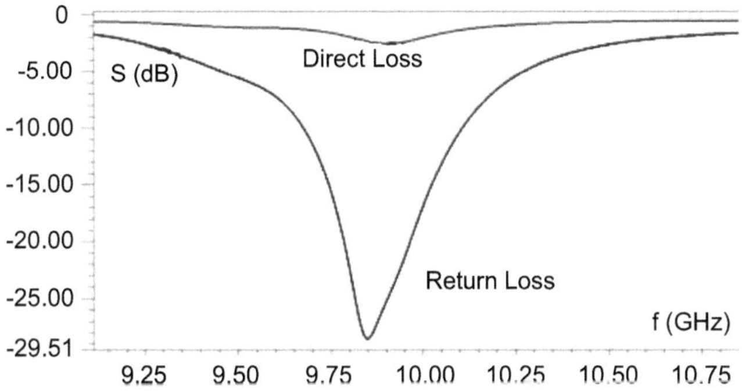

Figure 2.66 The coplanar transmission line: dependence of attenuation *vs.* frequency. Resonator dimensions are 0.6 × 4 × 0.1 mm^3, slot width is 0.4 mm, the center conductor width is 0.6 mm, ε of substrate is 40, substrate thickness is 1 mm, magnetizing field is 3125 Oe.

Figure 2.67 shows the experimental frequency dependence of the coplanar transmission line attenuation in the forward and reverse directions. The experimental investigation of the ME microwave properties of the bilayer structures were based on the measurements of the resonators' frequency responses for different values of external dc voltage and bias magnetic fields, for example, reflection spectra $S_{11}(f) = 10\log|P_{ref}(f)/P_{in}(f)|$, where $P_{in}(f)$ is the incident power, $P_{ref}(f)$ is the reflected power, and f is the excitation frequency. The frequency responses measured using out with Agilent network analyzer.

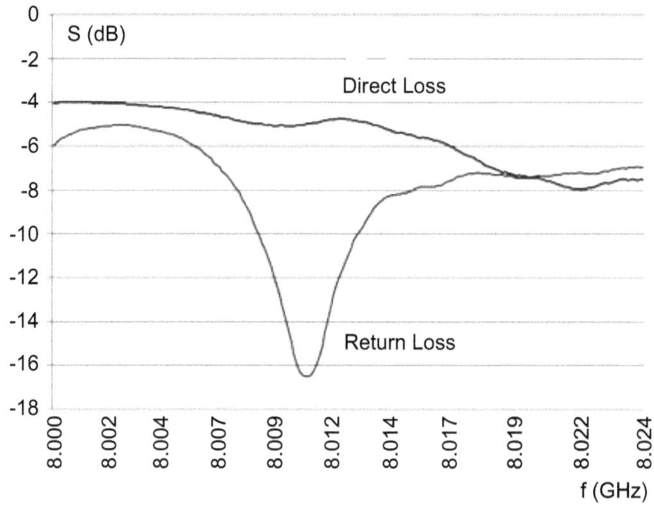

Figure 2.67 For comparison: coplanar waveguide and the experimental frequency dependence of attenuation. Magnetizing field is 2780 Oe.

Computation, design, and manufacturing technology of nonreciprocal microwave devices are currently of great interest. The main directions for further research are based on the use of modern computer design programs. The use of modern simulation software allows fast designing of various types of nonreciprocal microwave devices.

This simulation allows to select substrate parameters and the shape of the ME resonator. The ME resonator based on layered structure of YIG and PZT was used. To decrease the control voltage and increase the valve ratio it is necessary to reduce the thickness of the piezoelectric, and hence the thickness of the ferrite. The use of computer simulation for ME structures in nonreciprocal microwave devices opens promising opportunities for the creation of new devices.

2.5 ME Gyrator

The prospects for further development of microwave technology, including integrated technologies for the creation of transmit–receive modules for phased arrays involve the use of multifunctional and complex materials with significant advantages over traditional ones. Magnetoelectric materials which have been intensively studied in recent years all over the world allow to create devices with improved characteristics. Along with well-studied devices such as phase shifter, attenuator, and filter, there are a number of devices whose study is still ongoing. An ME gyrator is included in such devices to the present time. By definition, which was first given by B. Tellegen in the article [65], a gyrator is a basic element of an electric circuit, four-pole in the ideal case corresponding to the formula:

$$V_1 = -sI_2, \tag{2.35}$$

$$V_2 = sI_1, \tag{2.36}$$

where V is voltage, I is current, s is the coefficient of gyration.

Such gyrator converts the impedance of the circuit and changes the phase of the passing signal. A paper [65] described a case of gyrator that can be made using a hypothetical material that describes the general equations of the two-port network:

$$V_1 = Z_{11}I_1 + Z_{12}I_2, \tag{2.37}$$

$$V_2 = Z_{21}I_1 + Z_{22}I_2, \tag{2.38}$$

where V is voltage; I is current; Z_{11} is the input impedance of inductive character, tending to zero; Z_{22} is the output impedance of a capacitive character; Z_{12} and Z_{21} are the coefficients of gyration related to ME coefficient by formula [66]:

$$Z_{12} = \frac{\alpha_{\text{me}}}{\varepsilon_{\text{eff}}}\frac{d}{l},\ Z_{21} = \frac{\mu_{\text{eff}}}{\alpha_{\text{em}}}\frac{d}{l}, \tag{2.39}$$

where α_{me} and α_{em} is ME susceptibility related to ME coefficient by the proportion $\alpha_{me} = \alpha_{ME} \cdot \varepsilon_{eff}$ and $\alpha_{em} = \alpha_{em} \cdot \varepsilon_{eff}$ [67]; ε_{eff} and μ_{eff} is effective dielectric and magnetic constant, d and l are thickness and length of ME element, respectively.

ME voltage coefficient in the general case can be calculated as:

$$\alpha_{ME} = (dE/dH), \tag{2.40}$$

where H and E is the intensity of the alternating magnetic and electric fields, respectively. The coefficient α_{EM} is calculated similarly.

The use of gyrators in microwave circuits is necessary due to the difference in wave resistance of various sections of the microwave-integrated device. The gyrators can be used for impedance matching different values. The use of the integrated microwave ME gyrators will significantly improve the amplitude-frequency and phase-frequency characteristics of the transmit–receive modules, reduce the size and weight, and will allow to control characteristics of the microwave tract most efficiently. Currently, sufficiently effective ME materials that can be used to create ME gyrator have already been developed [67, 68].

In contrast to the well-known and widely used microwave technique gyrator based on Faraday effect [72] that rotates the phase by 180°, the gyrator using ME effect has the ability to invert the impedance of the transmission line and to convert the input voltage into current and vice versa [72]. The gyrator with ME effect has a compact shape due to the used of the lumped element. Overall dimensions of the proposed microwave gyrator can thus be significantly reduced and it is possible to apply the structure to micro- and nanoelectronics devices. By using this device in microwave technology for impedance inversion, the rotation phase of the electromagnetic wave converting the electrical current into an electrical voltage or, conversely, transforming electrical voltages into electric current becomes possible, too [69–71].

2.5.1 Gyrator's Element Design

The prototype of the device for carrying out experiments to study the gyration effect was developed.

Elements of ME gyrator were located on a substrate of thickness 1 mm and dielectric constant equal to 5 (Fig. 2.68). Two microstrip elements both with length 20 mm and width 0.5 mm and 6 mm, respectively, were located at a distance of 0.5 mm from each other. The input and output lines for better matching connected at some distance from the edge of the microstrip elements. The ME element is mounted between microstrip elements of the gyrator. The ME element represents, in this case, a layered structure; one layer is of yttrium iron garnet (YIG) with a thickness of 6.4 microns on a substrate of gadolinium gallium garnet (GGG) and thickness of 450 μm. The size of the ferrite element is the square of side 3 mm. The second layer is a piezoelectric PZT with thickness of 200 microns and diameter of 5 mm coated on both sides with metallic electrodes. The two layers (YIG and PZT) are glued together with a glue based on ethyl-2-cyanoacrylate.

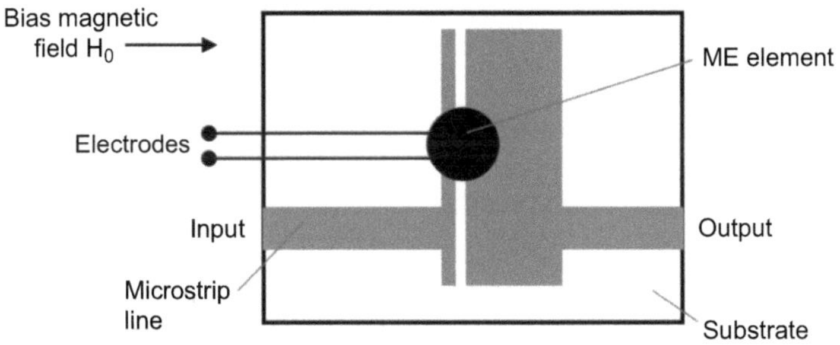

Figure 2.68 Design of magnetoelectric microwave gyrator.

ME effect in composites can be considered as the result of the interaction between piezoelectric and magnetostrictive properties. The mechanism of ME effect in such structures consists of the following: the external electric field causes deformation of the piezoelectric components, causing mechanical stress in the ferrite component. The magnetic properties of ferrite change as a result. As a result, the shift of the ferromagnetic resonance (FMR) line is observed in the microwave range. Thus, the parameters of the gyrator are changed, which is indicated by the measurement equipment. Expected changes in the parameters should occur for the transmission and reflection coefficients, and should change the phase and impedance of the transmission line. Experimental registration of changes in these parameters will be the evidence confirming the validity of assumptions about the possibility of the practical construction of gyrator on ME resonator.

The equivalent circuit of the magnetoelectric gyrator is shown in Fig. 2.69(a). Figure 2.69(b) shows a simplified equivalent circuit of the device.

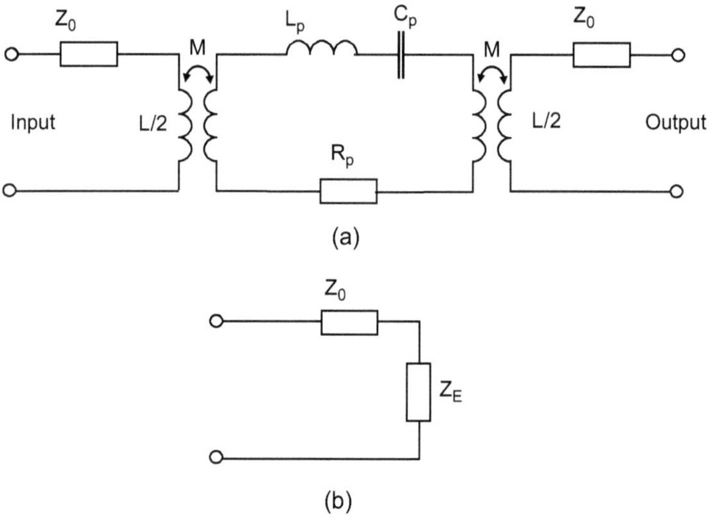

Figure 2.69 The equivalent circuit of the magnetoelectric resonator.

Z_E of the equivalent circuit is described by the formula:

$$\dot{Z}_E = |Z(\omega)| \cdot e^{j\phi(\omega)}, \tag{2.41}$$

where $Z(\omega)$ is the resistance and φ is the phase.

The transmission coefficient in the considered resonator is calculated according to the following formula:

$$K(\omega) = \frac{\dfrac{\omega^2 M^2}{Z_0}}{\dfrac{\omega^2 M^2}{Z_0} + j\omega L_p + R_p + j\omega L - \dfrac{j}{C_p \omega}}; \tag{2.42}$$

The prototype of the device on which measurements were carried out is depicted in Fig. 2.70.

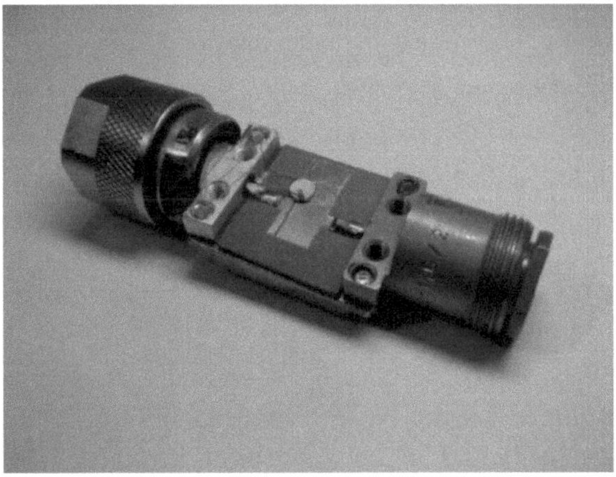

Figure 2.70 The prototype of the gyrator.

The impedance conversion in the transmission line and the phase change of the signal is the result of the described interaction.

2.5.2 Measurement Data

For measuring parameters of the gyrator, the PLANAR Obzor-804 network analyzer, the electromagnet, the power supply of the electromagnet, regulated power supply, and gauss meter were used. The setup used for parameters measuring of gyrator is shown in Fig. 2.71.

The power of the signal at the output of the analyzer was set to 10 mW.

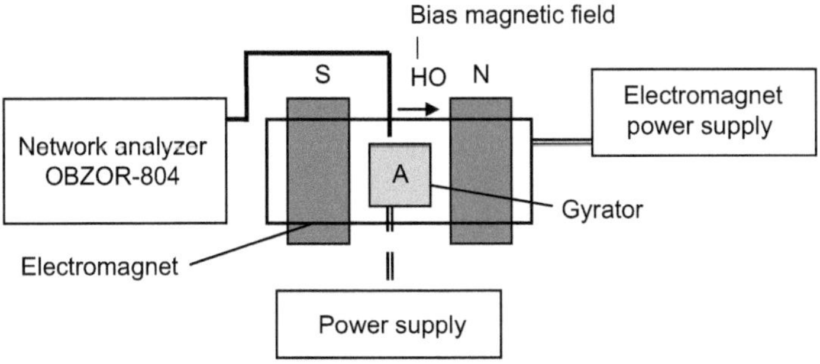

Figure 2.71 The setup used for measuring the parameters of gyrator.

Measurements were carried out in the frequency range of about 4.14 GHz. A constant magnetic field in the range from 0 up to 1000 Oe was applied to the ME element, and it was further found that a magnetic field of 830 Oe was optimal for changing the characteristics. The optimal parameters are close to the ferromagnetic resonance frequency in YIG film, with losses applied by the gyrator not exceeding 0.6 dB. The constant magnetic field acting on the ME element was fixed at 830 Oe. The constant electric voltage applied to the ME gyrator was in the range from 0 to 100 V.

The theoretical calculation of resonant frequency for the plate magnetized in the plane parallel to YIG shows satisfactory coincidence of the magnitude of the biasing field (830 Oe) with the ferromagnetic resonance line found with the empirically value of 4.12 GHz for the resonance frequency. The formula for the eigenmode frequencies of the precession of the gyrator with ferrite magnetized in a plane parallel to the sample plane is expressed as Eq. 2.33.

Initial data for calculation is equal to 2.8 MHz/Gauss; for YIG saturation magnetization is equal to 1780 Gauss, the magnetizing field is equal of 830 Oe, and the obtained resonance frequency f is equal to 4.12 GHz.

Losses in the studied frequency range are mainly due to losses in the device connectors at a frequency of 4.14 GHz; the losses in the connectors and transmission lines amounted to 1.8 dB. The change of the transmission coefficient S_{21} at a frequency of 4.14 GHz is at the magnetic bias field of 830 Oe and when the voltage is applied to the ME element at 0 V and 100 V, it was no more than 0.6 dB, which indicates the smallness of the gyrator insertion loss in a transmission line.

The reflection coefficient S_{11} was significantly changed when the magnetizing field was 830 Oe and the voltage applied to ME element was 100 V. Reflection coefficient S_{11} was –20.3 dB at 0 V and –16.4 dB at 100 V.

Active and reactive components of the device wave resistance were equal to 47 Ohm and –25 Ohm, correspondingly, at bias field of 830 Oe. Applied voltage of 100 V gave rise to an increase of 2 Oe in both active and reactive components of wave resistance.

The experimental data obtained for the gyrator coincided with the estimated data obtained by the formula in Ref. [72]. Comparison of the obtained experimental results with those calculated from the formula of the transmission coefficient S_{21} and the equivalent resistance (Eqs. 2.40 and2.41) led to satisfactory compliance.

2.6 ME Microwave Antenna Array

Radiolocation is an area of radio engineering that solves the problems associated with radar observation of various objects, i.e., their detection, measurement of coordinates and parameters of movement, and determination of some structural or physical properties by using radio waves reflected or re-emitted from objects or their own radiations.

Information obtained during the process of radar observation is called radar's information. The devices for radar observation are called radar station or radar. Objects of radar observation are called radar targets. When using reflected radio waves, radar targets are any non-uniformity of electrical parameters of the environment (permittivity and permeability, conductivity) in which radio waves propagate. To this include flying machines (airplanes, helicopters, weather balloons, etc.), meteorological objects (rain, snow, hail, clouds, etc.), river and seacrafts, terrestrial objects (buildings, cars, planes in airports, etc.). A special kind of radar targets are astronomical targets.

Irrespective of placement type, the radar system consists of antenna-feeder system, receiving and transmitting systems, processing devices (secondary processing of received signal), management system of protection and monitoring.

An important part of any radar is the antenna, which is designed for transmission or receiving of information by radio waves through environmental space.

Today, there are a large number of designs of antennas, but according to the construction types they can be differentiated as follows:

Antennas of the linear type (dipole, frame, slit type) characterize the fact that sizes of their transverse section are small compared with the wavelength. Usually such antennas are executed from a wire segment or several wires, or in the form of a rod.

Aperture antennas (horn, mirror, lens, open ends of waveguides, antennas surface waves) are antennasin which electromagnetic energy is radiated (or received) by some imaginary surface (the aperture of the antennas), are represented in the plane form, and have a size which is usually much larger than the wavelength.

An antennas array is a set of identical radiating elements (irradiators) that are arranged in a specific order and are excited by one or a set of the coherent sources.

Progress in the development of new types of aircraft and missiles, which became especially intensive in the middle of the 20th century, has led essentially to a growth in the speeds of radar targets and a decrease in their effective surface scattering. A considerable enhancement of radar stations is needed. Just in this period, phased antenna arrays have been widely implemented in designing radars for different application. However, their development has shown that changeover of the mirror antenna by the passive phased antenna array leads to increased energy losses in the high-frequency radar several times. To preserve the tactical characteristics of the radar, these losses have to be offset by increasing the transmitter power output, which would entail an increase in size and would simultaneously increase the power consumption of the radar. One of the solutions of this problem was to transition to the use of active phased array radar.

As mentioned in previous chapters, the magnetoelectric materials in the microwave range may experience a number of effects due to ME interaction. The resonant magnetoelectric effect concerns them. This effect is observed in the field of magnetic resonance and represents offset (or shift) resonance lines in an external constant electric field. This effect can be used for creation of a row of HF devices described above.

The possibility of control of radiation pattern of the antenna array by ME effect is assumed. To explore this issue, a number of models of antenna arrays was developed, irradiators of which represent multiferroic structures.

This contribution reports multiferroic magnetoelectric (ME) materials and their practical application for phased array antenna. Multiferroic structures are ideal for studies of wide-band ME interactions between magnetic and electric subsystems that are facilitated by interface strain produced by external magnetic and electric fields. The ME effects are manifested by the change of the electric polarization, *P*, of the material under the application of an external magnetic field *H* (the direct effect) or by the change in the magnetization of the material *M* under the influence of an electric field *E* (inverse effect). The ME effects can be used to regulate the microwave ME devices parameters. The inverse ME effects are often used at microwave frequencies. A significant decrease of energy required for the control of the microwave device is the main advantage of applying this effect. An earlier research [73] showed that ME effects in multiferroic materials are of great interest in terms of phased array antenna [74]. The proposed idea was a development of the phased array antenna with a control via the ME effect [81, 82]. The ME antenna element and ME phase shifter were used as prototypes to create multiferroic microwave patch antennas [76, 77]. ME effects in multiferroic bilayers consisting of lead zirconate titanate (PZT) and yttrium iron garnet (YIG) on gadolinium gallium garnet (GGG) substrates was studied earlier in Ref. [78].

2.6.1 Modeling and Results

The first model is the two-element antenna array (Fig. 2.72). Each irradiator of the antenna represents a multiferroic structure (Fig. 2.73) 20 × 20 mm with metallization deposited on top and bottom of substrate. In the metallization of the top layer, 12 cutouts of size 9 × 1.5 mm were made, with internal electrodes of size 8.5 × 0.5 mm. The multiferroic structure consists of a PZT layer with thickness of 5 μm on a YIG layer with a thickness of 5 μm deposited on a GGG substrate with a thickness of 500 μm. Electromagnetic field on the antenna surface in the slot zone at the resonance frequency has a circular polarization that can be used for efficient excitation of ferromagnetic resonance (FMR) in the YIG layer.

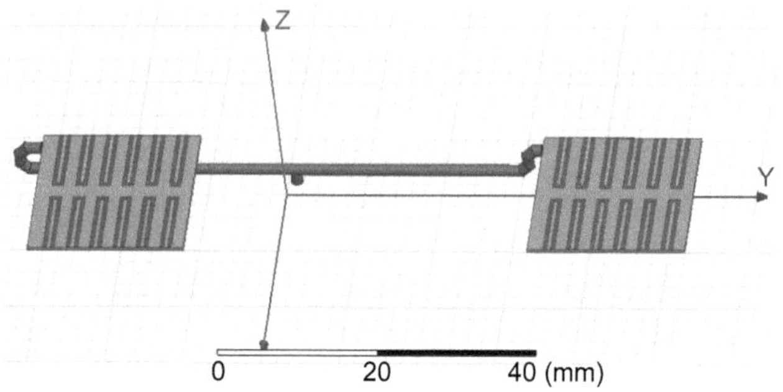

Figure 2.72 Two-element antenna array model.

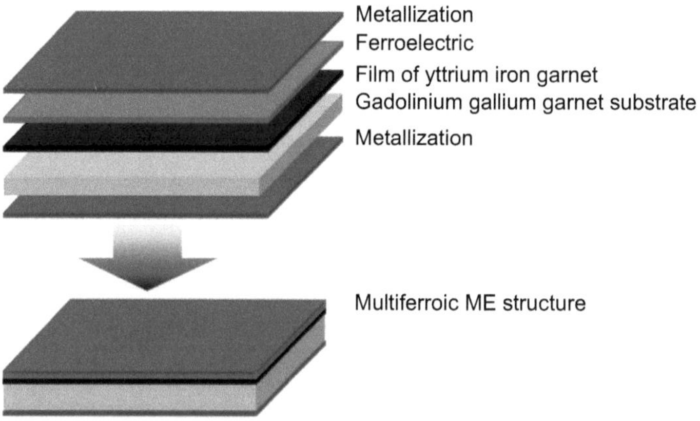

Figure 2.73 Multiferroic ME structure.

We used the following physical parameters for the simulation of the multiferroic structure. For YIG layer, the relative permittivity is 15.3, the tangent of angle of dielectric losses is 2×10^{-4}, saturation magnetization is 1780 Gs, and the width of ferromagnetic resonance line is 1 Oe. For PZT layer, the relative permittivity is 1870 and the tangent of angle of dielectric losses is 0.0168. For GGG substrate, the relative permittivity is 11.6 and the tangent of angle of dielectric losses is 0.0011.

Control of the antenna element properties was realized as follows: During metallization, a DC voltage of 0 to 0.5 kV was applied on the top and bottom layers of the antenna. The electric field leads to a deformation of the PZT layer owing to piezoelectric effects. Since the PZT layer is strongly connected with the ferrite layer, deformation is transferred to the YIG layer. Mechanical deformation of the ferrite brings about a shift of the FMR signal. If the ferrite is pre-magnetized so that the FMR signal is located in the region of resonance of the antenna, applying the control voltage brings about a shift of the FMR signal, resulting in a change of the antenna parameters and directional diagram. Characteristics of ME multiferroic devices obtained using the FMR line shift caused by ME effect in general correspond to the similar characteristics where the shift of FMR lines was caused by the change of the bias magnetic field, which was corroborated earlier [73]. This property is used for modeling multiferroic structures in which ME effect is simulated through the change of the bias magnetic field.

Figure 2.74 shows the dependence of the reflection coefficient S_{11} *vs.* f for different values of magnetic bias field for single irradiator without electrodes in top layer metallization. The natural resonance of the structure, which depends on its size at the frequency of 2.25, 3.3, and 4.45 GHz without magnetic bias field, is clearly visible in Fig. 2.74(a). Figure 2.754(b) shows the change of the coefficient S_{11} *vs.* f under the action of the bias magnetic field. The change of the resonance frequency, i.e., shift of FMR line occurs under the action of the bias magnetic field in the range from 64000 A/m to 110000 A/m in increments of 1600 A/m. This FMR shift may be used to control the antenna array radiation pattern. If one of the elements has a certain value of magnetic field while the other has a different value, the elements will have different frequency and phase response, and hence the radiation pattern of the antenna array may change when doing a magnetic adjustment of separate antenna elements.

The simulation of two-element antenna array characteristics was carried out in the frequency region from 1 to 5 GHz. The distance between the elements was $\lambda/2$. Figure 2.75 shows the dependence of the coefficient S_{11} *vs.* the frequency for two-element antenna array for values of the bias magnetic field from 54400 A/m to 57600 A/m. The frequency region, which is outside the natural resonance of the structure but has a maximum value of FMR, was selected for further studies. The value of the coefficient S_{11} is −40 dB when the bias magnetic field is 55200 A/m at the resonance frequency of 1.841 GHz.

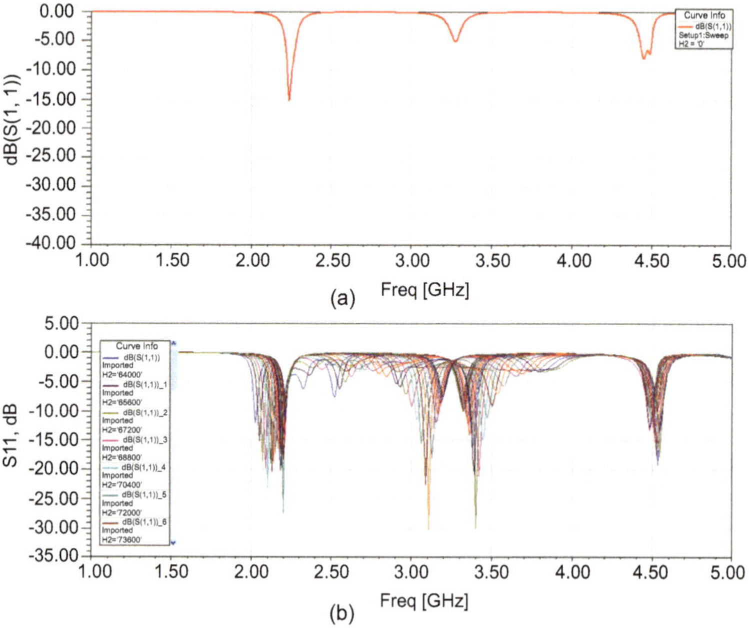

Figure 2.74 Dependence of the reflection coefficient S_{11} *vs.* frequency. (a) S_{11} *vs. f* without magnetic bias field, (b) S_{11} *vs. f* when the bias magnetic field is in the range from 64000 A/m to 110000 A/m, in increments of 1600 A/m.

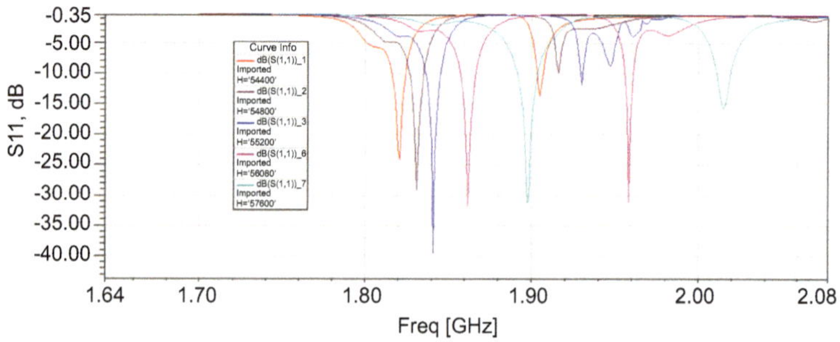

Figure 2.75 S_{11} *vs. f* when the bias magnetic field is in the range from 54400 A/m to 57600 A/m.

Modeling the radiation pattern of the antenna array was performed at a frequency of 1.841 GHz; the simulation results are shown in Fig. 2.76. The radiation patterns of the antenna array were obtained for three different cases with different values of the bias magnetic field on each element: (a, d) when the value of the bias on the first element is 54800 A/m and the second one is 55200 A/m; (b, e) when the value of the bias on the first element is 55600 A/m and the second one is 55200 A/m; (c, f) when the value of the bias on the first element is 55600 A/m and the second one is 55000 A/m.

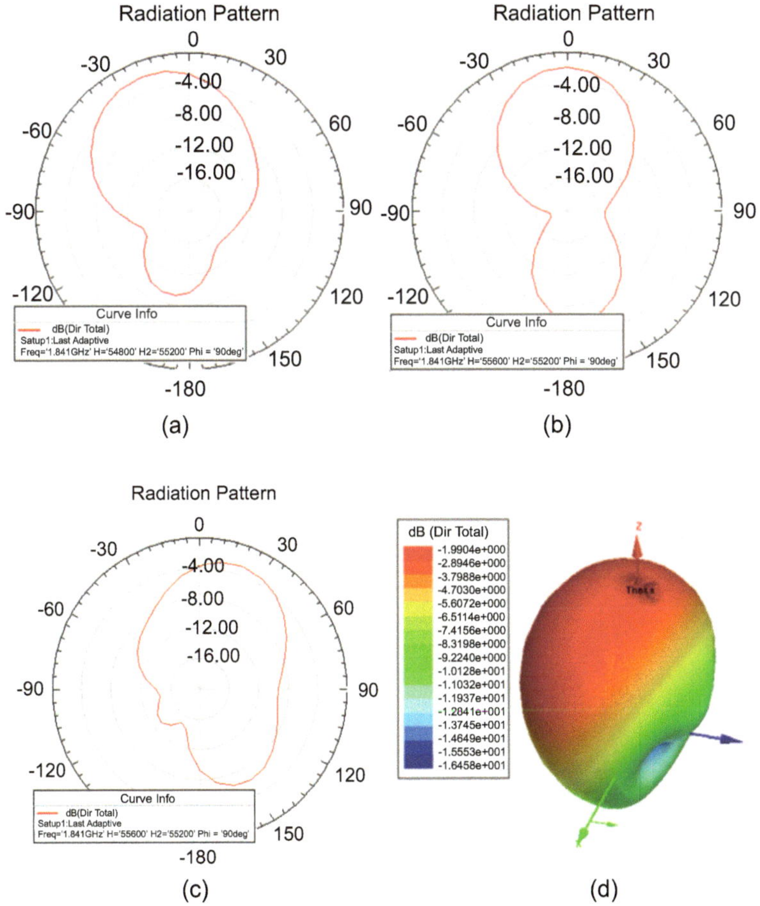

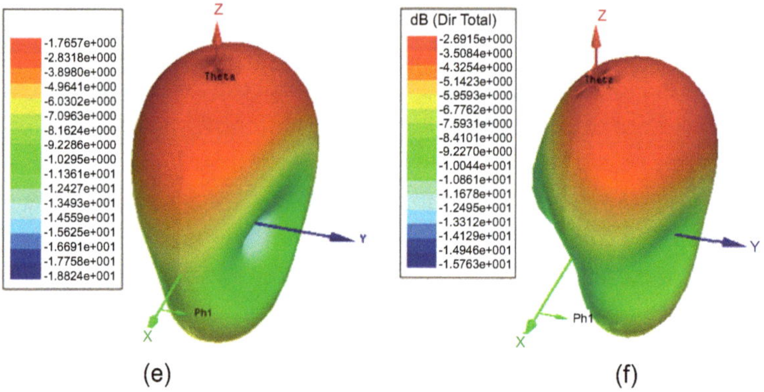

Figure 2.76 The radiation patterns of two-element antenna array (a–c) in the radial plane, (d–f) is 3D view.

Table 2.3 Topological dimensions of microwave antenna array

No.	Designator	Value (mm)
1	L	11.0
2	l	9.0
3	w	9.0
4	p	2.9
5	p_1	1.1
6	p_2	0.9
7	p_3	0.9
8	wl	0.5
9	Wp	1.3
10	$\lambda/2$	11.42

The second model is also the two-element antenna array. However, on it likenesses end. The model consists of a two-element antenna array (Fig. 2.77) on a PMN-PT substrate. Each separate element of the antenna array is radiating ME element, the structure of which was the size of 10 × 10 mm and consisted of piezoelectric PMN-PT 100 μm thick layer and YIG with thickness of 100 um. On the top and bottom of the structure, metallic copper (Cu) layers were deposited. Multiferroic ME structure is shown in Fig. 2.78. All dimensions are given in Table 2.3.

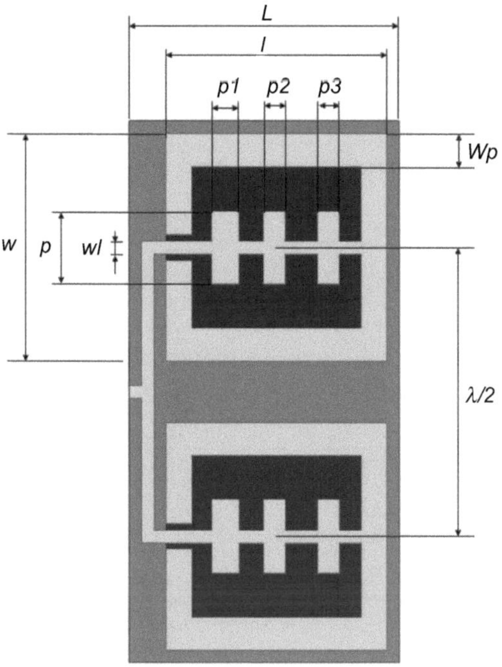

Figure 2.77 Model microwave antenna array.

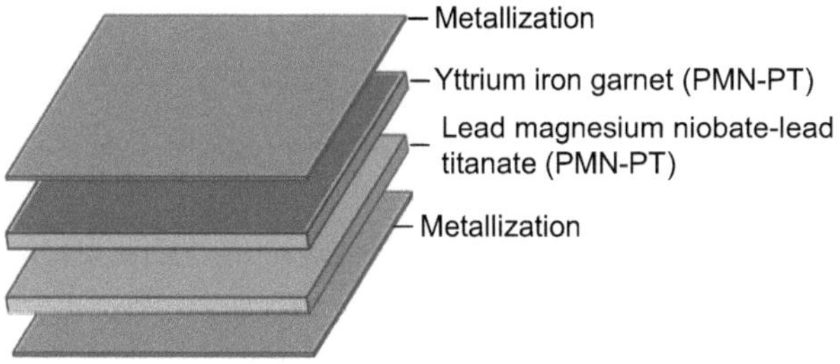

Figure 2.78 Multiferroic ME structure.

We used the following physical parameters for the simulation of multiferroic structure. For YIG layer, the relative permittivity is 15.3, the tangent of the angle of dielectric losses is 2×10^{-4}, and saturation magnetization is 1780 Gs. For PMN-PT layer, the relative permittivity is 1380 and the tangent of the angle of dielectric losses is 0.008.

Initially, the frequency dependence of the reflection coefficient S_{11} was researched for values of applied magnetics bias field on condition of the same value on elements of antenna array (Fig. 2.79). The frequency dependence is of interest in case of value of applied magnetics bias field equal to 54400 A/m. The value of the coefficient S_{11} is −28.8281 dB at a frequency of 2.935 GHz.

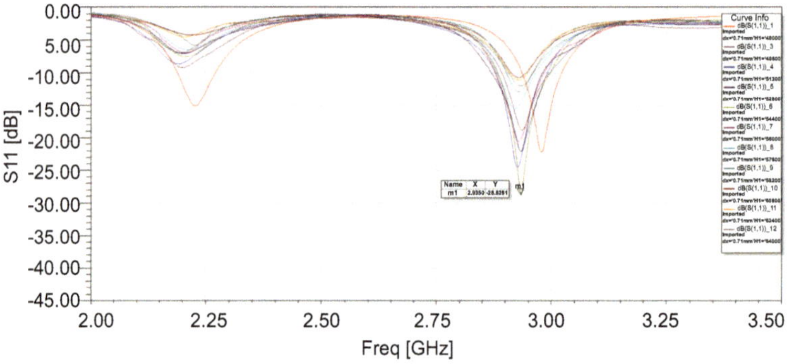

Figure 2.79 Dependence of the reflection coefficient S_{11} *vs.* f when the bias magnetic field is in the range from 48000 A/m to 64000 A/m in increments of 1600 A/m.

Figure 2.80 shows frequency- and phase-response characteristics of the coefficient S_{11} in case the value of applied bias field equals 54400 A/m. This characteristic gives the value of resonance frequency as 2.951 GHz.

The next modeling of the antenna array was performed at a frequency of 2.951 GHz.

Graphic representation of the radiation pattern of the antenna array in the rectangular plane is shown in Fig. 2.81 and in 3D view in Fig. 2.82 in case of different values of applied magnetic bias field of YIG layers on each element: (a) when the value of bias on the first element is 52000 A/m and the second one is 56800 A/m; (b) when the value of bias on the first element is 55200 A/m and the second one is 53600 A/m; (c) when the value of bias on the first element is 56000 A/m and the second one is 52800 A/m; and (d) when the value of bias on the first element is 57600 A/m and the second one is 51200 A/m.

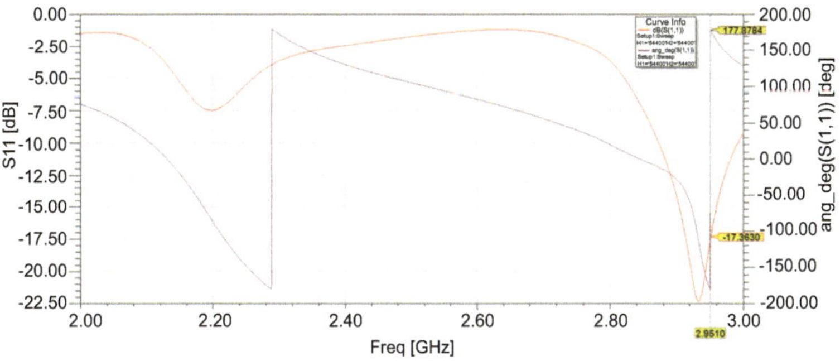

Figure 2.80 Frequency-response and phase-response characteristics of the coefficient S_{11} in case the value of applied bias field equals 54400 A/m.

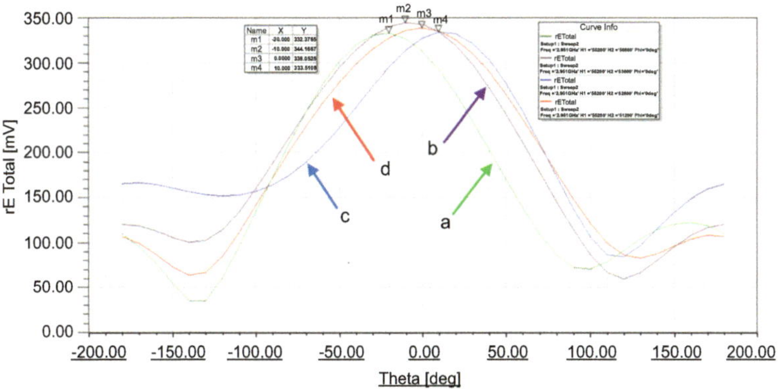

Figure 2.81 The radiation pattern of the antenna array in the rectangular plane in case of different values of the applied magnetic bias field of the YIG layers.

Modeling of several of two-element's antennas array has performed. The obtained values of the radiation pattern deviation rendered within limits ±30º. Thus, it is possible to control the shape and direction of the radiation pattern antenna array, if to give the different control voltage for each element separately. In the future research, it is possible to increase the range of adjustment of directional diagram of principal ray, and to adjust the transmission coefficient of the antenna array.

2.7 ME Microwave Modules and Radars

This content is devoted to the research on the designing of the receiving and transmitting microwave modules using devices based on magnetoelectric (ME) composites. Progress which has been made recently in the field of ME materials and microwave devices [73, 79–81] allows us to offer a microwave module, which would have certain advantages over traditional modules based on semiconductor and ferrite devices. Advantages of the proposed module are the electronic tuning of any of ME microwave devices, which allow to create the necessary amplitude–frequency response at any time, as well as low level of intrinsic noise, high radiation resistance, and low power consumption at the control circuit.

Proposed module can be used for the designing of the receiving and transmitting devices, including phased arrays system [63, 75]. The advantages of our module are small size and weight, the electronic control method, high radiation resistance, good thermal performance, the possibility of miniaturization and integration on a single substrate, and the possibility of creating a microantenna system. All this performance allows the use of the proposed modules to create a radar, including the cosmos system, as well as use the device for local computer networks and other communication systems.

2.7.1 Basic Element Design

ME device such as an antenna, isolator, filter, attenuator, phase shifter, gyrator, amplifier, modulator, and detector devices have been created by ME technology over the past twenty years, and the

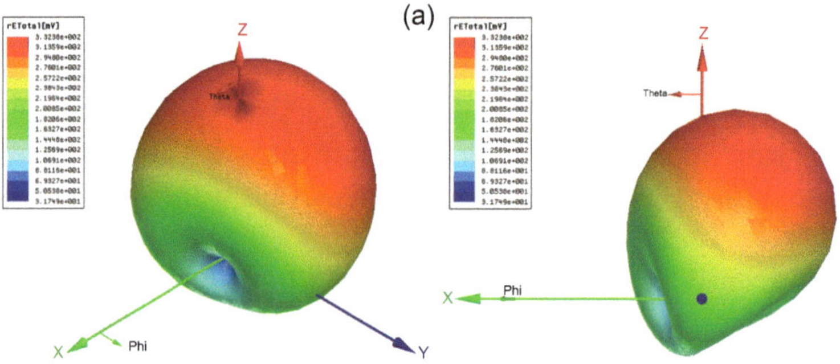

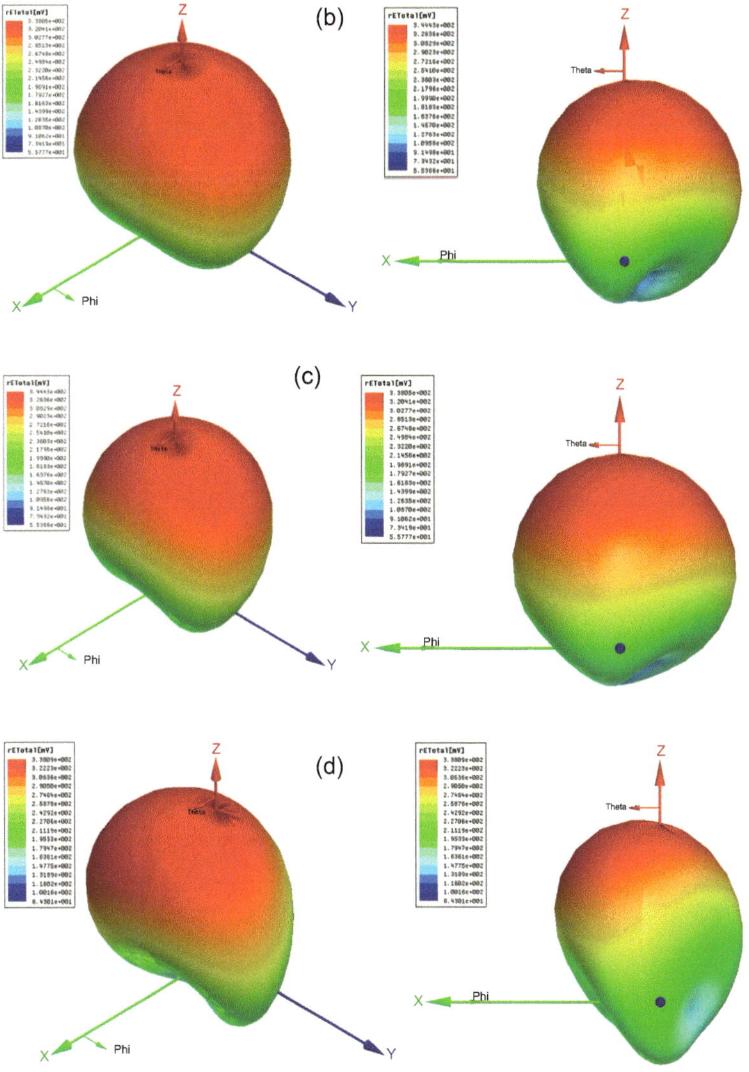

Figure 2.82 The radiation pattern of the two-element antenna array in 3D view

specifications are not inferior to similar competitive models [64, 80–85]. Consider the basic technical and technological possibilities of the proposed devices by the example of ME attenuator.

The main characteristics of studied microwave attenuators are: continuously tuned by electric field of attenuation that is up to 35 dB at a frequency of 4.478 GHz, VSWR is 1.1, and adjusting potential is 400 V. Characteristics of amplitude *vs.* frequency is shown in Fig. 2.83.

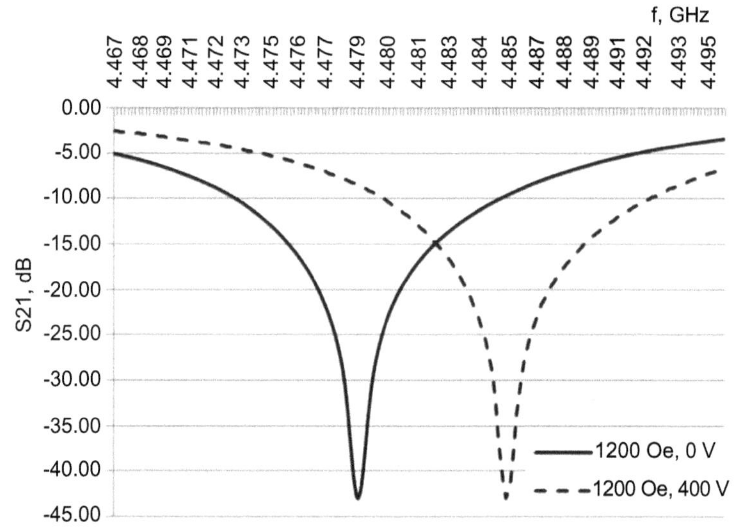

Figure 2.83 Amplitude *vs.* frequency of ME microwave attenuator.

Attenuator control characteristic is shown in Fig. 2.84.

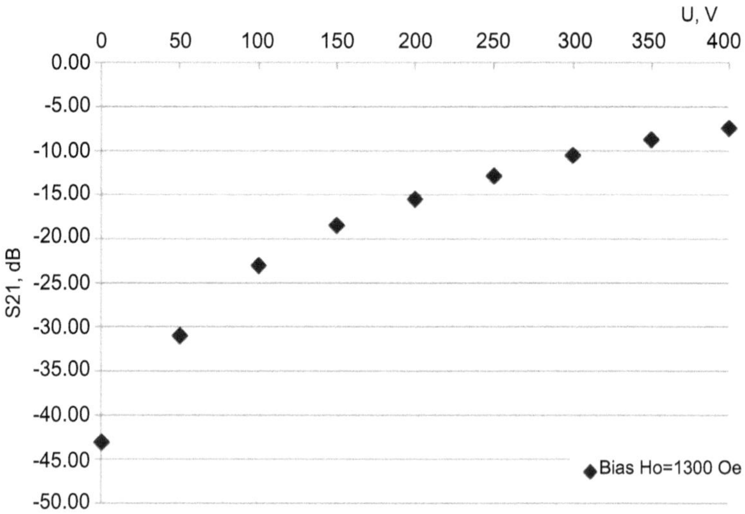

Figure 2.84 The control characteristic of ME attenuator.

The shift of the FMR line in Fig. 2.83 may be determined by the following formula:

$$\delta f = \frac{\gamma AU}{2\pi d} \tag{2.43}$$

where γ is the gyromagnetic ratio; A is magnetoelectric coefficient, in our case $A = 3.5 \times 10^{-6}$ m/V; U is the control voltage; and d is the thickness of the piezoelectric material.

A topology of ME microwave attenuator is shown in Fig. 2.85: The base of the attenuator is a microstrip transmission line with two stubs on dielectric substrate and ME resonator and two electrodes for applying control voltage. ME resonator can be made from single-phase or layered materials. The layered materials, which are mainly ferrite YIG film in conjunction with the piezoelectric PZT have an optimal use today. The YIG film has a thickness of 60 microns and the GGG substrate has a thickness of 400 microns and a diameter of 4 mm. PZT has a thickness of 0.5 mm and a diameter of 5 mm. Formula (2.43) shows that the thinner the piezoelectric layer, the bigger the shift of FMR line and, therefore, we have a more efficient control of the device. On the other side, from this formula, it follows that a decrease of the piezoelectric thickness can reduce the control voltage proportionately.

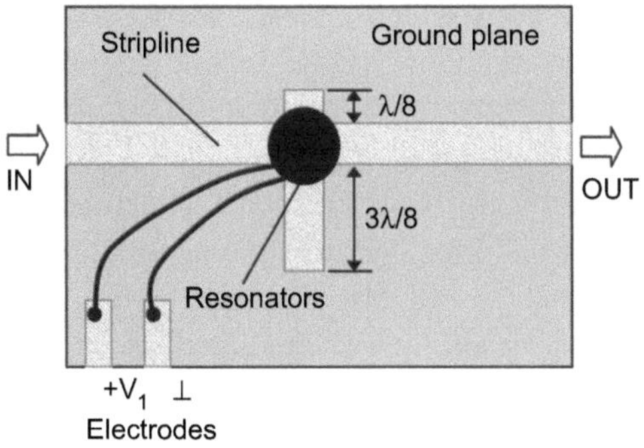

Figure 2.85 The topology of ME microwave attenuator.

Characteristic of attenuator calculated by the formula for the transmission coefficient by the matrix method as:

$$S = e^{-L+i\phi} \begin{vmatrix} 0 & 1 \\ 1 & 0 \end{vmatrix} \tag{2.44}$$

where $L = P_{in}/P_{out}$ is the ratio of the input and output power and φ is the phase shift.

P_{out} can be defined by the formula:

$$P_{out} = P_{in} \cdot e^{-0.5\left(\frac{AU}{d\sigma}\right)^2}, \tag{2.45}$$

where σ is half-width of the FMR line.

Similarly, S parameters can be calculated for all devices used in the module. Then, by the matrix method a general formula can be defined for the transmission coefficient of the module as a whole.

The general formula by which the characteristics of the module are calculated at the central frequency by the matrix method is

$$S = e^{-L+K+i\Phi}\begin{vmatrix} 0 & 1 \\ 1 & 0 \end{vmatrix} \tag{2.46}$$

where L is adjustable attenuation, K is adjustable amplification coefficient, and Φ is adjustable phase shift.

2.7.2 Microwave Magnetoelectric Module

Ferrite–piezoelectric structures are ideal for the design of ME modules. ME interaction in these structures allows to implement most of the functions which are necessary for work of receiving and transmitting modules. Hybrid technology allows to combine separately manufactured devices into a single module.

The basic ME elements of the modules for microwave phased arrays (Fig. 2.86) are devices such as antenna, isolator, filter, attenuator, phase shifter, gyrator, modulator, amplifier, detector, and circulator [62, 82–88]. ME effect can be implemented in these devices for a precise setting, or adjustment, etc.

The microwave magnetoelectric amplifier invented in Ref. [82] contains a microstrip dielectric substrate with an element of energy conversion (Fig. 2.87). It amplifies the microwave signal in the microstrip resonator topology that is placed in the magnetoelectric planar element, working in the field of magnetoacoustic resonance. ME element mounted in the microstrip substrate and located under microstrip resonator with a size of 1/2 wavelength of the amplified signal with two strips of 1/8 and 3/8 of the wavelength of the amplified signal. ME element is connected on the one end via pump frequency band-pass filter with the inlet of the pump generator, and on the other end via the amplified signal frequency band-pass filter and the circulator with the input and output of microwave signals. The permanent magnet is located near the magnetoelectric element. The invention provides the possibility of amplification of microwave signals, the most easy and cheap method, and also reduces the device dimensions to improve manufacturability.

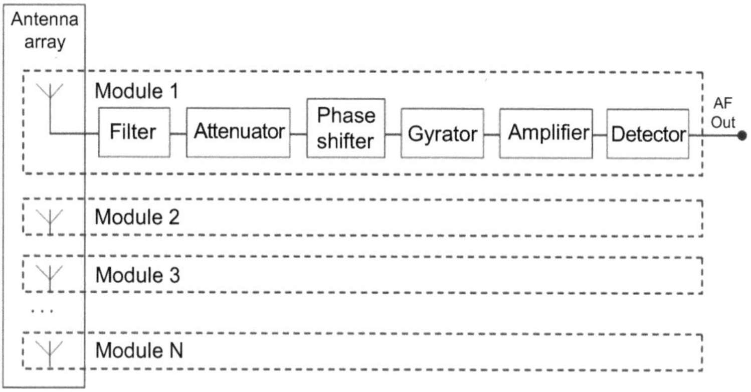

Figure 2.86 ME receiving module of phased array antenna.

Selective detector of microwave power [83] consist of a dielectric substrate, a sensing element placed in the field of a permanent magnet, and a semiconductor element that is made in the form of a disc from a composite structure of ferrite-ferroelectric-planar semiconductor diode. The design placed in the circular polarization of microwave magnetic field created use microstrip line with a length of $\lambda/2$ and $\lambda/4$. Thus biasing of the resonant field can be done by the permanent magnet in disc form that is mounted on the reverse side of the dielectric substrate and by the coupling capacitors, low-pass filters, and bus of the control and output signals that are mounted on the top of substrate. The technical result of this device is to increase the accuracy and performance of the frequency tuning of the detector.

For example, a module developed by Micran (a Russian company) is based on semiconductor devices and has the following features: dynamic range gain is 23 dB, phase adjustment range is 360° with steps of 5.6 and a noise figure of the receiver not more than 5 dB. We propose the module with comparable characteristics and the following parameters in the range of 10 GHz: the dynamic gain range is 25 dB, phase adjustment range is 360°(continuous), and noise figure of the receiver is not more than 0.9 dB, which exceeds the characteristics of "Micran" module in several times. The proposed solution is based on the ME devices, which improves sensitivity of the system and provides advantages for the module as a whole.

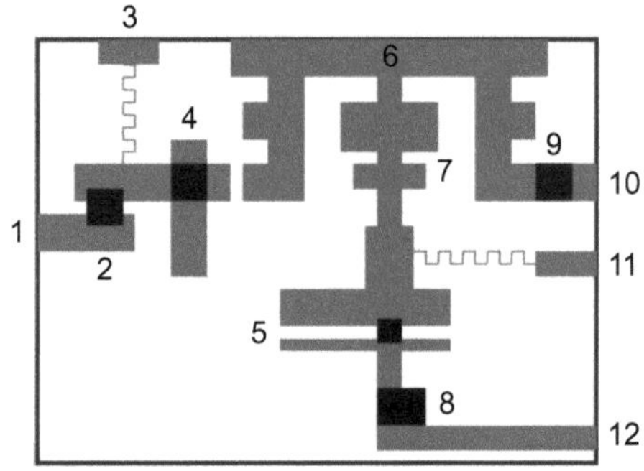

Figure 2.87 Topology of ME receiving module of phased array antenna. 1 is the microwave signal input, 2 is the ME filter, 3 is the control voltage input of ME phase shifter, 4 is the ME phase shifter, 5 is the ME gyrator-converter, 6 is the Y-circulator, 7 is the band-pass filter, 8 is the ME amplifier of pumping signal, 9 is the ME amplifier of output signal, 10 is the microwave signal output, 11 is the control voltage input of ME gyrator-converter, and 12 is the input of microwave pumping signal.

The implementation of transceiver modules is possible on a single technological substrate as well as separately by devices. Hybrid technology allows to separately combine manufactured devices in the single integrated module. Integrated technology, which combines the production of all ME devices in a single technological cycle can give the greatest advantages. The use of GGG substrate is possible with already applied YIG film and further deposition of the piezoelectric layer using the installation magnetron sputtering, molecular beam epitaxy, and deposition of this layer by a laser ablation method or a sol–gel method. Further procedures allow you to execute in a single cycle all of the devices that can make their subminiature. The reliability of such devices is large enough, in addition, they possess the property of radiation resistance and are suitable for space applications. Hybrid thick film technology can be used in addition to the integrated thin-film technology, because the the techniques is much cheaper and the statement of production can be carried out in a shorter period.

Advances in technology allow to develop and produce ME devices with competitive performance compared to devices based on other physical principles. ME microwave modules have advantages such as small size and light weight, electronic tuning, continuous phase adjustment, low noise of receiver, high radiation resistance, and miniaturization.

2.8 Conclusions

In this chapter we considered the main constructions, equivalent circuits, and characteristics of ME devices for application in the wide frequency range. It was showed that ME devices have some advantages over ferroelectric and semiconductor ones in the sensitivity, low price, and radiation resistance.

As for the ME inductance, it has been established that the most effective sample is a bimorph piezoelectric structure. It was also found that in a sample based on a piezoelectric bimorph structure with a magnetostrictive phase thickness of 20 μm, the inductance readjustment value is higher than in samples with a larger Metglas thickness. On the whole, the experiment successfully agrees with the theory, but it should be noted that the real values of the relative inductance tuning are much lower than the calculated ones. The discrepancy is explained by the fact that the technology of making samples does not have the proper technological elaboration and rigging to ensure reliable mechanical connection between the layers of Metglas and piezoceramics. The very nature of the theoretical dependence of the ME inductance and its relative tuning is in good agreement with the experimental data.

Although a large number of parameters of ME sensors have been attained, several important issues still remain to be solved including the following:

- Increasing the sensitivity and design of sensors to simultaneously measure the orientation and magnitude of dc and ac magnetic fields
- Designing contactless sensors and also for measuring the current at different range from leakage current up to 100 A
- Increasing the sensitivity and design of sensors for more accurate measurements of cogwheel magnetic field

As shown, ME effect can be used for designing of the energy harvesting devices. The output power of ME element in the resonance mode is much higher than out of resonance mode. The frequency of the excited resonant oscillations of ME element is influenced by the parameters of the element material, its length, thickness, and method of fixture. The prototype of ME generator including of eight ME elements with dimensions of one element $40 \times 10 \times 0.5$ mm at the frequency of the alternating magnetic field of 38 Hz at the output of the device showed a constant voltage of 1.12 V and a current of 3.8 μA. Alternating voltage until the rectifier bridge was of 1.7 V. Total generator power amounted to 4.28 μW. Designed generator can be applied in the construction of different power generating sets.

The characteristics obtained on the models show the possibility of creating ME microwave devices for a number of parameters that exceed the ferrite and semiconductor analogs. Despite significant progress in the development of ME microwave devices, we note three factors that inhibit this development: the required high-control electric field, the required high-resonant magnetic field, and the absence of active ME devices. In the future, these problems can be solved by using an electric field at the frequency of electromechanical resonance and gradient magnetic structures. To create active ME devices it is necessary to include in composite structures the piezoelectric semiconductors.

The obtained theoretical and experimental data confirm the possibility of using microwave gyrator for designing integrated devices. Besides, the device allows to solve an important problem of microwave technology which consists in matching the wave impedances of devices and transmission lines.

The use of a microwave antenna array with the substrate for radiating elements from the ME material is of interest in many practical cases. It can be mobile and stationary radars, telecommunications circuits, modems, means for targeting, local data networks, security systems, etc. The main advantages of such antenna arrays over antenna arrays with ferrite phase shifters consist in less power for control circuits, high processing speed, and capabilities of integrated technologies that will significantly reduce the size and weight of devices.

Application of ME film structures will significantly reduce the control voltage. Piezoelectric and ferrite materials with better properties than used now, will further significantly improve the parameters of devices, and application of microwave materials, such as, for example, hexaferrites will significantly expand the frequency range of modules up to 100 GHz.

References

1. R. V. Petrov, A. N. Soloviev, K. V. Lavrentyeva, *et al.*, Magnetoelectric transducers, *Progress In Electromagnetics Research Symposium Proceedings*, Stockholm, Sweden, Aug. 12–15, 2013, pp. 1271–1274.
2. M. I. Bichurin, V. M. Petrov, R. V. Petrov, *et al.*, Magnetoelectric microwave devices, *Ferroelectrics,* **280**, 211–218 (2002).
3. Y. Wang, J. Li, D. Viehland, Magnetoelectrics for magnetic sensor applications: status, challenges and perspectives, *Mater. Today*, **17**, 269–275 (2014).
4. J. Gao, Y. Wang, M. Li, *et al.*, Quasi-static (f<10−2 Hz) frequency response of magnetoelectric composites based magnetic sensor, *Mater. Lett.*, **85**, 84–87 (2012).
5. J. Clarke, R. H. Koch, The impact of high-temperature superconductivity on SQUID magnetometers, *Science*, **242**, 217–223 (1988).
6. Y. J. Wang, J. Q. Gao, M. H. Li, *et al.*, A review on equivalent magnetic noise of magnetoelectric laminate sensors, *Phil. Trans R. Soc. A*, **372**, 20120455 (2014).
7. Y. Wang, D. Gray, J. Gao, *et al.*, Improvement of magnetoelectric properties in Metglas/Pb(Mg1/3Nb2/3)O3–PbTiO3 laminates by poling optimization, *J. Alloy. Compd.*, 519, 1–3 (2012).
8. Y. Wang, D. Gray, D. Berry, *et al.*, Self-amplified magnetoelectric properties in a dumbbell-shaped magnetostrictive/piezoelectric composite, *IEEE Trans. Ultrason. Ferroelectr. Freq. Control*, **59**, 859–862 (2012).

9. Y. Wang, J. Gao, M. Li, *et al.*, Ultralow equivalent magnetic noise in a magnetoelectric Metglas/Mn-doped Pb(Mg1/3Nb2/3)O3-PbTiO3 heterostructure, *Appl. Phys. Lett.*, **101**, 022903 (2012).
10. X. Zhuang, S. Saez, M. Lam Chok Sing, *et al.*, *Sens. Lett.*, **10**, 961 (2012).
11. R. Jahns, H. Greve, E. Woltermann, *et al.*, Sensitivity enhancement of magnetoelectric sensors through frequency-conversion, *Sens. Actuators, A*, **183**, 16–21 (2012).
12. T-D. Onuta, Y. Wang, S. E. Lofland, *et al.*, Multiferroic operation of dynamic memory based on heterostructured cantilevers, *Adv. Mater*, **27**, 202–206 (2014).
13. S. Marauska, R. Jahns, C. Kirchhof, *et al.*, Highly sensitive wafer-level packaged MEMS magnetic field sensor based on magnetoelectric composites, *Sens. Actuators, A*, **189**, 321–327 (2013).
14. J. Petrie, D. Viehland, D. Gray, *et al.*, Shifting the operating frequency of magnetoelectric sensors, *J. Appl. Phys.* **111**, 07C714 (2012).
15. J. R. Petrie, J. Fine, S. Mandal, *et al.*, Enhanced sensitivity of magnetoelectric sensors by tuning the resonant frequency, *Appl. Phys. Lett.*, **99**, 043504 (2011).
16. Y. Wang, D. Gray, D. Berry, *et al.*, An extremely low equivalent magnetic noise magnetoelectric sensor, *Adv. Mater.*, **23**, 4111–4114 (2013).
17. E. Lage, C. Kirchhof, V. Hrkac, *et al.*, Exchange biasing of magnetoelectric composites, *Nat. Mater.*, **11**, 523–529 (2012).
18. C. Kirchhof, M. Krantz, I. Teliban, *et al.*, Giant magnetoelectric effect in vacuum, *Appl. Phys. Lett.*, **102**, 232905 (2013).
19. A. Piorra, R. Jahns, I. Teliban, *et al.*, Magnetoelectric thin film composites with interdigital electrodes, *Appl. Phys. Lett.*, **103**, 032902 (2013).
20. G. Sreenivasulu, V. M. Petrov, L. Y. Fetisov, *et al.*, Magnetoelectric interactions in layered composites of piezoelectric quartz and magnetostrictive alloys, *Phys. Rev. B*, **86**, 214405 (2012).
21. T. T. Nguyen, F. Bouillault, L. Daniel, *et al.*, Finite element modeling of magnetic field sensors based on nonlinear magnetoelectric effect, *J. Appl. Phys.*, **109**, 084904 (2011).
22. R. Pavel, T. Alois (Eds.), *Modern Sensors Handbook*, ISTE Ltd, 2007, p. 518.
23. Y. Ouyang, J. He, J. Hu, *et al.*, A current sensor based on the giant magnetoresistance effect: design and potential smart grid applications. *Sensors*, **12**, 15520–15541 (2012).

24. M. I. Bichurin, V. M. Petrov, R. V. Petrov, *et al.*, Magnetoelectric magnetometers, In *High Sensitivity Magnetometers*, Eds. A. Grosz, et al., pp. 127–166, © Springer International Publishing Switzerland (2016).
25. M. I. Bichurin, V. M. Petrov, R. V. Petrov, *et al.*, Magnetoelectric sensor of magnetic field. *Ferroelectrics*, 2002, *280*, pp. 199–202.
26. R. V. Petrov, N. V. Yegerev, M. I. Bichurin, *et al.*, Current sensor based on magnetoelectric effect, *18th International Symposium on Electrical Apparatus and Technologies (SIELA)* 2014, 29–31 May, Bourgas, Bulgaria.
27. I. N. Solovyev, A. N. Solovyev, R. V. Petrov, *et al.*, Sensitivity of magnetoelectric current sensor, *Proceedings of 11th International Conference on Applied Electromagnetics - ПЕС 2013*, September 01 – 04, 2013, Niš, Serbia, pp. 109–110.
28. R. V. Petrov, I. N. Solovyev, A. N. Soloviev *et al.*, Magnetoelectic current sensor. *PIERS Proceedings*, Stockholm, Sweden, Aug. 12–15, 2013, pp. 105–108.
29. N. Balabanian, *Fundamentals of Circuit Theory, Literary Licensing*, LLC, 2012, p. 570.
30. S. X. Dong, J. Y. Zhai, Equivalent circuit method for static and dynamic analysis of magnetoelectric laminated composites, *Chin. Sci. Bull.,* ***53***(14), 2113–2123 (2008).
31. M. I. Bichurin, V. M. Petrov, *Modeling of Magnetoelectric Effects in Composites* (Springer Series in Materials Science, Vol. 201), 2014, p. 108.
32. M. Bichurin, D. Viehland, *Magnetoelectricity in Composites, Pan Stanford Publishing*, 2011, Singapore, p. 286.
33. S. Bozhkov, I. Milenov, D. Danchev, *et al.*, Researching the signals of the automobile electromagnetic sensors, *12th International Conference on Applied Electromagnetics - PEC 2015*, ISBN 978-86-6125-145-0, O5-4, Niš, Serbia, 2015.
34. M. I. Bichurin, V. M Petrov., S. Priya, Magnetoelectric multiferroic composites. In: *Ferroelectrics–Physical Effects*, M. Lallart (Ed.), InTech, 2011, pp. 277–302.
35. M. I. Bichurin, V. M. Petrov, S. Dong, *et al.*, Dual magnetic field and mechanical vibrations energy harvesting system : Engineered multiferroics—magnetoelectric interactions, sensors and devices, *Materials Research Society. Symposia Proceedings*, Warrendale, PA, 2009, Vol. 1161E,.

36. R. V. Petrov, D. N. Ivanov, Energy harvesting system with magnetoelectric converter. *Vestnik NovSU, S.: Tech. Sci.*, **55**, 42–44 (2010) (in Russian).

37. V. Bedekar, M. Bichurin, I. Solovjev, *et al.*, Multimodal Energy Harvesting System. *Proc. SPIE 8035, Energy Harvesting and Storage: Materials, Devices, and Applications II*, 80350T (May 17, 2011); doi:10.1117/12.884747.

38. Y. Zhou, D. J. Apo, S. Priya, Dual-phase self-biased magnetoelectric energy harvester, *Appl. Phys. Lett.*, **103**, 192909 (2013); doi: 10.1063/1.4829151.

39. M. Nierla, M. Löffler, S. J. Rupitsch, Iterative finite element scheme for magnetoelectric energy harvesters, *Eurosensors 2015*, Procedia Engineering Volume 120, 2015, Pages 496–500 Eurosensors 2015 doi:10.1016/j.proeng.2015.08.682.

40. S. Priya, J. Ryu, C.-S. Park, *et al.*, Piezoelectric and magnetoelectric thick films for fabricating power sources in wireless sensor nodes. *Sensors*, **9**, 6362–6384 (2009); doi:10.3390/s90806362.

41. P. Finkel, R. P. Moyet, M. Wun-Fogle, *et al.*, Non-resonant magnetoelectric energy harvesting utilizing phase transformation in relaxor ferroelectric single crystals., *Actuators*, **5**, 2 (2016); doi:10.3390/act5010002.

42. R. V. Petrov, N. A. Kolesnikov, M. I. Bichurin, Magnetoelectric alternator. *Energy Harvesting and Systems*. Volume 3, Issue 2, Pages 173–180, ISSN (Online) 2329–8766, ISSN (Print) 2329–8774, DOI: 10.1515/ehs-2015-0024.

43. M. Bichurin, N. Kolesnikov, R. Petrov, *et al.*, Magnetoelectric energy source. *Proceeding of 12th International Conference on Applied Electromagnetics - ПЕС 2015*, August 31 – September 02, 2015, Niš, Serbia.

44. L. D. Landau, E. M. Lifshits, *Theoretical Physics, Vol. 7: Theory of Elasticity*. Pergamon Press, 1987, p. 248 (in Russian).

45. V. M. Petrov, M. I. Bichurin, G. Srinivasan, *et al.*, Bending modes and magnetoelectric effects in asymmetric ferromagnetic-ferroelectric structure. *Solid State Phenom.* **190,** 281–284 (2012).

46. M. I. Bichurin, V. M. Petrov, Magnetic resonance in layered ferrite-ferroelectric structures, *Sov. Phys. Tech. Phys.*, **33**, 1389 (1989).

47. M. I. Bichurin, V. M Petrov, Magnetic resonance in layered ferrite-ferroelectric structures, *Sov. Phys. JETP*, **58**, 2277 (1988).

48. M. I. Bichurin, I. A. Kornev, V. M. Petrov, *et al.*, Resonance magnetoelectric effect in multilayer composites, *Ferroelectrics*, **280**, 187–198 (2002).

49. M. I. Bichurin, V. M. Petrov, R. V. Petrov, *et al.*, Magnetoelectric microwave devices, *Ferroelectrics*, **280**, 213–220 (2002).

50. Yu. Poplavko, Yu. Prokopenko, V. Molchanov, *et al.*, Ferroelectrics study at microwaves, *Ferroelectrics: Characterization and Modeling*, M. Lallart (Ed.), InTech, (2011).

51. M. I. Bichurin, R. V. Petrov, Yu. V. Kiliba, Magnetoelectric microwave phase shifters, *Ferroelectrics* **204**, 311–319 (1997).

52. M. I. Bichurin, V. M. Petrov, R. V. Petrov, *et al.*, Magnetoelectric microwave devices, *Ferroelectrics* **280**, 211–218 (2002).

53. S. Shastry, G. Srinivasan, M. I. Bichurin, *et al.*, Microwave magnetoelectric effects in single crystal bilayers of yttrium iron garnet and lead magnesium niobate-lead titanate, *Phys. Rev. B* **70**, 064416 (2004).

54. G. Srinivasan, A. S. Tatarenko, M. I. Bichurin, Electrically tunable microwave filters based on ferromagnetic resonance in single crystal ferrite-ferroelectric bilayers, *Electronics Lett.* **41**, 596–597 (2005).

55. S. Tatarenko M. I. Bichurin, Microwave magnetoelectric devices, *Adv. Cond. Matter Phys.*, **2012**, 10 (2012).

56. A. S. Tatarenko, D. V. Snisarenko, M. I. Bichurin, Modeling of magnetoelectric microwave devices, *FACTA UNIVERSITATIS, Series: Electronics and Energetics*, Vol. 30, No 3, September 2017, pp. 285–293.

57. M. Perić, S. Ilić, S. Aleksić, *et al.*, Covered microstrip line with ground planes of finite width, *FACTA UNIVERSITATIS Series: Electronics and Energetics*, Vol. 27, No 4, December 2014, pp. 589–600.

58. M. I. Bichurin, A. S. Tatarenko, D. V. Lavrenteva, *et al.*, Magnetoelectric microwave devices. *Proceedings of 11th International Conference on Applied Electromagnetics - ПЕС'2013*, Niš, Serbia, September 01 – 04, 2013, pp. 77–78.

59. C. P. Wen, Coplanar waveguide: a surface strip transmission line suitable for nonreciprocal gyromagnetic device applications, *IEEE Transactions on Microwave Theory and Techniques*, **17**(12), 1087–1090 (1969).

60. S. B. Cohn, Slot line—an alternative transmission medium for integrated circuits, *1968 IEEE G-MTT International Microwave Symposium Digest*, 104–109 (1968).

61. E. A. Mariani, C. P. Heinzman, J. P. Agrios, *et al.*, Slot line characteristics, *IEEE Trans. Microw. Theory Tech.*, **17**, 1091–1096 (1969).

62. M .I. Bichurin, R. V. Petrov, Yu. V. Kiliba, Magnetoelectric microwave phase shifters, *Ferroelectrics*, **204**, 311 (1997).

63. M. I. Bichurin, V. M. Petrov, R. V. Petrov, *et al.* Magnetoelectric microwave devices, *Ferroelectrics* **280,** 213–220 (2002).

64. A. S. Tatarenko, M. I. Bichurin, Microwave magnetoelectric devices, *Adv. Cond. Matter Phys.*, **2012**, 10 (2012).

65. B. D. H. Tellegen, The gyrator, a new electric network element, *Philips Res. Rep.*, **3**, 81 (1948).

66. J. Zhai, J. Li, S. Dong, *et al.*, A quasi (unidirectional) Tellegen gyrator, *JAP*, **100**, 124509 (2006).

67. J. Zhai, J. Li, D. Viehland, *et al.*, Large magnetoelectric susceptibility: the fundamental property of piezoelectric and magnetostrictive laminated composites, *JAP*, 101, 014102 (2007).

68. M. I. Bichurin, V. M. Petrov, O. V. Ryabkov, *et al.*, Theory of magnetoelectric effects at magnetoacoustic resonance in single-crystal ferromagnetic-ferroelectric heterostructures, *Phys. Rev. B*, **72**, 060408(R) (2005).

69. M. I. Bichurin, R. V. Petrov, A. V. Filippov, Magnetoelectric microwave gyrator, Patent RU2357356. H03H011/42. 2009, p. 5.

70. R. V. Petrov, S. V. Averkin, A. V. Filippov, The study of the gyration effect in the magnetoelectric resonator, *Vestnik NovSU. Ser.: Techn. Sci.*, **60**, 85–89 (2010) (in Russian).

71. M. I. Bichurin, R. V. Petrov, I. N. Solovyev, *et al.* The study of the magnetoelectric microwave gyrator, Modern problems of science and education, *Electronic Scientific Journal*, **2** (2012); URL: www.science-education.ru/102-5370. www.rae.ru/snt/?section=content&op=show_article&article_id=6603. (in Russian).

72. A. G. Gurevich, *Ferrites at Microwave Frequencies*, Fizmatlit Publications, 1960, p. 408 (in Russian).

73. C.-W. Nan, M. I. Bichurin, S. X. Dong, *et al.*, Multiferroic magnetoelectric composites: historical perspective, status, and future directions, *J. Appl. Phys.* **103**, 031101, 2008.

74. M. Bichurin, V. Petrov, *Modeling of Magnetoelectric Effects in Composites* (Springer Series in Materials Science), 2014, p. 201.

75. M. I. Bichurin, R. V. Petrov, Magnetoelectric Phasers for PAS, In *Proceedings of the 2nd International Conference and Exhibition on Satellite Communications (ICSC'96)*, Moscow, Russia, 1996, pp. 236–241.

76. R. V. Petrov, D. V. B. Murthy, G. Sreenivasulu, *et al.*, A slot antenna with magnetoelectric elements, *Microw. Opt. Technol. Lett.*, **55**, 533–535 (2013).

77. R. V. Petrov, G. Srinivasan, *Design of Magnetoelectric Phased Antenna Array*, Vestn. Novg. Gos. Univ Ser.: Tehn. Nauki. 2009. No. **50**, pp. 61–65. (in Russian).

78. A. S. Tatarenko, R. V. Petrov, A. O. Nikitin, *et al.* Modelling of Multiferroic Microwave Patch Antenna, *Proceedings of International Conference Progress In Electromagnetics Research Symposium*, PIERS-2016, August 8–11, 2016, Shanghai, China.

79. M. I. Bichurin, D. Viehland, *Magnetoelectricity in composites*, Pan Stanford Publishing, Singapore, 2012, p. 273.

80. M. I. Bichurin, R. V. Petrov, A. S. Tatarenko, Magnetoelectric microwave devices for PAS, *Modern Problems in Science and Education*, **4** (2011), URL: www.science-education.ru/98-4727.

81. A. S. Tatarenko, M. I. Bichurin, Electrically tunable resonator for Microwave applications based on hexaferrite-piezoelectric layered structure, *Am. J. Cond. Matter Phys.*, **2**(5), 135–139 (2012) DOI: 10.5923/j.ajcmp.20120205.05.

82. R. V. Petrov, M. I. Bichurin, Yu. V. Kiliba, *et al.*, Magnetoelectric microwave amplifier, Patent RU 2439751 C1 (28.07.2010).

83. M. I. Bichurin, S. N. Ivanov, Microwave power selective detector, Patent RU 2451942 C1 (11.01.2011).

84. A. S. Tatarenko, M. I. Bichurin, D. V. Lavrentieva, Magnetoelectric microwave isolator, Patent RU 119940 U1 (05.03.2012).

85. M. I. Bichurin, A. S. Tatarenko, G. A. Semenov, *et al.*, Magnetoelectric attenuator, Patent RU 115125 U1 (26.10.2011).

86. V. M. Petrov, K. V. Lavrentyeva, Adjustable inductive element, Patent RU 123575 U1 (27.02.2012).

87. M. I. Bichurin, A. S. Tatarenko, R. V. Petrov, Active phased-array antenna receiving module, Patent RU 123586 U1 (03.05.2012).

88. M. I. Bichurin, V. M. Petrov, G. A. Semenov, Magnetoelectric material for components of radioelectronic devices, Patent No. RU 2363074 C1 (submitted 11.03.2008, published 27.07.2009).

Index